T0136664

FIELD GUIDE TO THE
ORCHIDS
OF **MADAGASCAR**

Dedicated to Jean Bosser
whose work on Madagascan orchids has been an inspiration.

ORCHIDS
OF MADAGASCAR

Phillip Cribb and Johan Hermans

Mapping by Mijoro Rakotoarinivo

Kew Publishing
Royal Botanic Gardens, Kew

PLANTS PEOPLE
POSSIBILITIES

Text: © The Board of Trustees of the Royal Botanic Gardens, Kew 2009
Line drawings: © the artist, as stated in the captions
Photographs: © the photographers as listed (pp. 430–431)

The authors have asserted their rights to be identified as the authors of this work in accordance
with the Copyright, Designs and Patents Act 1988.

All rights reserved. No part of this publication may be reproduced, stored in a retrieval system, or
transmitted, in any form, or by any means, electronic, mechanical, photocopying, recording or
otherwise, without written permission of the publisher unless in accordance with the provisions of
the Copyright Designs and Patents Act 1988.

Great care has been taken to maintain the accuracy of the information contained in this work.
However, neither the publisher, the editors nor the authors can be held responsible for any
consequences arising from use of the information contained herein.

First published in 2009 by
Royal Botanic Gardens, Kew
Richmond, Surrey, TW9 3AB, UK
www.kew.org

ISBN 978 184246 158 7

British Library Cataloguing in Publication Data
A catalogue record for this book is available from the British Library

Production Editor: Sharon Whitehead
Typesetting and page layout: Christine Beard
Design by Media Resources, Information Services Department,
Royal Botanic Gardens, Kew

Front cover photo: *Cymbidiella flabellata*, Eastern Madagascar.
Back cover photos, from left: *Angraecum protensum*, *Cynorkis cardiophylla*, *Cymbidiella falcigera*
Page 1 photo: *Bulbophyllum auriflorum*

Printed and bound in the United Kingdom by Henry Ling Limited

Mixed Sources
Product group from well-managed
forests and other controlled sources
www.fsc.org Cert no. SA-COC-001860
© 1996 Forest Stewardship Council
FSC

The paper used in this book contains material sourced from responsibly managed
and sustainable commercial forests, certified in accordance with the FSC
(Forestry Stewardship Council).

For information or to purchase all Kew titles please visit
www.kewbooks.com or email publishing@kew.org

*Kew's mission is to inspire and deliver science-based conservation worldwide, enhancing the quality of life.
All proceeds go to support Kew's work in saving the world's plants for life.*

Contents

Acknowledgements

The idea for this field guide originated from discussions with guides in Andasibe and Perinet, which took place during the joint Royal Botanic Gardens, Kew–Parc Tzimbazaza project on endangered Madagascan plants sponsored by the Kew Friends and Foundation. The need for a field guide was further emphasised during discussions with staff from the Missouri Botanical Garden and Conservation International in Madagascar. Many orchid enthusiasts have contributed photographs for this guide, either from the wild or from cultivated plants. We would particularly like to thank the following: Jean-Michel Hervouet, Marc Morel, Olaf Pronk, Dominique Karadjoff, Jean-Noel Labat, Mark Clements, Moritz Grubemann, Marcel Lecoufle, Eva Smrzova, Anton Sieder, Sven Buerki, Gunter Fischer, David Roberts, Jean Bosser, Marc Tessier, Sven Buerki, R. Hromniak, Martin Callmander, Jean-Claude Guérin of the French Orchid Society, Bill Baker, John Dransfield, David Du Puy and the late Karl-Heinz Senghas. Conservation International and the National Geographic Society supported the work by Martin Callmander.

We would like to thank the Friends and Foundation of the Royal Botanic Gardens, Kew, Dr Pamela le Couteur, Aliona Gavrilova and Roustan Tariko who have provided generous financial support for Kew's Madagascan orchid programme. Stuart Cable and Paul Smith who run Kew's Madagascan programme have been especially supportive.

We would also like to thank the Keepers and Librarians of the following institutes for access to their collections: Basel, Geneva, Royal Botanic Gardens, Kew, the Natural History Museum (London), Missouri, Paris, Tzimbabzaza, and Vienna.

The detailed line drawings of *Habenaria* and *Junellea* are the work of Juliet Beentje. We thank Justin Moat of Royal Botanic Gardens, Kew, and Mijoro Rakotoarinivo of Kew's Madagascan team for their work on maps, particularly Mijoro for providing all the distribution maps.

In Madagascar, we have been helped by many people, notably by the staff of the Royal Botanic Gardens, Kew–Parc Tzimbazaza project based in Antananarivo and the guides of many nature reserves and national parks. We are also grateful to the Ministry of Water and Forests for permits to work in areas under their control.

The inspiration for this work has been Jean Bosser of the Muséum d'Histoire Naturelle, Paris. We are pleased to dedicate this book to him for his life-time's work on Madagascar and its orchids, without which we would scarcely have known where to start.

Preface

Madagascar has one of the richest and most distinctive orchid floras anywhere in the world. The island boasts almost 1000 species in 57 genera; 17% of the genera and nearly 90% of the species are endemic. Visitors to the island are almost certain to see orchids on their visit. The well-known nature reserves and parks are rich in orchid species. Some orchids are spectacular but most are cryptic and grow in shady places under trees and shrubs or on the trunks and branches of trees in the woods and forests. A good pair of binoculars and a ×10 magnifying lens are particularly useful tools when trying to find and identify orchids.

The origin of this field guide can be traced to a joint project on Madagascar's threatened plants between the Parc Tzimbazaza in Antananarivo and the Royal Botanic Gardens, Kew, which was funded by the Friends and Foundation of Kew. Orchids were one of the three target groups, the others being palms and succulents. Detailed field and propagation studies were undertaken on a select number of threatened orchid species, with a view to their eventual re-introduction. Re-introducing orchids is not simple, especially when the number of plants in the wild has been reduced to less than 50, as was the case for several of the target species. With a narrow genetic base, it is important to breed with the right plants from the appropriate populations and to find appropriate means of raising the seedlings to maturity. It is also necessary to understand the ecology of the orchid well before re-introduction can occur. Finally, re-introduction must be followed by protection and monitoring to assess the success or otherwise of the project.

In the course of these studies, we visited many protected areas. Our guides were, almost to a man, keen to learn more about orchids but lacked any guide books with easily understood text and good photographs. The two parts of Henri Perrier de la Bâthie's seminal orchid account in the *Flore de Madagascar* were published in 1939 and 1941, and have long been out-of-print. Second-hand copies, when available, fetch high prices and are well out of the range of nearly everyone. The idea of producing a field guide sprang from the perceived need of the park rangers and guides and of tourists interested in Madagascar's unique plant life, of which orchids make up nearly 10% of the total species.

We are very grateful to a number of biologists and photographers who have photographed orchids in Madagascar for allowing us to use their images for this guide. Photographers are listed on pp. 430–431 and the copyright of their images remains with them.

Phillip Cribb
Johan Hermans
March 2009
Kew

How to use this field guide

This guide contains photographs of about a third of Madagascar's orchids. There are many gaps, but illustrations of most of the more obvious and showy orchids are included. The key to genera is necessarily technical in places, but the genus of most orchids can be identified more readily by a quick thumb through the photographs. Once the right genus has been determined and where several orchids are listed, the species are grouped together under headings to allow comparison of similar species. In many descriptions, technical terms are used because orchids have some unique structures that defy easy categorisation. A glossary is provided to help users understand the terminology. For each species, we provide a short description, distributional information by region and country outside of Madagascar, and information on ecology, altitudinal range and flowering season. Readers should appreciate that the information provided comes from the literature, herbarium material and personal observations. Some species undoubtedly occur outside of the described places, habitats and elevations, and have flowering times that differ from those provided. The photographs have been taken without a scale and users are referred to the sizes given in the text to judge if their plant fits the description provided. A few orchids are so poorly known that we have been unable to provide a full description, or to give distribution information, ecological or altitudinal data.

The distribution maps, which have been prepared by Mijoro Rakotoarinivo, are based upon geo-referenced herbarium collections at BM, E, K, MO, P, TAN, W (magenta areas) and non geo-referenced literature and other sources (pink areas), including sight records and unlocalized herbarium material. Dot maps have been eschewed because they can provide information that allows collectors to find rare species. The distribution maps should be used with care. An uncoloured area may merely reflect the lack of collection intensity in that area rather than the absence of the species from it.

The authors would appreciate any observations made by users so that future editions can be improved. Apart from the recently published *Field Guide to Madagascan Palms*, this is the only such guide to a group of plants in Madagascar.

Introduction

The Indian Ocean island of Madagascar lies about 400 km from the east coast of Africa, the tropic of Capricorn runs through the south, and the northern tip points towards the equator. At 587,000 km^2, it is the fourth largest island in the world. It is thought that the first human inhabitants arrived from Indonesia and Malaya about 2000 years ago, and that their arrival was followed by migration from mainland Africa and elsewhere.

Madagascar has a great variety of bedrock and soil types together with wide topographical variation, ranging from tropical coastal forest to several mountain ranges of over 2000 m. The island separated from the mainland about 165 million years ago during the continental drift, leading to the development of a largely endemic floral and fauna; there is an exceptional level of endemism. There are an estimated 10,000 different vascular plants, 90% of them endemic (Moat & Smith, 2007); at the last count, there were almost 1000 different orchids in 57 genera. One in ten vascular plants on Madagascar are orchids.

The Madagascan orchids have been catalogued in two recent publications (Du Puy *et al.*, 1999; Hermans *et al.*, 2007), the latter being an update of the first. In these books, data are provided on the distribution, ecology and nomenclature of each species. A limited selection of the species is illustrated in each book. Together, these publications also provide a comprehensive account of the literature on Madagascan orchids.

New species are being described every year, and we suspect that there are many more awaiting discovery. Sadly, this is often the result of previously inaccessible areas being opened up by forest destruction, mining or road-building.

References

Du Puy, D., Cribb, P.J., Bosser, J., Hermans, J. & Hermans, C. (1999). *The Orchids of Madagascar*. Royal Botanic Gardens, Kew.

Hermans, J., Hermans, C., Du Puy, D., Cribb, P.J. & Bosser, J. (2007). *Orchids of Madagascar*. Second Edition. Royal Botanic Gardens, Kew.

Moat, J. & Smith, P. (2007). *Atlas of the Vegetation of Madagascar*. Royal Botanic Gardens, Kew.

What is an orchid?

An understanding of the floral and vegetative structures of orchids provides the clues needed to identify orchids. Knowledge of floral morphology is crucial for naming orchids because they are, for the most part, classified into genera and species on the basis of the finer details of the structure of their sepals, petals, lip and column. Floral dissections provide the essential information for identification. For most species, the shape of the sepals, petals and especially the lip will provide all of the information the reader needs. For the more difficult taxa, however, details of the column, anther, pollinia and rostellum might be needed before accurate identification is possible. The vegetative features also provide crucial clues to the identities of orchid species.

Madagascan orchids come in all shapes and sizes. Some orchids live on the ground (called terrestrials) whereas others grow perched on trees (epiphytes) or rocks (lithophytes). The species of *Vanilla* and *Galeola* are vines that scramble or climb many metres up shrubs and trees, using their roots for support. The smallest orchids to be found are members of the *Angraecum* genus that are scarcely 2 cm tall and have flowers 2 mm across. What then unites these diverse plants into the family called the Orchidaceae? The distinctive features of orchids that separate them from other flowering plants lie primarily in their flowers but a number of vegetative features are also distinctive.

The flower

Orchid flowers are simple in structure and characteristically have their floral parts arranged in threes or multiples of three. This can be seen most easily in the two outer whorls of the flower. Let us take, as an example, the common Madagascan orchid *Eulophia plantaginea* (see p. 403), which is similar in general floral structure to the majority of orchids from these islands. Its floral parts are situated at the apex of the **ovary,** which itself can be seen to be tripartite in cross-section. The outermost whorl of the flower is the calyx, which consists of three **sepals** that are petal-like and coloured pale green. The two **lateral sepals** differ slightly from the third, called the **dorsal** or **median sepal**, which lies uppermost in the flower. In some orchids, such as bulbophyllums and polystachyas, the lateral sepals form a more-or-less conical chin at their base, called a **mentum**.

The corolla of *Eulophia plantaginea* comprises three showy **petals**. The two lateral white petals, resembling the dorsal sepal in shape, are uppermost in the flower and differ markedly from the third petal, which lies at the bottom of the flower. The third petal, called the **lip** (or **labellum**), is highly modified and 3-lobed. It has a callus of ridges on its upper surface and a short spur or nectary at the base. In other orchids, callus structures of various shapes can be present or absent and the spur, if present, can range from filiform and cylindrical to more-or-less saccate. The lip is an important adaptation of the orchid that facilitates cross-pollination. It can be imagined as a flag to attract potential and specific pollinators, which are then guided towards the pollen and stigmatic surface by the form of the callus and the perceived presence of nectar in the spur. The lip, therefore, can be supposed to act as a landing platform and the callus structure as a guidance system for the pollinator.

The central part of the orchid flower is a single structure called the **column**. In *Eulophia plantaginea*, a single **anther** lies at the apex of the column. The pollen in the anther is not powdery as in most plants, but is borne in four discrete masses, called **pollinia** (singular pollinium). The pollinia are attached to a sticky mass called a **viscidium**. In other species, the number of pollinia may be two, six or eight. These are attached to the viscidium either directly or by a stalk, called a **stipe** in most epiphytic orchids, and a **caudicle** in most terrestrial ones.

The **stigma**, the receptive surface on which pollen alights and germinates, is also positioned on the column in the centre of the orchid flower, on its ventral surface. The stigma is a sticky lobed depression that is situated below and behind the anther in most orchids; in some terrestrial genera, such as *Habenaria*, the stigma is bilobed with the receptive surfaces at the apex of each lobe. In many species, the pollen masses are transferred to the stigmatic surface by a modified lobe of the stigma called the **rostellum**. This is developed in *Eulophia plantaginea* as a projecting flap below the anther and above the stigma that catches the pollen masses as the pollinator passes beneath on its way out of the flower.

An interesting feature of the development of most orchid flowers is the phenomenon of **resupination**. In bud, the lip lies uppermost in the flower while the column lies lowermost. In species with a pendent inflorescence, the lip will, therefore, naturally lie lowermost in the flower when it opens. This should not be the case, however, in the many orchid species that have erect inflorescences, including *Eulophia plantaginea*. In these species, the opening of the flower should naturally lead to the lip assuming a place at the top of the flower above the column. In most species, however, the lip is lowermost in the flower. This position is achieved by means of a twisting of the flower stalk or ovary through 180 degrees as the bud develops, a process termed resupination.

The inflorescence

Orchids carry their flowers in a variety of ways. Even within the same genus, different species have different means of presenting the flowers. Most orchids in Madagascar have inflorescences that bear two or more flowers, usually on a more or less elongate floral axis comprising a stalk called the **peduncle** and a portion bearing the flowers, the **rachis**. In *Eulophia plantaginea* the flowers are borne in an elongate erect **raceme**, which is unbranched with the flowers arranged in a lax spiral around the rhachis. In a raceme, the individual flowers are attached to the floral axis by a stalk called the **pedicel**. In some species, pedicels are virtually absent and the flowers are sessile on the axis, such inflorescences are termed a **spike**.

In some terrestrial orchids, notably some *Satyrium* and *Disa* species, the inflorescence is borne on a separate shoot from the leaves.

In the genus *Bulbophyllum*, we find some interesting variations on the multi-flowered inflorescence. In several species, the flowers are borne all facing to the same side of the rhachis, this being called a secund inflorescence. The most spectacular group, however, are those in which the rhachis is so contracted that the flowers all appear to come from the top of the flower stalk in an umbel, with the inflorescence rather resembling the head of a daisy. For this reason, these bulbophyllums, such as *Bulbophyllum longifolium* Thou., were formerly considered to be in the separate genus *Cirrhopetalum*.

Compound inflorescences with many flowers are uncommon in island orchids, but where branching inflorescences are found, e.g. in *Graphorkis scripta*, they are termed **panicles**.

In some species, the flowers are borne one-at-a-time either sessile or on a shorter or longer peduncle. Solitary flowers can be found in many genera, including *Bulbophyllum* and *Disperis*.

Structure of the plant

The vegetative features of orchids are, if anything, more variable than their floral ones. This is scarcely surprising when the variety of habitats in which orchids are found is considered. Madagascan orchids grow in almost every situation: on the permanently moist floor of lowland tropical rain forests; in the uppermost branches of tall forest trees, where heavy rainfall is followed by scorching sun for hours on end; on rocks; on the summits of mountains; in the dry spiny forest; and in the grassy areas found on landslips and road-sides. The major adaptations seen in the vegetative morphology of orchids allow them to withstand adverse environmental conditions, in particular, the problems of water conservation on a daily and seasonal basis.

Tubers, rhizomes, stems and pseudobulbs

Madagascan orchids often suffer from periodic water deficits, which they overcome by various strategies. Rainfall is not continuous, even in the wettest habitats; in many places in the island, the rainfall patterns are markedly seasonal. Terrestrial orchids, such as *Cynorkis*, *Habenaria* and *Satyrium*, lack pseudobulbs but have underground **tubers** that survive drought. The new growth emerges from one tip of the tuber in suitable conditions, but the tuber then produces a new shoot when it starts to rain again. In other orchid genera, such as *Cheirostylis* and *Platylepis*, the stems are succulent but not swollen. The horizontal stem or **rhizome** creeps along the ground in the leaf litter and erect shoots bearing the leaves are sent up periodically. By contrast, epiphytic orchids, which grow on the trunks, branches and twigs of trees, and lithophytic ones, which grow on rocks, experience rapid water run-off; these orchids dry out quickly in the sunshine that follows rain. Many of these species have marked adaptations of the stem and leaves, some as dramatic as those encountered in the Cactaceae, which allow them to survive periodic droughts. The stem can become succulent and develop into a water-storage organ. This is so common in tropical orchids that the resulting structure has been given a technical name, a **pseudobulb**. In *Cymbidiella*, the pseudobulbs comprise several internodes, whereas in *Bulbophyllum*, they are of one internode only. Pseudobulbs are also found in many terrestrial orchids and can grow either above the ground, as in *Calanthe*, or underground, as in *Eulophia*.

Leaves

The leaf is another organ that has undergone dramatic modification in the orchids. Fleshy or leathery leaves with restricted stomata, such as those of *Angraecum* and *Bulbophyllum* species, are common.

A number of Madagascan orchids have scale-like leaves that lack any chlorophyll, the substance that gives leaves their green colouration. The epiphytic *Microcoelia* has leaves that have been reduced to **scales** so that photosynthesis takes place in its long green roots. Scale leaves are also present in four Madagascan *Vanilla* species, namely *V. decaryana*, *V. humblotii*, *V. madagascariensis* and *V. perrieri*, but these scales are green and the plants also have green stems. The vine-like *Galeola*, together with terrestrial species such as *Habenaria saprophytica*, *Eulophia hologlossa* and *E. mangenotiana*, lack chlorophyll altogether and cannot photosynthesize. They obtain all of their nutrients from the mycorrhizal fungus with which they are associated. All the other Madagascan orchids have green leaves of various shapes and sizes.

Plant species with photosynthetic green leaves are termed **autotrophic**. Terrestrial orchid species usually have much thinner and more textured leaves than their epiphytic cousins. In lowland forest, the perpetually moist atmosphere and lack of direct sunlight means that such leaves are not vulnerable to drought. Some of the terrestrial species of the forest floor have beautifully marked leaves. In *Goodyera*, *Zeuxine, Platylepis* and their relatives, the leaves can range from green to deep purple or black in colour and may be mottled or reticulately veined with silver.

Roots

The roots of most epiphytic orchids are much modified. They provide both attachment to the substrate and also uptake of water and nutrients in a periodically dry environment. These roots have an actively growing tip; the older parts are covered by an envelope of dead empty cells called a **velamen**. The velamen protects the inner conductive tissue of the roots and might also aid the uptake of moisture from the atmosphere, acting almost as blotting paper for the orchid.

Life in Madagascar can be inhospitable for orchids. In those regions with a more marked seasonality, conditions can be positively hostile for orchids at certain times of the year. Even tropical forests can have periods of relative drought during which the orchids have to survive days or even weeks without rain. In these conditions, tropical orchids that do not have water-storage capabilities in their stems or leaves can drop their leaves and survive on the moisture stored in their roots, which are protected by their cover of velamen.

Major orchid habitats

Orchids are found in almost every habitat in Madagascar, from the coast to the highest mountain summits and from the rain forest to the spiny forest. Of course, the distribution of orchids is not even throughout these habitats, but is dictated by soil, climate, elevation and aspect. Generally, orchids are most abundant in the wetter habitats and least frequent in the drier areas. They can form a major element of the vegetation in some places, whereas in others they can be difficult to locate. Many orchids are relatively widespread and tolerant of a number of habitats, whereas others are restricted to particular habitats. Even the driest places can have their own peculiar orchid flora.

Vegetation zones

Historically, Madagascar has been divided into a number of fairly well-defined areas of vegetation and climate; more recently, detailed mapping techniques such as satellite imaging have been used to re-define some of these zones. Moat and Smith (2007) have recently produced a detailed vegetation atlas of Madagascar. This shows that less than 10% of the island remains covered by natural vegetation. A summary of the most distinct areas of natural vegetation on Madagascar follows.

Humid evergreen forest at mid-elevation, Ankazobe

Moist evergreen forest near Andasibe, CE Madagascar

The eastern and northern rain forests

These forests, which include lowland evergreen and humid escarpment forests, cover a large part of the north and the east side of Madagascar from the coast to the crest of mountains lying parallel to the coast. This part of the country is under the influence of the south-east trade winds, which promote cloud formation and heavy rainfall throughout the year. Average annual rainfall is more than 2000 mm, it tops 6000 mm in parts of the Masoala peninsula in the north-east. There is no marked dry season at lower elevations and only a brief one higher up where the climate is cooler and able to support broad-leaf evergreen forest. Temperatures are generally high with mean winter readings in the coldest season ranging from 18°C at sea level to 10°C at the top of the escarpment. Primary vegetation is mostly lowland rain forest with tall forest trees (up to 25–30 m tall) in several distinct strata and a diffuse under-storey: this type of forest is rich in species. The forest is in better condition in the north and north-east than further south where just a patchwork remains. This type of vegetation continues in the north-west around the Sambirano river valley, where there is a short dry season and about 1800 mm of rain. Typical vegetation includes numerous palms, *Pandanus* (screw pines), *Canarium* and, in more degraded parts, *Ravenala*, the traveller's palm.

Littoral forest, near Tolagnaro, SE Madagascar

The rain forest is the home of numerous orchids; *Angraecum, Aerangis* and *Aeranthes* are well-represented, growing mainly epiphytically. *Bulbophyllum* is the largest genus, with about 200 different species growing in a variety of substrates, whereas polystachyas can be found in the lower strata of the forest and also on the ground, there are more than 20 different species on the island. Several *Liparis* species can be found in leaf litter on the forest floor, and there are some very colourful terrestrial orchids such as *Calanthe*, *Phaius* and *Gastrorchis* species. One of the highlights of the Madagascar orchid flora is the genus *Cynorkis*; there are about 120 different species, most of them being terrestrial but several growing in moss as true epiphytes.

Coastal forests are often rich in orchids, even where man has disturbed them. The best place to see the famous comet orchid (*Angraecum sesquipedale*) is on trees along the east coast. *Aeranthes grandiflora* and *Graphorkis concolor* are also found there, often planted on trees in the grounds of the hotels and resorts. *Angraecum sesquipedale* can also be found elsewhere in remnant coastal forest, such as the white sand forest near Tôlagnaro (Fort Dauphin).

Xerophytic scrub near Tolagnaro, SE Madagascar   Lichen-rich forest, Lakato, EC Madagascar

Some epiphytic species on the coast grow on rocks or even on sand-banks. The magnificent *Angraecum eburneum* is one such species, often forming large clumps in light shade. Some terrestrial species, especially of *Oeceoclades* and *Eulophia*, are found exclusively in the sandy coastal areas: *Oeceoclades calcarata* and *O. decaryana* resemble sanseverias when not in flower. Some typically terrestrial orchids, especially some *Cynorkis* species, can grow epiphytically in these forests.

Central highlands and mountains

Although there is some overlap with the eastern habitats, this area covers the central plateau plus a number of mountain ranges, altogether covering nearly 40% of Madagascar. The climate is more seasonal than that in the east, with an annual rainfall of 1200–2500 mm; there are frequent mists in some areas. Mean temperature in the coldest months ranges from over 13°C in the east and west down to 5°C in high mountains. The remaining forest is very much reduced in extent and is mainly composed of moist montane forest. The tree canopy is up to 25 m high with low-branched trees bearing numerous epiphytes, and there is a thick herbaceous understorey. At higher elevations or on exposed ridges, there are formations of sclerophyllous montane vegetation, also known as lichen forest, in which the smaller trees bear many epiphytes. The understorey of such forests is quite open, the soil being covered with moss and lichen. At lower elevation, the principal vegetation type is ericoid thicket, which forms a dwarf open habitat generally just a few metres in height.

Evergreen sclerophyllous 'tapia' forest, C Madagascar

Highland savannah dominated by inselberg, C Madagascar

The areas of open woodland are sometimes dominated by *Uapaca bojeri*, also known as 'tapia'. These trees have thick, fire-resistant bark allowing them to survive the fires that almost annually burn the coarse grassland around them. Tapia belongs in the family Euphorbiaceae; the fruit is edible and the tree is a host of the Malagasy silkworm, from which beautiful textiles are woven.

There is a very different ecosystem towards the summit of the mountains. Frosts and blizzards have been reported during the early morning hours of the dry season, which lasts approximately 7–8 months, and mist and dew are common. The mountain summits contain many endemic plants that have highly restricted distributions. Many of the massifs are not protected, although they contain numerous locally endemic species. Inselbergs (i.e. isolated monoliths of smooth rounded granite) occur in the central area and have characteristics similar to those of the mountains. The majority of the central region consists of a complex of hills and valleys, with vast expanses of species-poor grassland or pseudo-steppe composed of a few cosmopolitan fire-adapted grasses. There is evidence that some of this grassland was once thinly wooded. Apart from vast tracts of introduced trees, such as pine and *Eucalyptus*, these are also the main areas of human habitation and food production. The region rests on degraded laterite soils, and there are annual fires that produce new herbaceous growth for grazing. In localities protected from fire, such as erosion gullies, some forest survives.

Montane habitat, Mt Ibity, C Madagascar

The sandstone massif, Isalo, SW Madagascar

The Andringitra mountain range, SE Madagascar

The orchids of the central highlands and mountains are now becoming quite scarce, most relying on specific microhabitats to thrive. Angraecums have often adapted to an exposed lithophytic habitat although several species grow on the gnarled stems of tapia trees. From moister and more shaded forest comes the monotypic *Lemurorchis*. The terrestrial orchids of this area are also interesting, with the genera *Cynorkis, Eulophia, Disa* and *Satyrium* being very well represented with some large and colourful species.

Dry forests of the south and west

The land gradually slopes westwards from the central highlands to the coast, with remaining forest extending in a broad zone towards the coast from the northern tip of the island. The western zone receives rainfall only during the wet season with a dramatic reduction from the north to the south, with some southern-most areas receiving less than 400 mm of precipitation per year. The south and west support a variety of sclerophyllous deciduous woodland, open woodland and wooded savannah, and there are small patches of humid vegetation along rivers and in marshes dominated by *Raphia* palms. The main primary vegetation consists of western deciduous forest, which is hot and dry, rainfall averaging 500–1500 mm and with a long dry season of 5–10 months; this climate has a profound effect on the vegetation. The deciduous forest changes towards the south into xerophytic forest, where the canopy is 10–15 m tall. The extreme south and south-west have an average rainfall of 700 mm or less, with the driest parts not even getting 300 mm; the climate is arid and hot, the average annual maximum temperature is 30–33°C, the minimum 15–21°C. Many plants have adaptations to store what little moisture is available and to minimise water loss. The vegetation is mainly low-growing. This xerophytic forest is also known as spiny forest or bush, it is dominated by *Euphorbia* and Didieraceae, the world's greatest diversity of baobabs (*Adansonia*) and many other endemic succulents occur here. The area is interspersed with limestone formations, which have with their own unique vegetation. The northern part of the island is characterised by a wide variety of topography: there are high mountain peaks and coastal plains. The climate of the north is similar to that of the east except for a much drier pocket at the northern tip.

Orchids are not plentiful in these habitats but they make up for this by their variety of form and interesting adaptation to extreme conditions. Some local forms of *Grammangis, Aerangis* and *Aeranthes* can be found in pockets of forest. *Paralophia* is also restricted to the dry zone.

The genus *Sobennikoffia* is restricted to warm and dry areas, and there is little difference between the two main species of this genus. The ultimate in succulence comes in the leafless *Vanilla*. The genus *Oeceoclades* is mainly found in warm and dry areas and shows a fascinating array of shape and colour, many species have very intricately patterned leaves.

Adansonia grandidieri, Morandava, W Madagascar

Transition forest, Andohalela, SE Madagascar

Moist evergreen forest, Manjakatompo   Tsingy of Ankarana, NW Madagascar

Forest on karst ('tsingy')

Surprising numbers of orchids can be found in the karst formations that are known as 'tsingy'. They can grow there as epiphytes, lithophytes on the bare rock, or as terrestrials. The epiphytes found in such areas, especially bulbophyllums, usually have well-developed pseudobulbs and one or two leathery leaves. Likewise, the lithophytic species, especially *Oeceoclades* and *Eulophia*, also have pseudobulbs and leathery, often mottled leaves.

A few vanillas are also found in karst woodland. The strange *Vanilla madagascariensis* is a vine that climbs trees using its roots, which adhere to the vertical trunk in a step-like manner.

Cultivated fields

Another habitat generally considered devoid of orchids is cultivated land. However, *Eulophia plantaginea*, *Cynorkis angustifolia* and *C. flexuosa* are often found on field and ditch banks, especially around rice paddies.

Conservation

All orchids are protected in Madagascar and cannot be collected except under licence. A further protection is provided as no plants can be removed from protected areas. Orchids brought into nurseries can only be exported with the appropriate plant health and CITES (Control in Trade of Endangered Species of Fauna and Flora) export permits. In Europe, a CITES import permit is also required for orchids.

As far as we are aware, no Madagascan orchid is extinct. The orchid flora of Madagascar is still very poorly understood, however, and a significant number of species are known only from the original collection. Because of changes in many of the major habitats on the island, it is likely that some of these species will indeed be extinct or, at least, threatened with extinction. Recent surveys revealed only a few plants of *Grammangis spectabilis* in its only known habitat in south-east Madagascar, and a few tens of plants of both *Erasanthe henrici* var. *isaloensis* and *Angraecum longicalcar* in the wild in the plateau region.

Threats to orchids come from three sources, all man-made. The first of these is habitat destruction through logging, mining or agriculture. Orchids are very sensitive indicators of habitat change. The removal of shade, the addition of fertiliser or the use of pesticides is guaranteed to eliminate orchids. Logging will remove the trees on which the epiphytic species grow. Terrestrial species, of genera such as *Calanthe*, *Gastrorchis* and *Liparis*, are intolerant of direct sunlight and will not survive clear felling of their habitats.

The second threat to orchid habitats is the widespread use of fire to stimulate the growth of grass at the end of the dry season. This traditional method used by cattle farmers is extremely destructive because the fires often burn from the grassland into the surrounding forest and woodland. Epiphytic and many terrestrial orchids growing in the path of a fire do not survive.

The third factor affecting orchid numbers is the collection of plants for horticulture and medicine. Nearly all orchids are potential targets for collectors, although most are interested only in the dramatic species with attractive flowers. The showy orchids are particularly vulnerable, especially species of the genera *Aerangis*, *Angraecum* and their allies, *Calanthe*, *Cymbidiella*, *Eulophiella*, *Gastrorchis* and *Grammangis* have been collected in large numbers over many decades to feed the international orchid trade. Plants for which a few Malagasy Ariary are paid to the local collector can fetch tens of dollars in America, Japan or Europe. Little of the profit remains in Madagascar, most going to nurseries in developed countries. There is also a large internal trade in orchids for cultivation within Madagascar, and this is perhaps a greater threat to some species than international trade. The catholic taste of orchid growers is reflected in the diversity of orchids to be found in the orchid nurseries of Antananarivo and in markets and roadside stalls.

Some orchids feature in local medicine, one of the best-known being native *Vanilla* vines, the stems of which are chopped into lengths and sold as a constituent of a pick-me-up or aphrodisiac. The stems are boiled with the bark of two trees and the infusion is drunk once it has cooled. Its efficacy is debatable. *Angraecum eburneum* is also used medicinally.

The cardinal rule is to photograph or draw the plants that you see and to leave them for others to admire. Too often tourists, even in protected areas such as Andasibe (Perinet) and Ranomafana, succumb to the temptation to pick orchids or to dig them up. It is unnecessary and illegal. Nearly all of Madagascar's showy orchids are available in the horticultural trade as nursery-raised plants, both in Madagascar and abroad. Treat Madagascar's unique flora, including its orchids, with respect and leave the plants where you find them!

Maps

Map 1. Vegetation map of Madagascar. The major vegetation types are indicated in the key to the right.
Courtesy of Justin Moat & Paul Smith (2007).

Map 2. (p. 26) Major phytogeographical regions of Madagascar.
Courtesy of Mijoro Rakotoarinivo and Justin Moat, Royal Botanic Gardens, Kew.

Map 3. Map of Madagascar, showing the main towns, main roads and protected areas.
Courtesy of Mijoro Rakotoarinivo and Justin Moat, Royal Botanic Gardens, Kew.

Map 4. (p. 28) Madagascan political regions, used for the distribution maps of the species.
Courtesy of Mijoro Rakotoarinivo and Justin Moat, Royal Botanic Gardens, Kew.

Map 5. (p. 28) Madagascan major regions, used for the distributions of the species in the text.
Courtesy of Mijoro Rakotoarinivo and Justin Moat, Royal Botanic Gardens, Kew.

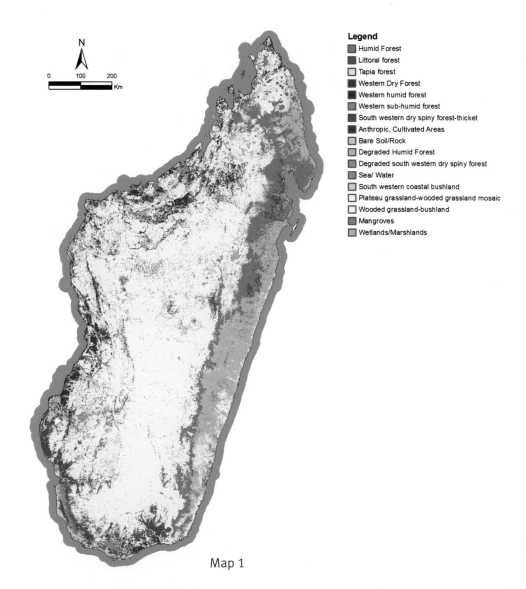

Legend
- Humid Forest
- Littoral forest
- Tapia forest
- Western Dry Forest
- Western humid forest
- Western sub-humid forest
- South western dry spiny forest-thicket
- Anthropic, Cultivated Areas
- Bare Soil/Rock
- Degraded Humid Forest
- Degraded south western dry spiny forest
- Sea/ Water
- South western coastal bushland
- Plateau grassland-wooded grassland mosaic
- Wooded grassland-bushland
- Mangroves
- Wetlands/Marshlands

Map 1

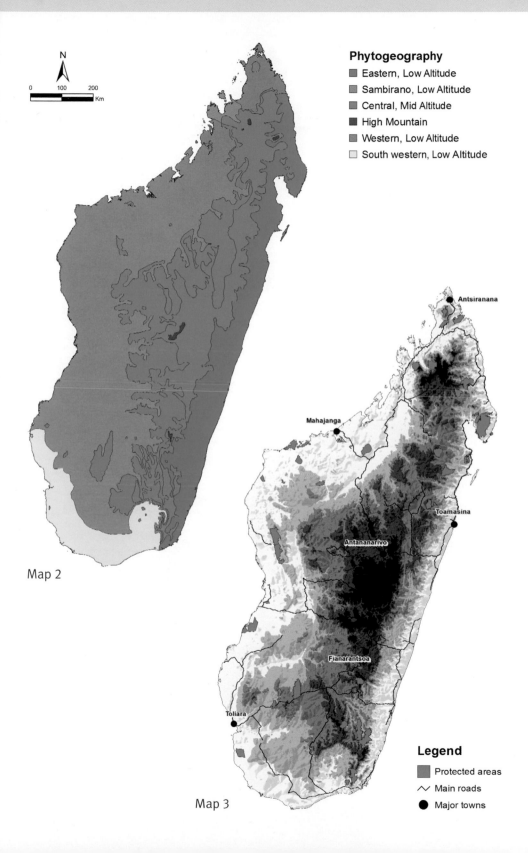

Phytogeography
- Eastern, Low Altitude
- Sambirano, Low Altitude
- Central, Mid Altitude
- High Mountain
- Western, Low Altitude
- South western, Low Altitude

N

0   100   200
Km

Antsiranana

Mahajanga

Toamasina

Antananarivo

Fianarantsoa

Toliara

Map 2

Map 3

Legend
- Protected areas
- Main roads
- Major towns

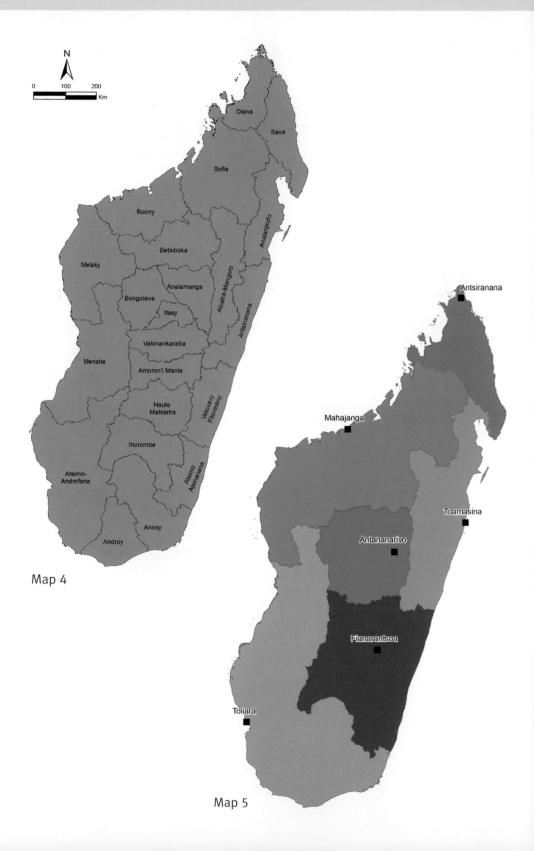

Map 4

Map 5

Keys to the genera

Key I. Madagascan orchid genera

1. Plants vines .... **Key II**
1. Plants not vines, either terrestrial, epiphytic or lithophytic 2
2. Plants saprophytic; leaves scale-like, not green; stems not green .... **Key III**
2. Plants green, usually with one or more well-developed green leaves 3
3. Column with a basally fixed anther, usually with two distinct anther
    sacs; terrestrial plants .... **Key IV**
3. Column with a terminal or dorsal anther with an easily removed
    anther cap; terrestrial, lithophytic or epiphytic plants 4
4. Plants growing on the ground 5
4. Plants growing on rocks or trees 6
5. Rhizomes fleshy, creeping; stems erect, leafy; leaves thin-textured;
    inflorescences terminal, unbranched .... **Key V**
5. Rhizomes obscure to elongate, not fleshy; pseudobulbs present or
    not; leaves thin-textured to leathery .... **Key VI**
6. Inflorescences terminal .... **Key VII**
6. Inflorescences lateral 7
7. Stems with distinct pseudobulbs .... **Key VIII**
7. Stems not pseudobulbous, short to long, rarely branching .... **Key IX**

Key II. Vine-like orchids

1. Plants green; leaves well-developed or scale-like but always green;
    flowers white or yellow, marked with apricot, red or maroon in lip .... **1. Vanilla**
1. Plants buff to brown, never green; leaves scale-like, brown to buff-
    coloured; flowers buff-brown .... **2. Galeola**

Key III. Saprophytic orchids, lacking chlorophyll and leaves

1. Sepals and petals fused for half their length and forming a tube;
    pedicel elongating after fertilisation of flower .... **22. Didymoplexis**
1. Sepals and petals free, not forming a tubular base; pedicels not
    elongating 2
2. Lip not spurred at base; sepals and petals fused into a tube in basal
    part .... **21. Auxopus**
2. Lip spurred or deeply saccate at base 5
3. Spur longer than the lip; lip ligulate, unlobed, lacking a callus ....
    .... **6. Habenaria saprophytica**
3. Spur shorter than the lip; lip 3-lobed, callose .... **51. Eulophia** (in part)

Key IV. Terrestrial orchids, usually with tubers and with a basally fixed anther with 2 more-or-less parallel anther loculi

1. Flower hood (lip uppermost in flower) with two spurs .... **13. Satyrium**
1. Flower hood with a single spur or spur-less 2
2. Flower hood with a distinct spur on its dorsal side; lip not spurred 3
2. Flower hood, if present, not spurred; lip often spurred or saccate at base 5
3. Lip complex, hidden inside hooded dorsal sepal; lateral sepals often spurred .... **14. Disperis**
3. Lip simple, usually pendent; lateral sepals not spurred 4
4. Petals free from the dorsal sepal; anther erect or horizontal .... **11. Disa**
4. Petals united to the dorsal sepal; anther horizontal .... **12. Brownleea**
5. Spur absent 6
5. Spur present, albeit sometimes short 7
6. Flowers purple; lip bilobed .... **3. Brachycorythis** (in part)
6. Flowers yellow or green, lip entire, often ornamented with depressions towards the base .... **5. Tylostigma**
7. Anther canals and lateral lobes of rostellum very short and obscure; spur short, often scrotiform or bifid at tip 8
7. Anther canals and lateral lobes of rostellum well-developed, often elongate 9
8. Flowers purple; lip entire with a two vertical calli at base; spur bifid at the tip .... **3. Brachycorythis** (in part)
8. Flowers green, yellow or white; spur globular or sometimes cylindrical; lip often 3-lobed or 3-toothed; lacking a callus .... **4. Benthamia**
9. Petals, bifid or rarely entire; lip deeply 3-lobed, the lateral lobes linear .... **6. Habenaria**
9. Petals entire; lip 3-lobed but lobes not linear 10
10. Flowers large, pure white; lateral sepals very oblique at base and forming a distinct mentum; lip with a broad keel in the basal part; lip midlobe very fleshy, tongue-shaped and recurved; stigma lobes fleshy, clavate .... **7. Megalorchis**
10. Flowers coloured, rarely white; lip not as above; stigma lobes not as above 11
11. Flowers bright orange, anther more than 3 times as tall as broad ....
    .... **8. Platycoryne**
11. Flowers not as above 12
12. Plants with a single leaf borne in the middle of the stem, persistent for two years or more; viscidia enclosed in a bursicle .... **10. Physoceras**
12. Plants with one or more basal or cauline, annual leaves; viscidia free, not enclosed in a bursicle .... **9. Cynorkis**

Key V. Terrestrial plants with creeping fleshy rhizomes and a removable anther-cap

1. Sepals and petals fused at the base .... **15. Cheirostylis**
1. Sepals and petals free at the base 2
2. Lip united to column for most of its length; lateral sepals and lip
   gibbose at base .... **18. Platylepis**
2. Lip free from column for nearly all its length; lateral sepals and lip
   not gibbose at base 3
3. Lip tapering to the tip; column well-developed .... **17. Goodyera**
3. Lip T-shaped with broadly spreading apical lobes; column reduced .... **16. Zeuxine**

Key VI. Terrestrial plants with short to elongate tough rhizomes and a removable anther-cap

1. Stems woody; inflorescences lateral and terminal .... **19. Corymborkis**
1. Stems not woody, fleshy and pseudobulbous or reduced 2
2. Leaf solitary, appearing after the plant has flowered; plant with an
   underground tuber .... **20. Nervilia**
2. Leaves 1–several, produced with the flowers; plant with fleshy stems
   or pseudobulbs, never tubers 3
3. Flowers large, turning blue when bruised 4
3. Flowers small, not turning blue when bruised 6
4. Lip 4-lobed with a basal warty callus; column short, fleshy, fused to
   the lip .... **26. Calanthe**
4. Lip 3-lobed or entire, lacking a warty callus; column elongate, free or
   fused at base only to the lip 5
5. Lip almost as long as broad, with a short fleshy, bilobed basal callus ....
   .... **28. Gastrorchis**
5. Lip longer than broad, with linear callus ridges .... **27. Phaius**
6. Lip lacking a spur; with or without pseudobulbs 7
6. Lip with a basal spur or mentum, sometimes very short; pseudobulbs
   or tuberous rhizomes present 9
7. Flowers large, yellow-green with a red lip spotted with black ....
   .... **54. Cymbidiella flabellata**
7. Flowers small, not coloured as above 8
8. Lip lowermost in the flower, elongate, often reflexed in middle and
   with a callus at the base; column slender, elongate .... **23. Liparis** (in part)
8. Lip often uppermost in the flower, flat; column very short .... **24. Malaxis**
9. Flowers non-resupinate, not opening widely, lip obovate with a
   terminal verrucose-papillose callus .... **31. Imerinaea**
9. Flowers resupinate, usually opening widely; lip usually 3–4-lobed
   with a basal or central callus 10
10. Rhizomes or pseudobulbs tuber-like, underground, many-noded;
    lip 3-lobed .... **51. Eulophia**
10. Pseudobulbs ovoid to fusiform, often 1-noded; lip 3-or-more
    lobed, usually 4-lobed .... **52. Oeceoclades**

Key VII. Epiphytic or lithophytic plants with terminal inflorescences

1. Stems pseudobulbous; leafy in apical part only 2
1. Stems not pseudobulbous, leafy along length 3
2. Flowers with a distinct mentum (chin) formed by the lateral sepals
   and column-foot . . . . **30. Polystachya**
2. Flowers lacking a mentum (chin) . . . . **23. Liparis** (in part)
3. Leaves bilaterally flattened, iris-like and imbricate; flowers minute,
   in whorls in a dense spike . . . . **25. Oberonia**
3. Leaves dorsi-ventrally flattened, thin-textured, twisted at base to
   lie in one plane; flowers larger, not in whorls . . . . **29. Agrostophyllum**

Key VIII. Epiphytic or lithophytic plants with lateral inflorescences and pseudobulbs

1. Leaves 1–2, leathery; pseudobulbs 1-noded . . . . **32. Bulbophyllum**
1. Leaves 3 or more, pleated; pseudobulbs usually several-noded 2
2. Lip with a distinct spur at the base 3
2. Lip lacking a spur at the base 4
3. Sepals and petals yellow, spotted with brown; roots branching,
   sometimes with erect, sharp branches . . . . **53. Graphorkis** (in part)
3. Sepals and petals not spotted; erect roots absent . . . . **56. Paralophia**
4. Pseudobulbs 4-angled in cross-section, silvery; sepals and petals
   yellow heavily blotched with brown . . . . **57. Grammangis**
4. Pseudobulbs not angular, green; flowers with concolorous sepals
   and petals 5
5. Pseudobulbs clustered; flowers small . . . . **53. Graphorkis** (in part)
5. Pseudobulbs well-spaced on an elongated rhizome 6
6. Flowers almost circular in outline, white to rose-purple, often very
   large . . . . **55. Eulophiella**
6. Flowers not circular in outline, green with a green or red lip, spotted
   with black . . . . **54. Cymbidiella**

Key IX. Epiphytic or lithophytic plants with lateral inflorescences and short to long stems, never pseudobulbous

1. Plants leafless; roots photosynthetic . . . . **34. Microcoelia**
1. Plants with green leaves, rarely leaves falling before flowering 2
2. Sepals and petals spotted; lip fleshy, with a very short basal spur . . . . **33. Acampe**
2. Sepals and petals concolorous; lip of similar texture to the sepals
   and petals, usually with a long spur at the base 3
3. Rostellum elongate, entire or deeply bifid or bilobed, longer than
   the auricles of the column-apex 4
3. Rostellum usually 3-lobed, the midlobe often obscure or tooth-like,
   shorter than the lateral auricles 7
4. Rachis not articulated; pedicel not articulated at the base; lip obscurely
   4-lobed, papillose, white with a green blotch in the throat . . . . **35. Beclardia**

4. Rachis articulated; pedicel articulated just above the base; lip entire
   or 3-lobed, not papillose 5
5. Lip trilobed; rostellum bifid; each pollinia with its own stalk and
   viscidium .... **37. Angraecopsis**
5. Lip entire or rarely obscurely trifid 6
6. Sepals less than 4 mm long; spur shorter than the pedicel and
   ovary .... **38. Microterangis**
6. Sepals 6 mm or more long; spur usually longer than the pedicel and
   ovary .... **36. Aerangis**
7. Column-foot greatly inflated and bearing a spur; lip articulated to
   apex of column-foot 8
7. Column-foot neither inflated nor spurred; lip spurred at base 9
8. Roots thin, wiry; flowers translucent pale yellow, off-white or green;
   lip not fimbriate on margin; spur less than twice the length of the
   lip, often much shorter .... **39. Aeranthes**
8. Roots thick, not wiry; flowers white with a green throat to the lip;
   lip margin fimbriate; spur more than 12 cm long .... **40. Erasanthe**
9. Inflorescence cylindrical, with many, densely arranged flowers ..... **41. Lemurorchis**
9. Inflorescence not as above, flowers usually well-spaced on rachis 10
10. Column elongated and thickened at the base into two elongated
    horizontal arms on either side of the mouth of the spur 11
10. Column not elongated at the base; sides of spur not thickened 12
11. Lip 3-lobed or 3-sinose, or rarely entire; lateral sepals and petals
    free at the base from the sides of the spur-mouth .... **42. Neobathiea**
11. Lip entire; lateral sepals and petals adnate with the lateral margins
    of the spur mouth .... **43. Jumellea**
12. Lip entire, concave at the base and sides enclosing the column ..... **44. Angraecum**
12. Lip 3–5-lobed 13
13. Lip uppermost in flower; mouth of the spur in front of the concave
    base of the lip 14
13. Lip lowermost in flower; stem often elongate and plant climbing or
    scrambling 18
14. Lip trifid at summit only; flowers white; sepals more than 1 cm long 15
14. Lip 3-lobed towards the base; flowers small and yellow 17
15. Lip hairy within, rolled into a tube, 3 cm or more long, the midlobe
    larger than the side lobes; inflorescence 1–3-flowered; plant stemless .... **45. Ambrella**
15. Lip glabrous; inflorescence 10-or-more-flowered; stem well-developed 16
16. Lip cornet-shaped, long-apiculate; stem slender, branching; leaves
    6 cm or less long .... **47. Oeoniella** (in part)
16. Lip more or less flat; stem robust, not branching; leaves 9 cm or
    more long (often much longer) .... **48. Sobennikoffia**
17. Lip margins erose or toothed, the side lobes attached to the column
    forming the sides of the spur-mouth .... **50. Oeonia** (in part)
17. Lip margins entire, the side lobes free from the column .... **46. Lemurella**
18. Lip 3-lobed, the midlobe much smaller than the side lobes .... **47. Oeoniella**
18. Lip 3–5-lobed but the midlobe as large as the side lobes 19
19. Petals lobed; lip often T-shaped, inserted on the sides of the spur-
    mouth .... **49. Cryptopus**
19. Petals simple; lip 3–5-lobed, the claw embracing the column .... **50. Oeonia** (in part)

1. VANILLA

A genus of about 100 species, pantropical. Seven species in Madagascar. Scrambling or climbing vines, with green or grey smooth or warty stems. Leaves present or reduced to scales (appearing leafless). Inflorescences lateral, short racemes, producing several flowers in succession. Flowers large, white or yellow, marked on lip with apricot, or dark red. Sepals and petals similar, oblanceolate. Lip funnel-shaped, entire or 3-lobed, often enclosing column; callus of hairs or papillae. Column elongate, clavate; pollinia powdery.

Key to species of *Vanilla*

1. Plants with green leaves 2
1. Plants leafless but with scales 4
2. Sepals and petals 4–7 cm long .... **1.3. V. planifolia**
2. Sepals and petals 2.5 cm long or less 3
3. Lip callus obsolete .... **1.1. V. coursii**
3. Lip callus hairy, yellow .... **1.2. V. françoisii**
4. Flowers white 5
4. Flowers yellow 6
5. Sepals 2.5–3 cm long .... **1.4. V. decaryana**
5. Sepals 5–6.5 cm long .... **1.5. V. madagascariensis**
6. Flowers with a deep red mark on lip; lip papillose; column 2 cm long .... **1.6. V. humblotii**
6. Flowers with an orange mark on the lip; lip not papillose; column 2.7–3 cm long .... **1.7. V. perrieri**

1.1. **V. coursii** H.Perrier

Flowers c. 2.5 cm long; lip with an obsolete callus at the base, carrying hairs. Toliara. In humid, low-elevation forest, on sandy soil; 200–300 m. Fl. January–February.

1.2. **V. françoisii** H.Perrier

Stem slender and carrying small leaves; leaves 2–2.5 × 0.6–0.9 cm; flowers solitary, small, whitish green, lip tinted rose and purple on side margins, the callus hairs yellow; sepals c. 2 cm long; lip 18–20 mm long, rolled and crenulate at the front edge. Toamasina. In coastal forest. Fl. February.

1.3

1.5

1.3. **V. planifolia** Andrews

(p. 35, top)

Plant a scrambling liana; leaves large, elliptic-oblong, acute to apiculate; flowers large, green with a yellowish lip; sepals and petals 4–7 cm long. Antsiranana, Toamasina, occasionally elsewhere in wet lowlands to 300 m. Commercial vanilla. Native of Mexico but widely cultivated for its seed-pods, especially in the NE of the island. Fl. October–December.

1.4. **V. decaryana** H.Perrier

Stem brownish green, warty; inflorescence 20–40-flowered; flowers white; sepals 2.5–3 cm long; lip 3–3.5 cm long, central crests carrying thick hairs from the base to the upper quarter; column 1.4–1.6 cm long; fruit up to 13 cm long. Mahajanga, Toliara. In dry, deciduous scrub, among *Didieraceae*; up to 700 m. Fl. October–January.
LOCAL NAME *Vahy amalona*, said to have aphrodisiac properties.

1.5. **V. madagascariensis** Rolfe

(p. 35, bottom)

Stems green, usually smooth; inflorescence 10–20-flowered; flowers white, the lip tinted red-pink; sepals and petals 5–6.5 cm long; lip 4–5 cm long, the disc carrying 2 crests that bear long fleshy hairs; column 2–2.5 cm long; fruits 15–20 cm long when mature. Antsiranana, Mahajanga, Toliara. In coastal forest and humid evergreen forest; in dry forests in the east; up to 800 m. Fl. June–October.
LOCAL NAME *Amalo*, said to be an aphrodisiac.

1.6. **V. humblotii** Rchb.f.

Flowers yellow with a deep red mark on lip; lip covered with large papillae and hairs, front edges fringed; column 2 cm long; staminodes with the margins crenulate. Antsiranana(?). Also in the Comoros. On rocks and low-elevation vegetation; up to 200 m. Fl. November–December.

1.7. **V. perrieri** Schltr.

(opposite, top)

Flowers bright yellow with an orange throat; lip not papillose, with just a scattering of hairs in the lower half; column 2.7–3 cm long. Mahajanga, Toliara, Morondava. In seasonally dry, deciduous woods on sandy soils; c. 200 m. Fl. October, January–February.

1.7

3.1

3.2

2. GALEOLA

A genus of some 10 species in tropical Asia across to E Australia and the Solomon Islands. A single species in Madagascar. Large scrambling saprophytic, brown to buff or off-white vines, with scales but lacking leaves. Roots adventitious. Inflorescences lateral, branching, many-flowered. Flowers brown to buff-coloured; sepals and petals similar; lip funnel-like with a crimped margin.

2.1. **G. humblotii** Rchb.f.

Toamasina, Toliara; also in the Comoros. In lowland forest; up to 300 m. Fl. October–February.

3. BRACHYCORYTHIS

A genus of about 23 species in tropical and South Africa and Madagascar. Two species in Madagascar. Terrestrials growing from fusiform or ellipsoidal tubers. Leaves many along stem, ovoid to lanceolate, acute to acuminate. Inflorescence terminal, densely many-flowered. Bracts leafy. Flowers white, purple or pink. Sepals and petals subsimilar; lateral sepals oblique. Petals adnate to sides of column. Lip bipartite; basal part shortly spurred or saccate; apical part flat, entire or 3-lobed. Column erect; stigma hollow; anther loculi parallel.

Key to species of *Brachycorythis*

1. Lip 3-lobed, with a short basal spur . . . . **3.1. B. disoides**
1. Lip bipartite, with a boat-shaped base . . . . **3.2. B. pleistophylla**

3.1. **B. disoides** (Ridl.) Kraenzl.   *(p. 37, bottom left)*

Plant stout, 15–30 cm tall; leaves ovate, acuminate, 2.5–5 cm long; flowers crimson, pink or white, the lip white, heavily spotted with purple; lip oblong, obtuse, obscurely 3-lobed, crenulate, with a ridge in the middle; spur short, 5 mm long and 2 mm wide, flattened. Antananarivo, Fianarantsoa. In grassland, seepages and on steep slopes amongst rocks; 1400–2000 m. Fl. December–July.

3.2. **B. pleistophylla** Rchb.f.   *(p. 37, bottom right)*
(syn. *Brachycorythis perrieri* Schltr.)

Plant up to 70 cm tall terrestrial; leaves lanceolate, up to 5 cm long; flowers mauve-purple to purple or crimson; lip bipartite, 10–15 mm long; basal part of lip boat-shaped; apical part bilobed. Antsiranana, Mahajanga, Toamasina; also in E and W Africa. In grassland, dry woods and woodland margins; 1400–1500 m. Fl. October–January.

4. BENTHAMIA

A genus of about 30 species, endemic to Madagascar and the Mascarene Islands, one species in E Zimbabwe. Terrestrial or rarely epiphytic herbs growing from ovoid tubers. Leaves 1–several, basal or cauline. Inflorescences spicate, laxly to densely many-flowered, sometimes second. Flowers small, white, yellow, green or reddish orange. Sepals and petals free, subsimilar. Lip 3-lobed or entire, sometimes with a basal callus, shortly spurred at base; spur cylindrical, clavate, saccate or scrotiform. Column with parallel anther locules; stigmas very short.

Key to species of *Benthamia*

1. Spur elongated, as long as the lip or longer 2
1. Spur saccate, globular or purse-shaped, very short 10
2. Leaves developing after flowering 3
2. Leaves developed at flowering 4
3. Lip oblong, very obscurely 3-lobed; spur 3 mm long, bifid at tip .... **4.8. B. exilis**
3. Lip 3-lobed with ligulate lobes; spur 2.5 mm long, acute .... **4.9. B. praecox**
4. Leaf solitary 5
4. Leaves 2 or more 7
5. Inflorescence second; sepals ligulate, 4–5 mm long; lip 3-lobed, the side lobes small, midlobe obate, acute; spur 6–8 mm long .... **4.1. B. glaberrima**
5. Inflorescence not second or only obscurely so; sepals 2–3 mm long, not ligulate; lip entire; spur 2.5 mm long 5
6. Leaf radical, petiolate, ovate, 1.3–3 cm long; lip concave, somewhat slipper-shaped, 2.5 mm long; rachis 2–4 cm long .... **4.2. B. longecalceata**
6. Leaf basal, oblong, 6–10 cm long; lip ovate-elliptic, flat, 3–3.5 mm long; rachis 8–15 cm long .... **4.3. B. monophylla**
7. Leaves 3–6 in lower part of stem .... **4.4. B. bathieana**
7. Leaves 2, basal or radical 8
8. Leaves radical; inflorescence slightly second; lip lozenge-shaped, expanded and 3-lobed near the tip .... **4.7. B. rostrata**
8. Leaves basal; inflorescence not second; lip entire or 3-lobed with midlobe twice the size of the side lobes 9
9. Plant up to 30 cm tall; leaves suborbicular, 6 × 5.5 cm; inflorescence up to 10 cm long; lip 3-lobed, side lobes obovate-cuneate; midlobe subquadrate, twice as large as the side lobes .... **4.5. B. leandriana**
9. Plant 45–65 cm tall; leaves oblong, 10–17 × 2.5–3.5 cm; inflorescence up to 25 cm long; lip rhombic, entire; spur subcylindrical, 3.0–3.5 mm long .... **4.6. B. perularioides**
10. Lip entire 11
10. Lip 3-lobed, sometimes obscurely so 15
11. Leaf solitary 12
11. Leaves 2 or more 13
12. Leaf grass-like, six times or more longer than broad; plant lacking separate sterile and fertile stems .... **4.10. B. calceolata**
12. Leaf oblong-lanceolate, 5 times longer than broad or less; plant with separate sterile and fertile stems .... **4.11. B. perfecunda**
13. Leaves ovate or narrowly ovate, cuspidate; lip ovate, entire .... **4.12. B. catatiana**
13. Leaves oblong-lanceolate, 13–15 cm long; lip concave, slipper-shaped 14

4.1. **B. glaberrima** (Ridl.) H.Perrier (*p. 45, top left*)

(syn. *Holothrix glaberrima* Ridl.; *Peristylus glaberrima* (Ridl.) Rolfe; *Platanthera glaberrima* (Ridl.) Kraenzl.; *Habenaria glaberrima* (Ridl.) Schltr.; *Rolfeella glaberrima* (Ridl.) Schltr.)

Plant 30–60 cm tall; leaf narrowly linear-lanceolate; spike densely many-flowered; flowers sulphur-yellow, somewhat secund; sepals ligulate, 4–5 mm long; lip ovate, 3-lobed; side lobes small; midlobe ovate, acute; spur cylindrical, 6–8 mm long. Antananarivo, Fianarantsoa; fairly common in the whole central highlands in grassland, rocky outcrops and dry scrub; 1500–2200 m. Fl. February–May.

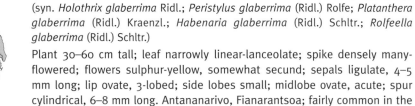

4.2. **B. longecalceata** H.Perrier

Plant 25–40 cm tall; leaf radical, ovate, acute or acuminate, 1.3–3 × 1–1.5 cm long, petiolate; spike densely 10–20-flowered, 2–4 cm long; sepals ovate, 2–3 mm long; petals subfalcate, obtuse; lip 2.5 mm long, slipper-shaped; spur cylindrical, 2.5 mm long. Antsiranana. On rocks and in peaty depressions; 1500–2000 m. Fl. March–April.

4.3. **B. monophylla** Schltr.   (p. 41)

Plant 15–30 cm tall; leaf basal, oblong, 6–10 × 2–3.5 cm; spike subdensely many-flowered, subsecund, 8–15 cm long; flowers yellow; sepals oblong, 3 mm long; lip entire, ovate-elliptic, 3.3 × 1.75 mm; spur narrowly conical, 2.5 mm long. Fianarantsoa, Toliara. Montane *Philippia* scrub and grassland, moss- and lichen-rich forest, and rock crevices; 1800–2600 m. Fl. March.

4.4. **B. bathieana** Schltr.   (p. 45, top right)

Plant up to 70 cm tall; leaves 3–6 in lower part of stem, ovate or elliptic, up to 16 × 7.5 cm; spike up to 30 cm long, densely many-flowered; flowers yellow; sepals oblong, 3.5–4 mm long; lip 4 × 3 mm, clearly 3-lobed in middle; side lobes triangular, obtuse; midlobe linear-ligulate; spur cylindrical, 4 mm long. Antananarivo, Fianarantsoa; also in the Mascarenes. Near shaded, wet rocks; c. 2000 m. Fl. February–April.

4.5. **B. leandriana** H.Perrier

Plant up to 30 cm tall; leaves 2, basal, subopposite, sessile, suborbicular, 6 × 5–5.5 cm long; spike sublaxly many-flowered, up to 10 cm long; sepals 3–3.5 mm long; petals obovate, dilated at apex; lip 3-lobed; side lobes obovate-cuneate; midlobe subquadrate, twice as large as the side lobes; spur clavate, 5 mm long. Toliara (Bemaraha), Morondava. On calcareous rocks in tsingy area; 150–200 m. Fl. December–January.

4.6. **B. perularioides** Schltr.

Plant 45–65 cm tall; leaves 2, near base, with several sheaths above, oblong, 10–17 × 2.5–3.5 cm; spike up to 25 cm long, densely many-flowered; flowers greenish; sepals narrowly oblong, 3.5 mm long; lip rhombic, entire; spur subcylindrical, 3–3.5 mm long. Antananarivo, Mahajanga. Grassland; 1200–2500 m. Fl. March–April, November.

4.7. **B. rostrata** Schltr.

Similar to *B. perularioides* but differs by the 2 radical leaves, the spike which is slightly unilateral, the 4 mm long perianth segments, the lozenge-shaped lip that is expanded near the middle and 3-lobed, and the longer (5 mm long) filiform spur. Antananarivo, Fianarantsoa. Shaded rocky outcrops, grassland; 1500–2000 m. Fl. February–March.

4.8. **B. exilis** Schltr.

Plant 13–28 cm tall; leaves not developed when flowering; spike up to 8 cm long, many-flowered; flowers yellowish or greenish; sepals oblong, 3.5 mm long; lip narrowly ovate, 4 mm long, obscurely 3-lobed; spur 3 mm long, cylindrical and bifid at the apex. Fianarantsoa. Montane, ericaceous scrub; 2000–2400 m. Fl. January–April.

4.8a. **B. exilis** var. **tenuissima** Schltr.

Differs from the typical form by its more slender, elongated, 60 cm long stem and yellow flowers with brown sepals. Antananarivo, Fianarantsoa. Montane, ericaceous scrub; 2000–2500 m. Fl. February–March.

4.9. **B. praecox** Schltr.　　　　　　　　　*(p. 45, bottom left)*

Slender terrestrial or lithophyte, up to 30 cm tall; leaves acuminate-aristate, not developed at flowering time; spike 15 cm long, subsecund, many-flowered; flowers brownish-green; sepals elliptic-oblong, 3 mm long; lip broadly cuneiform, 3-lobed above the middle; side lobes ligulate; midlobe similar; spur narrowly cylindrical, 2.5 mm long. Fianarantsoa. Rocky outcrops of granite and gneiss; high-elevation moss- and lichen-rich forest; ericaceous bush, with *Philippia* dominant; 1000–1950 m. Fl. September–January.

4.10. **B. calceolata** H.Perrier

Plant 30–60 cm tall with separate sterile and flowering stems; leaf oblong-lanceolate, acute, up to 14 × 3.2 cm; spike slender, up to 17 cm long; flowers very small; sepals 1.8–2 mm long, dorsally keeled; lip slipper-shaped, entire, slightly warty on outer side; spur scrotiform, 1 mm long, contracted at the base. Toamasina, Toliara. Lowland marshland; sea level to 500 m. Fl. June–August.

4.11. **B. perfecunda** H.Perrier

Plant 30–70 cm tall; leaf in middle of stem, grass-like, 5–15 × 0.8 cm; spike 6–15 cm long; flowers similar to *B. macra* (p. 46) and to *B. verecunda* (p. 50), but lip slipper-shaped; sepals 1.5 mm long; lip entire; spur purse-shaped, 0.6 mm long, slightly bilobed. Antsiranana. Mossy, montane forest in peaty soil; 1800–2000 m. Fl. April.

4.12. **B. catatiana** H.Perrier

Epiphyte, 30–40 cm tall; leaves 2–3, oblong-lanceolate, 13–15 × 2–4 cm; spike up to 12 cm long; flowers yellow; sepals 4.5 mm long; lip slipper-shaped, obtuse, 4.5–5 mm long; spur scrotiform, 1 mm long, obtuse; ovary 3 mm long. Toliara. Lowland marshland. Fl. unknown.

4.13. **B. misera** (Ridl.) Schltr.

(syn. *Habenaria misera* Ridl.)

Close to *B. spiralis* (p. 47). Leaves 2, ovate or ovate-lanceolate, cuspidate; flowers very small; sepals ovate-lanceolate, obtuse; lip entire, ovate; spur saccate. Antananarivo, Antsiranana, Fianarantsoa. In grassland; amongst mosses; in marshy areas; 900–1200 m. Fl. September, March–April.

4.14. **B. nigrescens** Schltr.

Plant up to 30 cm tall; leaves 2–3, basal, oblong-lanceolate, up to 9 × 2.8 cm; spike up to 18 cm long, very densely many-flowered; flowers greenish; sepals 5–5.5 cm long; lip ligulate, obtuse, 5.2–5.7 mm long; spur scrotiform; spike obscurely unilateral; lip ligulate, obtuse, c. 5 × 1.5 mm, thickened and contracted at the tip; spur scrotiform, a little wider than long. Antananarivo, Toamasina, Antsiranana. Shaded and wet rocks, grassland and as an epiphyte on moss- and lichen-covered trees in evergreen forest; 2000–2400 m. Fl. October–April.

4.14a. **B. nigrescens** subsp. **decaryana** H.Perrier

Differs by its more unilateral spike, the narrower lip, and the size of the anther. Mahajanga. Humid, highland forest; c. 1700 m. Fl. April.

4.1

4.4

4.9

4.16

4.14b. B. nigrescens subsp. humblotiana H.Perrier

Differs by its obtuse leaves, the denser spike and the flowers facing all directions; lip is thickened from the base; spur smaller. Antsiranana. Mossy forest; c. 2000 m. Fl. April.

4.15. B. dauphinensis (Rolfe) Schltr.

(syn. *Habenaria dauphinensis* Rolfe)

Plant up to 50 cm tall; leaf linear, almost 10 times longer than wide, 15 × 0.6–0.7 cm; spike many-flowered; sepals broadly elliptic-oblong, 1.5 mm long; lip very concave and fleshy, obscurely 3-lobed; spur 1.5 mm long, flattened and almost bidentate at the tip. Toliara. Lowland marshes. Fl. June.

4.16. B. cinnabarina (Rolfe) H.Perrier   *(p. 45, bottom right)*

(syn. *Habenaria cinnabarina* Rolfe; *Benthamia flavida* Schltr.; *B. perrieri* Schltr.)

Plant 15–35 cm tall; leaves 4–5, linear to linear-lanceolate, 6–12 × 0.6–1.2 cm; spike 5–16 cm long; flowers cinnabar-orange; sepals 4 mm long; lip oblong, 3-lobed, the lateral lobes very obtuse; spur scrotiform and a little flattened, 0.8 mm long. Antananarivo, Antsiranana, Mahajanga, Fianarantsoa. Marshes, peat-bogs and wet rocks; 1500–2200 m. Fl. February–May.

4.17. B. elata Schltr.

Plant 40–75 cm tall; leaves 2–6, ligulate or oblanceolate, up to 16 × 2 cm; spike cylindrical, up to 20 cm long; sepals 2–3 mm long; lip oblong, 3-lobed in the upper part, the side lobes narrowly lanceolate-ligulate, the middle one ligulate-obtuse, almost twice as big as the side lobes; spur semi-oblong, saccate. Fianarantsoa. Wet embankments; 1000–1200 m. Fl. June–August.

4.18. B. macra Schltr.

Plant up to 45 cm tall; leaves 3 along stem, narrowly linear, acute, up to 13 × 0.8 cm; spike subsecund, many-flowered; flowers brownish green; sepals 4 mm long; lip oblong, 3-lobed in upper part, 4 mm long; spur oblong-saccate. Fianarantsoa. Montane, ericaceous scrub; 1100–2500 m. Fl. February.

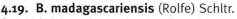

4.19. **B. madagascariensis** (Rolfe) Schltr.

(syn. *Holothrix madagascariensis* Rolfe; *Habenaria madagascariensis* (Rolfe) Schltr.; *Platanthera madagascariensis* (Rolfe) Kraenzl.; *Peristylus madagascariensis* (Rolfe) Rolfe; *P. macropetalus* Finet)

Plant 25–50 cm tall; leaves 3–5, linear, 8–15 × 0.3–0.5 cm; spike 5–10 cm long, subsecund; flowers white; sepals 7 mm long; lip 8–9 mm long, narrow, 3-lobed in apical part; lobes equal in length, 3 mm long; spur saccate, 1 mm long. Toamasina, Toliara. Marshes; sea level–500 m. Fl. December–July. LOCAL NAME *Soazombitra*.

4.20. **B. melanopoda** Schltr.

Plant 60–78 cm tall; leaves 3–4 in middle of stem, linear, up to 23 × 1.3 cm long; spike 10 cm long, cylindrical; flowers greenish yellow; sepals 3 mm long; lip ovate, 3 mm long; midlobe very fleshy; lateral lobes are also larger; spur saccate. Antananarivo, Fianarantoa. Grassland on peaty soil; 1500–2000 m. Fl. March–April.

4.21. **B. procera** Schltr.

Plant up to 50 cm tall; leaves 4–5, linear, acute, up to 18 × 1 cm; spike densely many-flowered, up to 20 cm long; flowers yellowish; sepals 2.5 mm long; lip oblong-ligulate, 3 × 1.8 mm, 3-lobed in the upper part, the middle lobe thickened, longer than the laterals which are obscure and obtuse; spur wider at the apex than long, 0.8 × 1.2 mm, purse-shaped. Toamasina. Shaded rocks; cool, shady spots in humid evergreen forest; river margins; 250–800 m. Fl. November–December.

4.22. **B. spiralis** (Thouars) A.Rich.

(syn. *Satyrium spirale* Thouars; *Spiranthes africana* Lindl.; *Habenaria spiralis* (Thouars) A.Rich.; *Herminium spirale* (Thouars) Rchb.f.; *Peristylus spiralis* (Thouars) S.Moore; *Habenaria minutiflora* Ridl.; *H. dissimulata* Schltr.; *Benthamia minutiflora* (Ridl.) Schltr.; *B. spiralis* var. *dissimulata* (Schltr.) H.Perrier)

A common and variable species. Plant up to 60 cm tall; leaves 3–5, ligulate or linear-lanceolate, 7–17 × 0.5–1.5 cm; spike 7–30 cm long; flowers greenish with yellow tips; sepals 2–3 mm long; lip 2–3 × 1–1.8 mm, 3-lobed at the apex, the middle lobe thicker and longer than the laterals, which are obscure and sometimes even missing; spur subrectangular, purse-shaped, entire or a little emarginate at the tip, c. 1 mm long. Antananarivo, Fianarantsoa, Toamasina, Toliara; also in the Mascarenes. Marshland, forest, woodland margins, shaded wet rocks; 100–1500 m. Fl. January–December.

4.23. **B. cuspidata** H.Perrier

Plant up to 45 cm tall; leaves 4 along stem, broadly obovate, cuspidate-obtuse, largest 14 × 6 cm; spike up to 25 cm long, cylindrical and densely flowered; sepals 4.5–6 mm long; lip 3-lobed; side lobes a rounded tooth; midlobe 5–6 times longer, obtuse; spur scrotiform, 1 mm long; staminodes very long. Toamasina. Lowland marshes; sea level–200 m. Fl. September.

4.24. **B. herminioides** Schltr.

Plant up to 20 cm tall; leaves 3, elliptic-lanceolate or lanceolate, up to 5 × 2.5 cm; spike 5–8 cm long; flowers whitish; sepals 2.75 mm long; lip rhombic, 2.75 × 2.5 mm, 3-lobed in the upper third; spur very short, subglobular. Antananarivo, Antsiranana. Mainly found at high elevation amongst rocky outcrops, in *Philippia* scrub and in lichen-rich forest; 1000–2000 m. Fl. February–March.

4.24a. **B. herminioides** subsp. **angustifolia** H.Perrier

Differs from the other subspecies by the more slender shape and the shorter, linear-acute leaves, c. 5 × 0.5 mm, and the distinct stigma. Antsiranana. Shaded, montane, ericaceous scrub; c. 2600 m. Fl. April.

4.24b. **B. herminioides** subsp. **arcuata** H.Perrier

Floral bracts at most equalling the ovary in length; lateral lobes of the lip narrow and acute; anther with the apicule higher, with central lobe of the rostellum rounded. Antsiranana. In recently burnt, montane, *Philippia* scrub; 2600–2800 m. Fl. April.

4.24c. **B. herminioides** subsp. **intermedia** H.Perrier

Flowers smaller than those of the other subspecies; lateral lobes of the lip obtuse and shorter, equalling half the size of the midlobe; rostellum with the central tooth narrower and obtuse; ovary 4 mm long. Antsiranana. Lichen-rich forest; c. 2000 m. Fl. April.

4.25. **B. humbertii** H.Perrier

Plant 50–70 cm tall, with two black sheaths at base of stem; leaves broadly ovate-lanceolate, 6–8 × 2–4 cm; spike 15–18 cm long, densely flowered; flowers yellow-green; sepals 4 mm long; lip broadly obovate, 5 × 4.5 mm, thick, 3-lobed in the upper third, the lobes short, obtuse, almost equal, the middle one a little wider; spur compressed, 1 × 2 mm. Fianarantsoa, Toliara. Humid, highland forest, amongst rocks; 1100–2000 m. Fl. November.

4.26. **B. majoriflora** H.Perrier   *(below, left)*

Robust epiphyte or terrestrial up to 60 cm tall; leaves 4–5 in lower half and middle of stem, ovate-lanceolate, acute, 15–25 × 3–5.5 cm; spike densely cylindrical, many-flowered, 10–18 cm long; sepals narrowly elliptic, 4.5–5 cm long; petals dilated at apex; lip 3-lobed in apical half, widely ventricose; side lobes falcate, acute; midlobe fleshy; spur purse-shaped, 1–1.5 mm long. Toamasina. In mossy forest; 500–1500 m. Fl. December.

4.26

4.28

4.27. B. nigrovaginata H.Perrier

Plant up to 80 cm tall, with two long black sheaths at base of stem; leaves 4, oblong-lanceolate, acute, 10–19 × 2–3 cm; spike 35–40 cm long; flowers greenish; sepals 4.5 mm long; lip very concave, 5 mm long, 3-lobed in apical part; midlobe 3 times longer than side lobes; spur saccate, 0.7 mm long. Fianarantsoa. River margins and riverine forest. Fl. October.

4.28. B. nivea Schltr.   (p. 49, right)

Epiphyte, 11–25 cm tall; leaves 3–4, oblong-lanceolate, acuminate, 5–14 × 2–3.5 cm; spike 6–13 cm long; flowers white; sepals 7–8 mm long; lip narrowly oblong, 8–9 × 3.5–4 mm; side lobes obtuse, 2.5 mm long; midlobe 3.5 mm long; spur saccate. Antananarivo, Toamasina. Mossy, montane forest; 2000–2200 m. Fl. February–March, September.

4.28a. B. nivea subsp. parviflora H.Perrier

Flowers smaller than typical variety; lip with narrow side lobes. Antsiranana. Mossy, montane forest; c. 2000 m. Fl. April.

4.29. B. verecunda Schltr.

Plant 35–50 cm tall; leaves 2 in lower part of stem, linear-lanceolate, 10–17 × 1.7–2.3 cm; spike elongate, up to 30 cm long, laxly many-flowered; flowers greenish, virtually secund; sepals 3.5 mm long; lip oblong, 3.5 × 2 mm, a little concave at the base, 3-lobed in the upper third, the middle lobe oval, subobtuse; spur almost globular, 1 mm across. Antsiranana. Montane, humid, evergreen forest, forested mountain-tops, amongst mosses; c. 1200 m. Fl. April.

5.2

5. TYLOSTIGMA

A genus of some 8 species, endemic to Madagascar. Terrestrial or rarely lithophytic plants growing from a bundle of fusiform roots. Leaves several, suberect, in lower part of stem, filiform to lanceolate, acute to acuminate. Inflorescence terminal, a narrowly cylindrical spike, many flowers; bracts as long as or longer than the flowers. Flowers minute, less than 3 mm long. Sepals and petals subsimilar, not opening widely. Lip 3-lobed; lobes triangular; spur very short, conical. Column very short.

Key to species of *Tylostigma*

1. Leaves towards base of plant ....**5.1. T. tenellum**
1. Leaves in middle of stem 2
2. Leaves linear-lanceolate ....**5.2. T. foliosum**
2. Leaves linear or almost filiform 3
3. Lip wrinkled at base 4
3. Lip not wrinkled, smooth at the base 5
4. Leaves 3–4, 2.5–3.5 mm broad ....**5.3. T. herminioides**
4. Leaf solitary in middle of stem, 2–2.5 cm broad ....**5.4. T. madagascariense**
5. Sepals keeled on reverse; petals 3-nerved ....**5.5. T . hildebrandtii**
5. Sepals not keeled; petals 2-nerved or less 6
6. Lip with a small basal callus; flowers yellow ....**5.6. T. filiforme**
6. Lip lacking a basal callus; flowers white or green 7
7. Leaves 1–2; flowers white; lip broadly ovate-cordate ....**5.7. T. nigrescens**
7. Leaves 3; flowers green; lip broadly triangular ....**5.8. T. perrieri**

5.1. T. tenellum Schltr.

Lithophyte; plant small, 6–11 cm tall, with 3–4 linear, acuminate leaves, close together at the base; flowers 1 mm long, yellow-green; lip broadly ovate, obtuse, with a semi-rectangular speculum at the base. Antananarivo. Shaded and humid rocks; c. 1600 m. Fl. May, December.

5.2. T. foliosum Schltr.　　　　　　　　　　　　　*(p. 51 and opposite, left)*

Plant 20–40 cm tall; leaves linear-lanceolate, up to 18 × 0.6–1.2 cm wide; raceme densely many-flowered; floral bracts as long as the flowers; flowers nutant, 2 mm long, green, much shorter than the bracts; sepals 3 mm long; lip broadly ovate, obtuse, 2.5 mm long, with a speculum near the base. Fianarantsoa. Marshes in peaty and marshy soil, by rocks; 2000–2500 m. Fl. January–February.

5.3. T. herminioides Schltr.

Plant 15–25 cm tall; leaves 3–4 towards the middle of the stem, linear, acute, 7–15 × 0.25–0.35 cm; rachis 4–10 cm long; flowers green; sepals 2 mm long; lip broadly ovate, obtuse, obscurely 3-lobed in front, lip narrowed in the middle. Antananarivo. Marshes; c. 2200 m. Fl. March.

5.4. T. madagascariense Schltr.

Plant up to 15 cm tall; leaf borne in middle of stem, linear-lanceolate, acute, up to 16 × 2–2.5 cm; rachis up to 9 cm long; flowers greenish-yellow; sepals and petals suborbicular, 1.5–1.8 mm long; lip triangular, 1.75 mm long. Antananarivo. Marshes in peaty and marshy soil; c. 1500 m. Fl. January.

5.2

5.7

5.5. T. hildebrandtii (Ridl.) Schltr.
(syn. *Habenaria hildebrandtii* Ridl.)

Plant slender, 20–40 cm tall; leaves grass-like, basal; rachis 5–6 cm long; flowers yellowish, longer than the bracts; sepals 1 mm long; lip ovate-lanceolate, glabrous, without ridges. Antananarivo, Fianarantsoa. Marshes; 1500–2000 m. Fl. January–March.

5.6. T. filiforme H.Perrier

Leaves 3–4, subfiliform, inserted in upper part of stem, 4–5 × 0.15–0.2 cm. Rachis 15–20-flowered, 2–3 cm long. Flowers yellow; sepals 2–2.2 mm long. Lip obtusely slipper-shaped, 2 mm long, with a small callus at the base and with two specula ending in a transverse callus. Similar to *T. perrieri*, but with the lower sheaths without a blade and with a different lip structure. Antsiranana (Marojejy). Marshes and in marshy depressions; 1800–2000 m. Fl. March–April.

5.6a. T. filiforme subsp. bursiferum H.Perrier

Differs from the typical form in its smaller habit, white flowers and the base of the lip, which is transversely purse-shaped. Antsiranana (Mt Beandroko). Marshes; c. 1400 m. Fl. March–April.

5.7. T. nigrescens Schltr.   (p. 53, right)

Plant slender up to 60 cm tall; leaves 1–2, linear, up to 23 × 0.2–0.3 cm; flowers off-white; sepals 2 mm long; petals oblong; lip broadly ovate, cordate at base, 2 mm long, with an obsolete speculum at the base. Antananarivo. Marshes in peaty and marshy soil; in cypress plantations; 1500–2000 m. Fl. December–February.

5.8. T. perrieri Schltr.

Plant slender, 20–35 cm tall; leaves 3, linear, 5–10 × 0.1–0.25 cm; bracts as long as flowers; flowers green; sepals 2 mm long; petals broadly ovate; lip broadly triangular, obtuse, 2 mm long, with a kidney-shaped speculum at the base. Antananarivo, Fianarantsoa, Mahajanga. Marshes, in peaty and marshy soil; 1500–2100 m. Fl. April–July.

6. HABENARIA

A large cosmopolitan genus of perhaps 600 species; 29 species in Madagascar. Small to large terrestrials growing from ellipsoid to fusiform tubers. Stems leafy, sometimes at base or all along, rarely lacking leaves. Leaves linear, lanceolate or elliptic, not pleated. Inflorescence terminal, spicate or racemose, few–many-flowered, white, yellow or green. Dorsal sepal forming a hood, often with the petals. Lateral sepals similar and spreading. Petal simple or bilobed. Lip simple or 3-lobed, spurred at base; side lobes entire or lacerate-fimbriate; spur elongate, often clavate. Column erect; anther loculi parallel; stigmas stalked; pollinia 2, sectile.

Key to species of *Habenaria*

1. Plants saprophytic, lacking green leaves; lip not lobed . . . . **6.1. H. saprophytica**
1. Plants autotrophic, with green leaves; lip 3-lobed, rarely obscurely lobed 2
2. Petals entire 3
2. Petals bifid at base, sometimes only slightly 11
3. Spur 5 cm or more long 4
3. Spur 3.5 cm or less long 5
4. Sepals 6–7 mm long; spur 5.5 cm long . . . . **6.2. H. bathiei**
4. Sepals 15–20 mm long; spur 15–20 cm long . . . . **6.3. H. beharensis**
5. Plant up to 1.5 m tall, rhizomatous . . . . **6.9. H. praealta**
5. Plants 1 m or less tall; tuber-bearing 6
6. Flowers white 7
6. Flowers pale yellow, yellow or green 8
7. Leaves lanceolate, 9–15 cm long; sepals 9 mm long; spur incurved, cylindrical, 7–12 mm long . . . . **6.7. H. monadenioides**
7. Leaves linear-lanceolate, up to 5 cm long; sepals 5.5–6 mm long; spur filiform, 30 mm long . . . . **6.8. H. quartziticola**
8. Bracts and sepals ciliate-papillate; plant up to 1 m tall . . . . **6.5. H. decaryana**
8. Bracts and sepals glabrous; plant 60 cm or less tall 9
9. Flowers pale yellow; leaves more or less basal, narrowly linear, 3–5 mm broad . . . . **6.4. H. ambositrana**
9. Flower green or yellowish green; leaves arranged in lower part of stem, 7 mm or more broad 10
10. Leaves linear-lanceolate, up to 18 mm broad; sepals 4–5 mm long; lip side lobes broadest at tip and somewhat bifid-dentate . . . . **6.6. H. lastelleana**
10. Leaves tapering to tip, up to 10 mm broad; sepals 6–7 mm long; lip side lobes linear, entire . . . . **6.10. H. simplex**
11. Spur 4 cm or more long 12
11. Spur 3 cm or less long 15
12. Spur 7–12 cm long 13
12. Spur 6 cm or less long 14
13. Dorsal sepal 5–7 mm long; flowers 15 or more; anther canals 1.3 mm long . . . . **6.12. H. clareae**
13. Dorsal sepal 13–16 mm long; flowers 8–10; anther canals 4 or more mm long . . . . **6.13. H. leandriana**
14. Petal lobes and side lobes of lip upcurved like cow-horns; anther canals 12 mm long . . . . **6.11. H. cirrhata**
14. Petal lobes and side lobes of lip somewhat deflexed, upcurved at tip; anther canals 6–6.5 mm long . . . . **6.14. H. tianae**

15. Flowers gnat-like; dorsal sepal reflexed; lateral sepals very oblique with
    apex on dorsal margin . . . . **6.15. H. incarnata**
15. Flowers not gnat-like; dorsal sepal erect; lateral sepals not noticeably
    oblique with a terminal apex 16
16. Leaves 3, prostrate on ground . . . . **6.16. H. tropophila**
16. Leaves 4 or more, spreading to suberect along stem 17
17. Spur spirally twisted in the middle; flower stalk pubescent papillose . . . .
    . . . . **6.21. H. cochleicalcar**
17. Spur pendent to upcurved; flower stalk glabrous 18
18. Sepals pubescent or ciliate, 10 mm long 19
18. Sepals 8 mm or less long, glabrous 20
19. Flowers white; inflorescence densely up to 45-flowered . . . . **6.29. H. tsaratananensis**
18. Flowers green; inflorescence 7–8-flowered . . . . **6.27. H. nautiloides**
20. Leaves obovate, oblanceolate or broadly elliptic 21
20. Leaves lanceolate to broadly lanceolate 23
21. Leaves less than 2.5 times as long as broad; anterior petal lobe longer
    than the posterior lobe, filiform, 7 mm long; lip side lobes filiform . . . . . **6.23. H. demissa**
21. Leaves 3 time longer than broad or more; anterior petal lobe, smaller than
    the posterior lobe, often reduced to a tooth; lip side lobes shorter than
    the midlobe, sometimes tooth-like 22
22. Lip side lobes longer than the midlobe; petal anterior lobe narrowly lanceolate . . . .
    . . . . **6.18. H. alta**
22. Lip side lobes often dentate; petal anterior lobe dentate . . . . **6.20. H. boiviniana**
23. Plant 60 cm or more tall; leaves 10–13 24
23. Plant 50 cm or less tall; leaves 4–8 25
24. Leaves 2–3 cm broad; petal anterior lobe tooth-like, much shorter than the
    posterior lobe; spur 12–15 mm long . . . . **6.17. H. acuticalcar**
24. Leaves up to 1 cm broad; petals deeply bipartite, the lobes 4.5 mm or more
    long; spur subfiliform, 6 mm long . . . . **6.24. H. ferkoana**
25. Stigma lobes 10 mm long . . . . **6.22. H. conopodes**
25. Stigma lobes 3 mm or less long 26
26. Leaves sheathing at the base; spur 12 mm or more long 27
26. Leaves not sheathing at the base; spur 12 mm or less long 28
27. Dorsal sepals 3 mm long . . . . **6.25. H. foxii**
27. Dorsal; sepal 4–4.5 mm long . . . . **6.26. H. hilsenbergii**
28. Leaves ovate-lanceolate, 2 cm broad, grading into the bracts; spur pendent . . . .
    . . . . **6.28. H. truncata**
28. Leaves lanceolate, 1–1.5 cm broad, not grading into the bracts; spur upcurved
    over the ovary . . . . **6.19. H. arachnoides**

6.1. **H. saprophytica** Bosser & P.J.Cribb

Plant 8–30 cm tall, achlorophyllous, aphyllous, glabrous, whitish; inflorescence 3–14-flowered; dorsal sepal 2.5–3.5 mm long; laterals 3.5–5.5 mm long; lip ligulate, 4.5–6 mm long; spur 8–9 mm long. Antsiranana. Lower montane forest; 610 m. Fl. February.

6.2

6.2. H. bathiei Schltr. (p. 57)

Plant up to 45 cm tall; leaves several, linear-lanceolate, suberect, gradually shorter up the stem, 14–16 × 1–1.3 cm; inflorescence many-flowered, cylindrical; flowers white with green tips to lip lobes; dorsal sepal 6–7 mm long; petals ovoid-deltoid; lip deeply 3-lobed; side lobes upcurved; spur filiform, 5–5.5 cm long, pale green. Fianarantsoa, Toamasina, Toliara. Humid, evergreen forest, highland savannah and secondary grassland; 800–2000 m. Fl. January–February.

6.3. H. beharensis Bosser

Plant to 50 cm tall; leaves 4–5, linear-oblong, 10–20 × 2–3.5 cm; flowers few, very large, white, pubescent; ovary glandular; sepals 1.5–2 cm long; petals entire; lip 3-lobed, with two small, obtuse swellings at the base, side lobes much longer than midlobe; spur pendent, 15–20 cm long. Toliara. Dry, deciduous scrubland; up to 500 m. Fl. March–April.

6.4. H. ambositrana Schltr. (opposite)

Leaves several near base of stem, narrowly linear, incurved-tubular at base; flowers with a long pedicel and ovary, pale yellow or yellow-green; petals elliptic, obtuse; dorsal sepal 6–7 mm long; lip 3-lobed; side lobes shorter than midlobe; spur slenderly clavate, 2.5–3 cm long; anther not apiculate. Antananarivo, Fianarantsoa. River margins, marshes, wet grassland amongst rocks; 1500–2000 m. Fl. February–June.

6.5. H. decaryana H.Perrier

Plant up to 1 m tall; leaves linear, acute, 20–25 × 0.8–1 cm; inflorescence unevenly laxly flowered, up to 60 cm long; bracts ciliate; flowers whitish yellow; sepals ciliate-papillose, 6 mm long; petals ovate, obtuse; lip 3-lobed; spur incurved, 13–15 mm long, slightly apically dilated. Antananarivo, Fianarantsoa. Open woodland and open humid forests; 1000–1200 m. Fl. October–November.

6.6. H. lastelleana Kraenzl.

Plant 40–60 cm tall; leaves suberect, several, linear-lanceolate, acute, up to 17 × 1.8 cm; raceme laxly many-flowered, the flowers 10–15 mm apart; flowers 8–10, green; sepals 4–5 mm long; lip 3-lobed; side lobes at right angle to midlobe; spur 1.5–3.2 cm long. Antsiranana, Toamasina. Semi-deciduous, lowland forest; 400–600 m. Fl. April.

6.7. **H. monadenioides** Schltr.

Plant 35–60 cm tall; leaves several, suberect, lanceolate, acute, 9–15 × 1.5–2.3 cm; raceme uniformly densely flowered; flowers white; lip 3-lobed, with a conical callus at the base; side lobes shorter than midlobe; spur slightly incurved, cylindrical, 7–12 mm long. Antananarivo, Fianarantsoa. On rocky quartzite outcrops; 1400–2200 m. Fl. February–March.

6.8. **H. quartziticola** Schltr.

Plant slender, 25–35 cm tall; leaved conduplicate, linear-lanceolate, acute, 5 × 1.8 cm; inflorescence cylindrical, densely up to 10-flowered; flowers white; sepals 5.5–6.5 mm long; petals suborbicular-rhombic; lip 3-lobed; side lobes filiform and longer than midlobe; spur filiform, narrowly acute, 3 cm long. Fianarantsoa. On rocky quartzite outcrops; 800–1600 m. Fl. February.

6.9. **H. praealta** (Thouars) Spreng.   *(opposite, top left)*
(syn. *Satyrium praealtum* Thouars)

Plant up to 1.5 m tall, rhizomatous; leaves lanceolate, acute; flowers yellow-green; sepals 6 mm long; lip midlobe slightly longer than side lobes; spur slightly dilated towards the apex, 5–6 mm long. Antananarivo; also in the Mascarenes. In riverine forest and marshy areas in evergreen forest, under *Pandanus*; 1300–1500 m. Fl. November–December.

6.10. **H. simplex** Kraenzl.   *(opposite, top right)*
(syn. *Habenaria stricta* Ridl.; *H. ichneumoniformis* Ridl.)

Plant 15–30 cm tall; leaves suberect, 2–7 cm long, tapering; flowers yellow-green; dorsal sepal narrow, 6–7 mm long; petals acute; lip 3-lobed, midlobe curved upwards; spur cylindrical, 2.5–3 cm long; anther apiculate. Antananarivo, Fianarantsoa. Grassy rifts amongst granite rocks; 1000–1500 m. Fl. October–January.

6.11. **H. cirrhata** (Lindl.) Rchb.f.   *(p. 63)*

Plant up to 70 cm tall; leaves several, ovate-lanceolate; flowers large; sepals 1–2 cm long; petals lobes curved upwards like cow-horns; lip 3-lobed with side lobes curling up; spur 7–12 cm long. Antananarivo, Fianarantsoa, Mahajanga; also in the Comoros and tropical Africa. Dry meadows and grassland. 100–1200 m. Fl. October–March.

6.9

6.10

6.12

6.14

6.12. **H. clareae** Hermans *(p. 61, bottom left)*

(syn. *Habenaria elliotii* Rolfe)

Plant up to 80 cm tall; leaves oblanceolate, up to 4 cm broad; inflorescence laxly many-flowered; flowers green; dorsal sepal 5–7 mm long; petals bilobed; lip 3-lobed, side lobes longer than midlobe; spur slender, 4–5 cm, a little dilated towards the apex. Antananarivo, Antsiranana, Fianarantsoa, Mahajanga, Toamasina; also in E Africa. Coastal forest; humid, evergreen forest; marshes; rocky outcrops; up to 1800 m. Fl. February.

LOCAL NAME *Sinananga*.

6.13. **H. leandriana** Bosser

Plant 25–40 cm tall; leaves 4–7, oblong to oblong-lanceolate, 6–13 × 3–5 cm; inflorescence 12–15 cm long, 8–10-flowered; flowers pale green; sepals 1.3–1.6 cm long; lip 3-lobed, side lobes twice as long as midlobe; spur 4–6 cm long. Similar to *H. clarae* but differs by its fewer and larger flowers, its longer stigmatic processes and by its more attenuate, stalked leaves. Antsiranana, Toliara. Seasonally dry, deciduous forest, woodland or dry deciduous scrub on karst; up to 500 m. Fl. March.

6.14. **H. tianae** P.J.Cribb & D.A.Roberts *(p. 61, bottom right)*

Plant up to 50 cm tall; leaves obovate, up to 20 × 5 cm, glossy bright green. Inflorescence 2–7-flowered. Flowers green and white. Sepals 1.9–2.2 cm long. Petals with posterior lobe 19 mm long, anterior lobe 29 mm long, upcurved. Lip 3-lobed; side lobes upcurved, filiform, 25–30 mm long; midlobe 22–23 mm long; spur spiralled, 10–12 cm long. Fianarantsoa, Toliara. In deciduous woodland on sand; 500–800 m. Fl. February.

6.15. **H. incarnata** (Lindl.) Rchb.f. *(p. 65, top)*

(syn. *Habenaria rutenbergiana* Kraenzl.; *H. humblotii* Rchb.f.; *H. diptera* Schltr.)

Plant to 70 cm tall; leaves several, lanceolate; flowers small, gnat- or fly-like; lateral sepals strongly reflexed, almost 2 times longer than the dorsal sepal; petals bifid; lip side lobes shorter than midlobe; spur 18–25 mm long, parallel to ovary. Antananarivo, Antsiranana, Fianarantsoa, Toliara; also in the Comoros. Grassland and rocky places, forest and woodland margins, margins of cultivated fields; sea level–1550 m. Fl. January–June.

6.16. **H. tropophila** H.Perrier

Leaves fleshy, broadly ovate; flowers pale green; lip 3-lobed, the lobes filiform; spur hooked, c. 12 mm long, slightly thickened. Toliara. Humid, evergreen forest; rocky outcrops; 100–300 m. Fl. January–February.

6.11

6.17. H. acuticalcar H.Perrier

Plant 6–100 cm tall; leaves 10–12, lanceolate, acute, 7–20 × 2–3 cm; raceme 30–40 cm long, densely flowered; bracts as long as the flowers, 3-nerved; flowers green; sepals 5 mm long; petals bifid, anterior lobe tooth-like; lip 3-lobed; lobes cylindrical, 12–15 mm long, somewhat incurved; spur pendent, 12–15 mm long, inflated towards tip, acutely narrowed. Antsiranana. Humid, montane forest. 500–2050 m. Fl. December–January, March–April.

6.18. H. alta Ridl.   *(opposite, bottom left)*

Plant up to 70 cm tall; leaves obovate, acute; flowers numerous, green; sepals 6 mm long; lip midlobe slightly longer than side lobes; spur slightly dilated towards the apex, 1–1.5 cm long. Antananarivo, Fianarantsoa, Mahajanga. Humid, mossy, evergreen forest; shaded and humid granitic rocks. 1500–2000 m. Fl. February–April.

6.19. H. arachnoides Thouars   *(p. 67)*

Plant 30–50 cm tall, with 5–6 broadly lanceolate leaves, rounded at the base, 5–10 × 1–1.5 cm broad; inflorescence short, laxly 8–15-flowered; flowers small, green; sepals 3–5 mm long; petals bilobed; lip papillose, divided into 3 narrow lobes; spur curving upwards over ovary, scarcely inflated, 8.5–12 mm long. Antananarivo, Fianarantsoa, Toamasina; also in the Mascarenes. Humid, hill and montane forest; 750–2000 m. Fl. March, July, October.

6.20. H. boiviniana Kraenzl.

(syn. *Habenaria nigricans* Schltr.; *H. perrieri* Schltr.)

Plant to 70 cm tall; leaves 5, obovate, 7–11 × 2.5–3.5 cm broad; inflorescence 10–20 cm long; flowers green; ovary pubescent; dorsal sepal 4–6 mm long; petals ciliate, bilobed but anterior lobe often much reduced; lip 3-lobed but side lobes often vestigial, basal teeth; spur filiform, 1–1.6 cm long; stigmatic processes broadened and flattened at the front. Antsiranana, Fianarantsoa, Mahajanga; also in the Comoros and E Africa. Semi-deciduous forest; humid, evergreen forest; on rocks; in woods; sea level–1000 m. Fl. March–April, August.

6.15

6.18

6.26

6.21. **H. cochleicalcar** Bosser

Leaves 7–8, linear-lanceolate, 7–17 × 2–3 cm; inflorescence 15 cm long, 10–12-flowered; stalk pubescent-papillose; sepals 9–10 mm long; lip 3-lobed, side lobes more than twice as long as midlobe; spur spirally twisted in middle, 15 mm long, dilated and a little flattened at the tip. Toliara. Dry, deciduous scrub; up to 200 m. Fl. March.

6.22. **H. conopodes** Ridl.

Plant 30–45 cm tall; leaves lanceolate, acuminate, up to 7.5 × 1.2 cm; raceme laxly many-flowered; flowers yellow-green; sepals 6–7.5 mm long; lip 3-lobed; spur incurved, slightly inflated at the apex, 8 mm long; stigma lobes 10 mm long. Antananarivo. Savannah; on rocks in the highlands. c. 1000 m. Fl. January–March.

6.23. **H. demissa** Schltr.

Plant 35–50 cm tall; leaves 4–5, broadly elliptic, up to 8 × 3.7 cm; inflorescence narrowly cylindrical, many-flowered; flowers green. Sepals 5–6 mm long; lip 3-lobed, 7–8 mm long; spur 15 mm long. Antananarivo, Toliara. Mossy forest; 1200–2400 m. Fl. December, March.

6.24. **H. ferkoana** Schltr.

Plant up to 60 cm tall; leaves up to 13, lanceolate, acuminate, up to 8 × 1 cm; inflorescence many-flowered, up to 15 cm long; flowers pale green; sepals 5–6.5 mm long; lip deeply 3-lobed; spur subfiliform, slightly sigmoid, 6 mm long, pointed; stigma lobes short; close to *H. arachnoides* (p. 64) but with larger and more leaves, slightly larger flowers; petals bilobed with more pointed upcurved side lobes; lip midlobe longer than the upcurved side lobes; spur slightly longer than the lateral sepals. Madagascar, exact locality and flowering time unknown.

6.19

6.25. H. foxii Ridl.

Plants 25–45 cm tall; leaves 5–6, obovate, up to 11 × 4.5 cm; flowers green; dorsal sepal 5–6 mm long; petal lobes upcurved; lip 3-lobed; spur curving above ovary, dilated in apical third, acute, 12–14 mm long; stigma lobed decurved, 3.5–4 mm long. Antananarivo, Fianarantsoa. In shady woods and forest. Fl. March, June.

6.26. H. hilsenbergii Ridl.   *(p. 65, bottom right)*

(syn. *Habenaria atra* Schltr.; *H. ankaratrana* Schltr.)

Plant 25–45 cm tall; leaves suberect, lanceolate, sheathing at base; inflorescence cylindrical, densely many-flowered; flowers green; sepals 4–6 mm long; lip 3-lobed, the lobes 4–6 mm long, the middle one shorter and broader; spur filiform, 1.2–3 cm long. Antananarivo, Fianarantsoa, Toamasina. Grassland and rocky outcrops; 1500–2000 m. Fl. February–March, June.

6.27. H. nautiloides H.Perrier

Plant 40–50 cm tall; leaves 7–8, narrowly linear, attenuate, 2–9 × 0.5–2 cm; raceme 7–8-flowered, 5–7 cm long; flowers green; sepals 10 mm long, ciliate; petals bipartite, lobes filiform, 12 mm long; lip 3-lobed, lobes filiform, the 12 mm long side lobes twice as long as the midlobe; spur 12–15 mm long, attenuate-acute at the tip; differs from *H. hilsenbergii* in having narrow oblanceolate leaves not enveloping the stem, differently shaped sepals and lip, and hairs on the keels of the sepals; flowers green; spur c. 15 mm long. Toliara. Humid, lower montane forest; c. 600 m. Fl. March.

6.28. H. truncata Lindl.   *(opposite)*

Plant 30–45 cm tall; leaves ovate-lanceolate, 6–8 × 2 cm, grading into bracts; raceme densely flowered, cylindrical; flowers green; lip yellow-green, of 3 narrowly linear lobes; spur clavate, 10–12 mm long. Antananarivo, Fianarantsoa, Toamasina. Grassland and rocky outcrops; 1200–2000 m. Fl. January–April.

6.29. H. tsaratananensis H.Perrier

Plant 30–60 cm tall, pubescent-scabrid; leaves linear-lanceolate, 7–13 × 1.1–1.7 cm; inflorescence cylindrical, very densely up to 45-flowered; flowers white; ovary papillose-pubescent; sepals 5–6 mm long; lip 3-lobed; spur 7–9 mm long, cylindrical, a little dilated. Antsiranana. Mossy montane forest; 1200–1500 m. Fl. April.

6.28

7. MEGALORCHIS

A monotypic genus, endemic to northern Madgascar. Large terrestrials, growing from stout hairy, fleshy roots. Stem leafy. Leaves 12–15, elliptic-ligulate, acuminate, stalked. Inflorescence terminal, densely many-flowered, up to 40 cm long, 11–12 cm in diameter. Flowers large, fleshy, white, glabrous. Dorsal sepal ovate; lateral sepals not spreading widely, obliquely ovate. Petals very obliquely ovate. Lip clawed at base, 3-lobed in middle, spurred at base; side lobes erect, oblong; midlobe shortly clawed, deflexed, ovate; spur cylindrical, pendent. Column dorsi-ventrally compressed; anther loculi horizontal; stigmas stalked; pollinia sectile.

7.1. M. regalis (Schltr.) H.Perrier
(syn. *Habenaria regalis* Schltr.)

Plant robust, up to 1.5 m tall; leaves up to 60 × 8 cm; flowers large, white; lip 3 cm long; spur 5.5–6 cm long. Antsiranana, Mahajanga. Mossy, montane forest; 1800–2000 m. Fl. October–November.

8. PLATYCORYNE

A genus of about 20 species in Africa, one species extending its range to Madagascar. Small to medium-sized terrestrials, growing from fleshy tubers. Leaves lanceolate to ligulate, acute. Inflorescences terminal, racemose, usually several-flowered. Flowers yellow or orange. Dorsal sepal forming a hood with the petals. Lateral sepals deflexed. Lip entire or obscurely auriculate at base, spurred; spur pendent, clavate. Column erect; stigmas decurved, short; anther loculi parallel; pollinia 2, sectile.

8.1. P. pervillei Rchb.f.   *(opposite, top and bottom)*
(syn. *Habenaria depauperata* Kraenzl.; *H. pervillei* (Rchb.f.) Kraenzl.)

Plant slender; leaves linear-lanceolate; flowers several in a head, bright orange; dorsal sepal hooded; lip ligulate, spurred at base; spur 12–19 mm long. Mahajanga; also in E Africa. Grassland and marshes; up to 500 m. Fl. January–April.

8.1

8.1

6.4

9. CYNORKIS

About 120 species in tropical and South Africa, Madagascar, the Comores and Mascarene Islands. In Madagascar, 97 mostly endemic species. Terrestrial or epiphytic herbs with fleshy or tuberous roots. Stem, ovary and calyx often with glandular hairs. Leaves 1–several, radical, with a few sheath-like and cauline. Flowers few to many in a terminal raceme, usually resupinate. Sepals free or slightly adnate to the lip, the dorsal forming a hood with the petals, the lateral sepals spreading. Lip entire or 3–5-lobed, spurred at the base. Column short and broad; androclinium erect or sloping; anther loculi parallel, the canals short or long and slender; viscidia 2, rarely 1, auricles distinct; stigmatic processes oblong, papillose, usually joined to the lobes of the rostellum. side lobes of rostellum usually elongated; midlobe often large, projecting forwards.

Key to the sections of genus *Cynorkis*

1. Caudicles of the pollinia united by a common viscidium placed below the extremity of the rostellum; lip and rostellum entire .... **V. Sect. Monadeniorchis**
1. Caudicle of each pollinium ending in a distinct viscidium; viscidia exceptionally fused, but in that case the lip 4-lobed 2
2. Rostellum (including the cliandrium, i.e. from the anther chambers to the front extremity) in a horizontal blade, toothed at the front, not divided in clearly defined arms nor deeply indented in the middle down to the anther filament .... **I. Sect. Parviflorae**
2. Rostellum with clearly defined arms or deeply indented in the middle down to the anther filament 3
3. Rostellum divided into 2 arms for about half its total length, entire in the basal part (cliandrium) 4
3. Rostellum deeply notched in the middle down to the anther filament, the middle lobe reduced to a small projection or developed into a blade placed against the filament in-between the 2 anther chambers 5
4. Rostellum with 2 middle lobes, the upper one reduced to a small swelling or developed into a lobule which is obliquely ascending at the front, the lower one is opposite to the first, inflexed into a lip-shape .... **II. Sect. Lowiorchis**
4. Middle lobe of the rostellum simple, with an obliquely ascending lobule at the front .... **III. Sect. Cynorkis**
5. Rostellar arms longer than the anther chambers; anther canals elongate .... **IV. Sect. Imerinorchis**
5. Rostellar arms below the anther chambers, hidden by them and not longer than them; canals and caudicles hardly there .... **VI. Sect. Lemuranthe**

I. Section Parviflorae (syn. Sect. *Hemiperis* Perrier)

Plants normally slender. Inflorescence often laxly or few-flowered. Flowers small or medium in size, often non-resupinate. Lateral sepals and petals often fused at their base to the base of the lip. Rostellum in a flat blade, horizontal, only toothed or lobed at the front, with teeth (normally 3 but some variation exists), or lobed on the same level, always very short; anther canals on the rostellar blade, the viscidia resting in-between the teeth or anterior lobes.

Key to groups of species in Section Parviflorae

1. Lip uppermost in the flower **9.1–9.12**
1. Lip lowermost in the flower 2
2. Plants stoloniferous or with bundles of roots, lacking tubers; leaves 3–6 .... **9.13–9.14**
2. Plants with tubers; leaves 1–3 3
3. Lip 5-lobed, basal lobes often small and tooth-like **9.15–9.22**
3. Lip entire, bifid or trilobed 4
4. Plants lacking leaves at flowering time 5
4. Plants with well-developed leaves at flowering time 6
5. Spur saccate to conical, 1 mm long or less **9.23–9.26**
5. Spur cylindrical, more than 3 mm long .... **9.27. C. filiformis**
6. Leaves 3–7 .... **9.28–9.31**
6. Leaves 1–2 7
7. Inflorescence glandular or hairy **9.32–9.43**
7. Inflorescence glabrous **9.44–9.47**

9.1. C. bathiei Schltr.

(syn. *Cynosorchis inversa* Schltr.)

Terrestrial, 7–30 cm tall; leaves 2, oblong or elliptic, oblong, 2–8 × 0.5–1.7 cm; inflorescence densely many-flowered; flowers rose-purple; sepals 4–6 mm long; lip deeply 3-lobed, 4–6 × 4–5 mm, the side lobes oblong; spur cylindrical, 1.5–3 mm long; anther 0.5 mm tall. Fianarantsoa. Montane, ericaceous scrub; rocky outcrops; c. 2400 m. Fl. February–April.

9.2. C. coccinelloides (Frapp. ex Cordem.) Schltr.

(syn. *Camilleugenia coccinelloides* Frapp. ex Cordem.)

Terrestrial; leaves 1–2, oblanceolate; inflorescence 3–10-flowered; flowers violet or white with violet spots; ovary not glandular; sepals 3–3.5 mm long; lip 5-lobed, 3.3 × 2.5 mm; spur scrotiform; the plants from Madagascar differ from those from la Réunion by their white flowers. Antananarivo, Antsiranana; also in the Mascarenes. Montane, ericaceous scrub amongst sphagnum and peat moss; amongst rocks; 1600–2400 m. Fl. March–April.

9.3. C. cuneilabia Schltr.

Terrestrial, 20–45 cm tall; leaves 2, elliptic-lanceolate, 9–18 × 0.5–2.5 cm; inflorescence 4–15-flowered; flowers pale pink; sepals 5–6.5 mm long; lip 8 × 5.5 mm, obovate, very shortly 3-lobulate at the tip, the middle lobule tri-crenulate; spur cylindrical, 3 mm long. Antananarivo. In the lighter parts of mossy forest; c. 2000 m. Fl. March.

9.4. **C. flabellifera** H.Perrier

Plant 20–25 cm tall; leaf solitary, radical, narrowly lanceolate, 4–9 × 0.7–1.5 cm; raceme 2–7 cm long, subdense, slightly glandular-hairy; sepals 4.5 mm long; lip 3-lobed, 5 mm long; side lobes rounded, 1 mm long; midlobe flabellate, 4.5–5 mm broad; spur 3–3.5 mm long, inflated at tip. Antsiranana (Marojejy). Montane, ericaceous scrub; 1800–2000 m. Fl. December.

9.5. **C. formosa** Bosser

Terrestrial 12–12.5 cm tall; leaves 2, lanceolate or linear-lanceolate, 2–5 × 0.5–1 cm; inflorescence densely many-flowered; flowers whitish-rose; sepals 2.5–3 mm long; lip 5-lobed, 3.5–4.5 mm long; spur cylindrical, 1.5–2 mm long; close to *C. bathiei* (p. 73) but differs by its much smaller flowers and sepals and in having petals and lip of a different shape; flower pinkish white with red spots. Antsiranana. Montane, ericaceous scrub on granite and gneiss; lichen-rich forest; 2000–2500 m. Fl. March.

9.6. **C. henricii** Schltr. *(opposite)*

Plant 40–50 cm tall; leaves 3–4, elliptic or elliptic-lanceolate, 7–11 × 3–4 cm; inflorescence 10–15-flowered; flowers white or lilac, the ovary glandular; sepals 6–7 mm long; lip 3-lobed, 12–14 × 6.5–8 mm, the side lobes triangular, the midlobe fan-shaped; spur 7.5–8 mm long, slightly recurved. Toamasina. Humid evergreen forest; c. 500 m. Fl. August.

9.7. **C. peyrotii** Bosser *(p. 77, top left)*

Epiphyte, 6–11 cm tall; leaves 2–3, lanceolate or linear-lanceolate, 2.8–4.5 × 0.7–1.3 cm, reticulately marbled with white; inflorescence 3–15-flowered, glandular; flowers rose-pink; sepals 3.5–4.5 mm long; lip 3-lobed, broadly ovate, 4–4.5 × 3 mm, side lobes subtriangular, midlobe broadly ovate, obtuse; spur sausage-shaped, 4–5 mm long, another form has a sub-entire lip and short, 1–2.5 mm long, cylindrical spur. The two forms coexist in the same populations. Antananarivo, Fianarantsoa, Toamasina. Humid, evergreen forest on moss- and lichen-covered trees; 1000–1500 m. Fl. July–August.

9.8. **C. quinqueloba** H.Perrier ex Hermans

(syn. *Cynosorchis quinqueloba* H.Perrier)

Slender, terrestrial, with 3–4 basal leaves, elliptical or oblong, cuspidate-acute, narrowed towards the base; spike with 7–10 flowers; floral bracts hairy, about a third the length of the flower; flowers very finely papillose on the outside, glandular on ovary; sepal broadly ovate, 5 mm long; petals broadly ovate, very obtuse; lip obtusely 5-lobed, 6 × 4.5 mm, the lateral lobes at least half the size of the middle one; spur scrotiform. Fianarantsoa, Toliara. Ericaceous scrub, with *Philippia* dominant, and moss- and lichen-rich forest; 1400–1900 m. Fl. November, January.

9.6

9.9. C. quinquepartita H.Perrier ex Hermans

(syn. *Cynosorchis quinquepartita* H.Perrier)

Terrestrial herb slender, 30–60 cm tall, lacking tubers but with bunched, downy, slightly fleshy roots; leaves radical, 1 or rarely 2, linear-lanceolate, acute, 10–20 × 0.7–2.8 cm; flowers violet, with a few glands on the outside; sepal 2.5–3.5 mm long; lip 5-lobed, 5 × 4.5 mm; the basal lobes bigger than the intermediate ones, but a little smaller than middle one; spur cylindrical, but broad, 4 mm long. Antsiranana. Mossy, montane forest; c. 1200 m. Fl. September.

9.10. C. ridleyi T.Durand & Schinz   *(opposite, top right and bottom)*

(syn. *Amphorchis lilacina* Ridl.; *Cynosorchis heterochroma* Schltr.; *C. moramangana* Schltr.; *C. rhomboglossa* Schltr.)

Terrestrial or rarely epiphyte; leaves 1–2, obovate-elliptic, often darker coloured underneath; flowers purple-pink; sepals 3–4 mm long; lip spotted with purple, with small lateral lobes, the midlobe very variable; spur a little recurved and slightly swollen at the tip, slightly shorter than the ovary; anther 1 mm tall. Antananarivo, Antsiranana, Fianarantsoa, Toamasina; also in the Comoros. Humid, mossy, evergreen forest on plateau amongst mosses; 600–2000 m. Fl. September–April.

9.11. C. schlechteri H.Perrier

(syn. *Cynosorchis exilis* Schltr. non (Frapp.) Schltr.)

Terrestrial; leaves 2, oblong, 3–4.5 × 0.8–1.3 cm; inflorescence 7–20-flowered; flowers small, lilac or with darker spots; sepals 2.5–3 mm long; lip 5-lobed, 5–5.5 mm long, the basal lobes very small; spur straight, 3 mm long. Toamasina. Humid, evergreen forest; rocky outcrops; sea level–800 m. Fl. September.

9.12. C. sororia Schltr.

Terrestrial or lithophyte, 13–20 cm tall; leaf 1, elliptic, acute, 4.5–8 × 1.3–2.2 cm; inflorescence 4–20-flowered; flowers pink; sepals 4–8 mm long; lip deeply 3-lobed, 9–15 mm long, the lateral lobes small, triangular and obtuse, the midlobe very big, the margins almost crenulate; spur apex distinctly inflated, 7–8 mm long. Related to *C. ridleyi* but differs in the shape of the rostellum. Antsiranana, Toamasina. Mossy forest, by side of streams in sand; 800–1700 m. Fl. May, September.

9.7

9.10

9.10

9.13. C. graminea (Thouars) Schltr. (opposite, top)

(syn. *Satyrium gramineum* Thouars; *Habenaria graminea* (Thouars) Spreng.; *Platanthera graminea* (Thouars) Lindl.; *Bicornella longifolia* Lindl.; *Bicornella parviflora* Ridl.; *B. similis* Schltr.; *Benthamia graminea* (Thouars) Schltr.; *Cynosorchis longifolia* (Lindl.) Schltr.; *C. similis* (Schltr.) Schltr.)

Terrestrial, 15–60 cm tall; roots fleshy; leaves grass-like, 6–30 cm long; flowers small, pink or purple with darker spots; sepals 3–4 mm long; lip entire, a little expanded in the lower part, 4.5–5 mm long; spur c. 4 mm long, a little contracted in the middle and widened at the tip. Antananarivo, Fianarantsoa, Mahajanga, Toamasina, Toliara; also in the Mascarenes. Marshland in peaty soil; wet grassland, beside streams, a common species in wet areas; sea level–2000 m. Fl. August–March.

9.14. C. stolonifera (Schltr.) Schltr. (opposite, bottom)

(syn. *Bicornella stolonifera* Schltr.)

Terrestrial, 4–60 cm tall, lacking tubers but sending out stolons; leaves long and grass-like, up to 15 × 0.5 cm; flowers reddish purple; sepals 3 mm long; lip ligulate, 3.5 mm long; spur cylindrical, c. 1.5 mm long. Antananarivo, Fianarantsoa. Marshes in peaty soil; 700–1600 m. Fl. January–March.

9.15. C. cardiophylla Schltr. (p. 81)

Terrestrial; leaf 1, radical spreading on the ground, suborbicular or broadly oval, 3–4.5 × 2.5–3.5 cm; flowers purple, glandular; sepals 7–9 mm long; lip 5-lobed, 8–10 mm long; spur slender, cylindrical, slightly apically dilated, 8–10 mm long. Antananarivo, Fianarantsoa. Montane, ericaceous scrub; rocky outcrops, in scree; 1500–2400 m. Fl. January–April.

9.16. C. elegans Rchb.f.

(syn. *Gymnadenia muricata* Brongn. ex Kraenzl.)

Leaves 2, oblong-lanceolate, acute, violet-purple beneath, spotted with violet above; flowers 3–7, small, white or rose-purple; lip 5-lobed; spur filiform, as long as the ovary. Fianarantsoa.

9.17. C. pinguicularioides H.Perrier ex Hermans

(syn. *Cynosorchis pinguicularioides* H.Perrier)

Plant up to 10 cm tall; leaves 3–4, radical, lanceolate, acute, 1.5–3 × 0.4–0.6 cm; raceme rather laxly 4–10-flowered; flowers small, purplish, papillose on the outside; sepals 4 mm long; lip 5-lobed, 5.5 × 4.2 mm, the lobes obtuse, the 3 front ones more or less equal, much larger than the 2 lower ones; spur scrotiform, 1.2 mm long. Toliara. Rocky outcrops; 200–500 m. Fl. May–July.

9.13

9.14

9.18. C. pseudorolfei H.Perrier

Plant 12–25 cm tall; leaves 2, narrowly oblong-lanceolate, petiolate, 4–8 × 0.6–2 cm; raceme laxly 3–12-flowered; flowers purple; sepals 6–7 mm long; lip 5-lobed, 7 mm long; side lobes subovate; apical lobe obovate, apiculate; spur slightly incurved, 3 mm long, slightly diclated at tip. Similar to *C. rolfei*, but the inflorescence is not hairy, the perianth is twice as long and clearly gibbose at the base, the lip is different and the spur is curved and cylindrical. Toliara. Coastal forest; 100–150 m. Fl. March.

9.19. C. raymondiana H.Perrier ex Hermans

(syn. *Cynosorchis raymondiana* H.Perrier)

Terrestrial herb glabrous; leaves radical, 2, rarely 3, oblong-lanceolate, cuspidate-acute, 4–8 × 1.2–2 cm; inflorescence slender, laxly 6–12-flowered; flowers very small, pinkish-white; sepals 3.2–6 mm long; lip narrowed at the base, very unequally 5-lobed, 5 × 1.5 mm; lower lobes small, triangular-acute and diverging, the intermediate ones larger and obovate, the middle one as long as the intermediates, but oblong-acute and narrower; spur club-shaped, 5 mm long. Toliara. Coastal and montane forest, in sand; sea level–1200 m. Fl. April, July.

9.20. C. rolfei Hochr.

Terrestrial, 15–25 cm tall; leaves 2, radical, ovate-lanceolate, 3.5–7.5 × 1.2–2.3 cm; flowers small; sepals 2.2–3 mm long; lip 5-lobed, 4–5 mm long; spur cylindrical, 1.6–2 mm long. Toamasina. Coastal forest in sandy soil; sea level–200 m. Fl. July, December.

9.21. C. saxicola Schltr.

Terrestrial, 12–25 cm tall; leaves 3, linear-ligulate, 3.5–6.5 × 0.4–0.7 cm; flowers 3–5, small, purple, glandular; sepals 3–5 mm long; lip 5–6 × 3 mm, dentate at base, expanded above the base; spur cylindrical, curved at the front and thickened at the tip, 6–8 mm long. Fianarantsoa. Shaded and wet rocks; c. 2000 m. Fl. January.

9.22. C. verrucosa Bosser

Terrestrial, 25–40 cm tall; leaves 2–4, lanceolate 3–10 × 1.5–3.5 cm; rachis 6–12 cm long, many-flowered, glandular; flowers small; sepals 2.5–4 mm long; lip 5-lobed, flabellate, 3.5–5 × 4.5–5 mm, with small warts on the surface; spur straight, clavate, 2.5–3 mm long. Toamasina. Humid, evergreen forest; 1000–1500 m. Fl. June.

9.15

9.23. C. andohahelensis H.Perrier

Terrestrial, 25–45 cm tall; flowers 10–15, white with a yellow lip; sepals 3–4.5 mm long; lip 3-lobed, 5 × 4 mm; spur 1 mm long. Toliara. Montane, ericaceous scrub in humus and in peat; 1800–2000 m. Fl. January.

9.24. C. ochroglossa Schltr.

Terrestrial, 30–40 cm tall, lacking radical leaves; raceme up to 15-flowered; flowers small, white with an orange-yellow lip; sepals 3–4 mm long; lip 3-lobed, 5 mm long; spur conical and very short, 1 mm long. Fianarantsoa. Rocky outcrops; c. 2400 m. Fl. February.

9.25. C. papillosa (Ridl.) Summerh. *(p. 85, top left)*

(syn. *Peristylus filiformis* Kraenzl.; *Habenaria filiformis* (Kraenzl.) Ridl.; *H. papillosa* Ridl.; *Helorchis filiformis* (Kraenzl.) Schltr.; *Cynorkis filiformis* (Kraenzl.) H.Perrier non Schltr.)

Slender terrestrial plant, 25–40 cm tall; leaves not developed; flowers 10–30, white with a purple lip; lip oblong, entire; spur conical. Toamasina, Fianarantsoa. Marshes in peaty soil and ericaceous scrub; 1500–2400 m. Fl. December–April.

9.26. C. sacculata Schltr.

Terrestrial, 12–25 cm tall; leaves absent; flowers 8–15, brownish, glandular; sepals 3.5–4 mm long; lip clawed, 3-lobed above, 3.5–4 × 4 mm; spur conical, obtuse, 2–3 mm long; related to *C. tryphioides* (p. 89) but differs in its 2–3 mm long spur which is conical saccate and obtuse, as wide at the base as long. Antananarivo, Fianarantsoa. Montane, ericaceous scrub; 1600–2300 m. Fl. February–March.

9.27. C. filiformis Schltr. *(p. 85, top right)*

(syn. *C. hirtula* H.Perrier)

Slender terrestrial; flowers small, c. 14 mm long, white with a purple lip; lip 4 × 5 mm, the base narrow, the lateral lobes very small and short; spur narrowly cylindrical, 4.5 mm long. Antananarivo, Fianarantsoa. Montane, ericaceous scrub, in sphagnum moss; 1100–2400 m. Fl. January–March.

9.28. **C. brevicornu** Ridl.

(syn. *Cynosorchis brevicornu* (Ridl.) T.Durand & Schinz)

Terrestrial, 15–35 cm tall; leaves 3–7, linear-lanceolate, acuminate, 7.5–10 × 1.2 cm; flowers small; lip linear, narrow; spur 2.2 mm long, cylindrical. Fianarantsoa. Damp places amongst rocks. Fl. unknown.

9.29. **C. decaryana** H.Perrier ex Hermans

Terrestrial, up to 25 cm tall; leaves 5, radical, ovate-lanceolate, 3–6 × 1.2–1.8 cm; raceme, elongate, laxly 15-flowered; flowers small, lilac; sepals 2.8–3.2 mm long; lip base narrow and concave, 3-lobed above the middle, 3.5 × 3.5 mm, the lateral lobes narrowly triangular, acute, a little shorter than the midlobe, which is very broadly transverse; spur recurved, 2–3 mm long, very inflated towards the tip; anther retuse; viscidium as long as the caudicles. Toamasina. Lowland forest; sea level–500 m. Fl. September.

9.30. **C. globosa** Schltr.

Terrestrial, 8–9 cm tall; leaves 4–5, radical, linear, 3–4 × 0.6–0.8 cm; inflorescence subglobular, densely 15–25-flowered; flowers small; sepals 4 mm long; lip 3-lobed, 4.5 × 5 mm; spur 5 mm long, obtuse. Toamasina(?). Fl. unknown.

9.31. **C. tenella** Ridl.

Small, slender terrestrial, 10–15 cm tall; leaves 4, radical, ovate-oblong, 2.5–4 × 0.6 cm; inflorescence 4–5-flowered; flowers small; sepals ovate; lip 3-lobed, the lateral lobes curved and subacute; spur 5 mm long, shortly expanded. Fianarantsoa; also in the Comoros. Fl. May.

9.32. **C. alborubra** Schltr.

Slender terrestrial with 2 basal spreading leaves, 2.5–7 × 0.6–1.7 cm; rachis 6–9-flowered, glandular; flowers white with purple spots on the lip; sepals 6–7 mm long; lip 3-lobed, 6.5 × 4.75 mm, the lateral lobes lanceolate, subacute, the midlobe oblong-ligulate, obtuse, almost 3 times longer than the laterals; spur 2.5–3 mm long, conical and acute. Fianarantsoa. Montane, ericaceous scrub; c. 2400 m. Fl. January.

9.33. C. ampullacea H.Perrier ex Hermans

(syn. *Cynorkis ampullacea* (H.Perrier) H.Perrier; *Cynosorchis cuneilabia* subsp. *ampullacea* H.Perrier)

Terrestrial with 1 or rarely 2, radical, linear-lanceolate leaves, 7–11.5 × 1–1.7 cm; raceme glandular-hairy, laxly 4–12-flowered; flowers violet; sepals 5–6 mm long; lip obtriangular-flabellate, 8 mm long, the base thickened and hollowed into a gutter, gradually expanded above; spur 4 mm long, swollen into a very obtuse flask-shape. Antananarivo, Fianarantsoa. Rocky outcrops; c. 1400 m. Fl. January–March.

9.34. C. ampullifera H.Perrier

Terrestrial, 15–30 cm tall; leaf ovate, apiculate, 2–3 × 1.5–1.8 cm; raceme corymbiform, few-flowered, glandular; flowers purple; sepals 3.5–4 mm long. Lip 3-lobed, 4.5 mm long and wide; side lobes subrectangular, 2 mm long; midlobe flabelliform, 2 × 3 mm, toothed on front margin; spur 2 mm long, slightly dilate-ampulliform. Close to *C. cardiophylla* (p. 78) but the leaves shortly petiolate and the lip 3-lobed. Antsiranana, Toamasina. Humid evergreen forest; lichen-rich forest; boggy ground; 400–1450 m. Fl. March, September.

9.35. C. aurantiaca Ridl.   *(opposite, bottom left)*

Slender terrestrial, 20–40 cm tall; leaf radical, grass-like, 8–20 × 0.3–0.4 cm; inflorescence glandular; flowers yellowish-white with the lip yellow-orange, glandular; sepals 3.5 mm long; lip 6 mm long, 3-lobed, the midlobe finely papillose; spur 4–4.5 mm long, slightly narrowed, obtuse. Antananarivo. Humid, evergreen forest on plateau, epiphyte in moss; marshes on plateau; 1200–1800 m. Fl. January–May.

9.36. C. baronii Rolfe   *(opposite, bottom right)*

(syn. *Cynosorchis pauciflora* Rolfe; *C. baronii* (Rolfe) Durand & Schinz; *C. nigrescens* Schltr.; *C. nigrescens* var. *jumelleana* Schltr.)

Terrestrial or lithophyte, 10–40 cm tall; leaves 2, spreading on the soil, ovate-lanceolate, 2.5–10 × 1–3.5 cm; flowers small, white, pink or pink spotted with red, glandular; sepals 3.5–4.5 mm long; lip obscurely expanded-sublobulate above the narrow base, then 3-lobed, 4.5–6 mm long; spur 3–3.3 mm long. Antananarivo, Fianarantsoa, Toliara. Montane, ericaceous scrub, a relatively common species in the highlands; rocky outcrops in debris; high-elevation grassland; 1500–2500 m. Fl. January–April.

9.37. C. glandulosa Ridl.

Allied to *C. lilacina* (p. 86) but distinguished by its 2 lanceolate leaves, 15 × 2.5 cm, smaller glandular flowers, bifid lip, and short points to the anther. Antananarivo. Rocky outcrops. Fl. March.

9.25

9.27

9.35

9.36

9.38. C. globifera H.Perrier

Terrestrial 15–40 cm tall; leaves 2–3, petiolate, narrowly oblong-lanceolate or oblanceolate, 4–12 × 1–2.2 cm; scape robust; raceme 2–13 cm long; ovary and veins of sepals hairy; sepals 4 mm long; lip 3-lobed, lobes equal; spur globular, 1 mm in diam. Toliara. Humid, lowland forest. Fl. March.

9.39. C. jumelleana Schltr.

Terrestrial, 20–40 cm tall; leaf ligulate or oblong-ligulate, 6–14 × 1–1.8 cm; inflorescence 10–30-flowered; flowers white or violet, glandular; sepals 5–6 mm long; lip 3-lobed, 6 × 5.5 mm; midlobe flabelliform; spur conical, 2.5–3 mm long; related to *C. tenerrima* (p. 89) but the lip is expanded above the narrow base and deeply 3-lobed. Fianarantsoa. In ericaceous shrub; c. 2500 m. Fl. February.

9.39a. C. jumelleana var. gracillima Schltr.

Terrestrial; more slender, 8-or-less-flowered; flowers smaller; lip wider; the staminodes narrower. Fianarantsoa. Montane, ericaceous scrub; c. 2400 m. Fl. January.

9.40. C. lilacina Ridl.   (*opposite, top*)
(syn. *Cynosorchis lilacina* (Ridl.) T.Durand & Schinz)

Terrestrial, 25–40 cm tall; leaf single, lanceolate, 6–18 × 1.2–2.5 cm, with a distinct petiole; inflorescence glandular; flowers violet or lilac, small; sepals 6–8 mm long; lip 3-lobed, 10 mm long; side lobes obtuse; midlobe oblong-lanceolate, obtuse; spur broad at the base then narrowed and inflated and obtuse-truncate at the tip, 7 mm long. Antananarivo, Fianarantsoa, Tomasina. Grassland; marshes; on inselbergs; 1300–1800 m. Fl. November–April.

9.40a. C. lilacina var. curvicalcar (H.Perrier) H.Perrier
(syn. *C. andringitrana* subsp. *curvicalcar* H.Perrier)

Differs from the species type in having petals with the outer edges often expanded-angular, a short, 5 mm long spur and a much shorter (1 mm long) rostellum. Antsiranana. In 'Savoka' (*Philippia*) scrub; c. 2400 m. Fl. January–April.

9.40

9.41

9.40b. C. lilacina var. laxiflora (Schltr.) H.Perrier

Terrestrial; flowers much smaller than those of the typical variety; raceme more lax; anther canals short. Fianarantsoa. Mossy forest; c. 1600 m. Fl. February.

9.40c. C. lilacina var. pulchra (Schltr.) H.Perrier

(syn. *Bicornella pulchra* Kraenzl. ex Schltr.; *Cynosorchis pulchra* Kraenzl. ex Schltr.)

Terrestrial; spur cylindrical then obtuse, 5–6.5 mm long; anther erect, small, the anther canals slender; stigmatic processes shorter than the anther canals. Antananarivo, Antsiranana, Fianarantsoa, Mahajanga. In *Philippia* scrub; humid, evergreen forest on plateau; 1600–2200 m. Fl. January–April.

9.40d. C. lilacina Ridl. var. tereticalcar H.Perrier ex Hermans

Terrestrial; differs from the typical variety in its spur, which is subcylindrical from the base to the apex, more or less elongate, and at times as long as the pedicellate ovary. Antsiranana, Fianarantsoa. Humid, evergreen forest on plateau; in *Philippia* scrub; 1500–2000 m. Fl. February–March.

9.41. C. melinantha Schltr.   (p. 87, bottom)

Terrestrial, 20–40 cm tall; leaf radical, linear, 8–20 × 0.3–0.4 cm; flowers orange spotted red on lip, glandular; sepals 6–7 mm long; lip 3-lobed, 6 × 6 mm; spur 3.5–4.5 mm long, cylindrical; related to *C. jumelleana* (p. 86) but differs in petal shape and lip shape, the longer spur and anther canals, and the orange-yellow flowers spotted red on lip. Fianarantsoa. Rocky outcrops on quartz soil; 1500–1700 m. Fl. February.

9.42. C. perrieri Schltr.

Terrestrial or lithophyte, 7–12 cm tall; leaves elliptic-lanceolate, 2.5–4 × 0.3–0.5 cm; inflorescence glandular, 3–7-flowered; flowers white, glandular; sepals 4–4.5 mm long; lip shortly unguiculate at the base then 3-lobed, 5.5–6 mm long; midlobe obcordate; spur cylindrical, obtuse, 4.5–5 mm long; close to *C. hispidula* (p. 94). Antananarivo. Rocky outcrops, in moss; 1600–2000 m. Fl. February–May.

9.43. **C. sagittata** H.Perrier

Terrestrial; close to *C. andringitrana* but has hairs on the inflorescence, smaller flowers, and an arrow-shaped lip. Antsiranana. Coastal forest; sea level–100 m. Fl. July.

9.44. **C. andringitrana** Schltr.

Terrestrial, 15–30 cm tall; leaf basal, elliptic-oblong, 5–17 × 1–2.5 cm; inflorescence glabrous; flowers pale purple; sepals 4–6 mm long; lip 3-lobed in the middle, 6–6.5 × 4.5 mm; spur narrowly cylindrical, 5–5.5 mm long; lip similar in shape to that of *C. tryphioides*, but the plant is taller, has only one basal leaf and much larger flowers. Antananarivo, Fianarantsoa. Mossy forest and wet areas; 1600–2200 m. Fl. February–April.

9.45. **C. bimaculata** (Ridl.) H.Perrier

(syn. *Habenaria bimaculata* Ridl.)

Plant 45–60 cm tall; leaves radical, 2–3, narrowly lanceolate, 12–16 × 1.5–2.2 cm; inflorescence many-flowered; flowers lilac with two darker marks on the lip; sepals 7 mm long; lip obovate-cuneiform, 5 × 3.5 mm; spur cylindrical, 4 mm long, obscurely bilobed at tip. Antananarivo. Humid, highland forest. Fl. March.

9.46. **C. tenerrima** (Ridl.) Kraenzl.

(syn. *Habenaria tenerrima* Ridl.)

Small terrestrial, 10–15 cm tall; leaves 2–3, radical, linear-lanceolate, 1.8–2 cm long; inflorescence 5–7-flowered; flowers small, white-pink or lilac; sepals 3 mm long; lip 3.5 mm long, broadly obcordate, the base narrow; spur cylindrical and straight, 2.2–2.8 mm long. Antananarivo. Marshes in peaty soil; 1400–2300 m. Fl. November–December.

9.47. **C. tryphioides** Schltr.

Very small terrestrial or lithophyte, 5 cm tall; leaves 2, ovate, prostrate on substrate, 3–3.2 × 2–2.3 cm; flowers many, very small, lilac, glabrous; sepals 3 mm long; lip angular, 3-lobed, 3 × 3 mm; spur cylindrical, acute, c. 3 mm long. Antsiranana, Mahajanga. On limestone rocks in open forest; in grassland; dry forest; 250–2000 m. Fl. December–March.

LOCAL NAME *Felantrandraka*.

9.47a. C. tryphioides var. leandriana (H.Perrier) Bosser
(syn. *Benthamia leandriana* H.Perrier)

Terrestrial or lithophyte; differs from the typical variety by the shape of its short, 3-lobed lip, the shape of the rostellum, and its larger leaves. Antsiranana, Mahajanga. Seasonally dry, deciduous forest or woodland on limestone; sea level–500 m. Fl. December–January.

II. Section Lowiorchis

Lip always 4-lobed, i.e. 3-lobed with the middle lobe more or less deeply divided. Rostellum with long or short arms, always entire for a certain length and in front of the anther, with the double middle lobe bilobed or almost so. The upper lobule reduced to 1 or 2 small obtuse protrusions or developed into an obliquely ascending blade; the lower lobule in a membranous piece forming a gutter or inflexed lobe, divided.

Key to groups of species in Section *Lowiorchis*

1. Leaves absent at flowering time ....**9.48. C. aphylla**
1. Leaves present at flowering time 2
2. Lip 6-lobed, fimbriate on margins of 4 apical lobes ....**9.49. C. fimbriata**
2. Lip 3- or 4-lobed, lip margins entire 3
3. Spur 40–45 mm long; lip 22 mm long ....**9.50 C. lowiana**
3. Spur 4–5 mm long; lip 10 mm long 4
4. Inflorescence 20–25-flowered; spur cylindrical ....**9.51. C. orchioides**
4. Inflorescence less than 20-flowered; spur clavate ....**9.52. C. sambiranoensis**

9.48. C. aphylla Schltr.

Terrestrial or lithophyte; leaves not developed at flowering time; flower large, dark pink; lip 4-lobed, wider than long, 3 × 2.4 cm; spur filiform, 4–5 cm long. Antsiranana, Mahajanga. Rocky outcrops; on sandstone and sand; sea level–200 m. Fl. September–November, January.

9.49. C. fimbriata H.Perrier ex Hermans
(syn. *Cynosorchis fimbriata* H.Perrier)

Terrestrial; stem and spike enclosed at the base by a long sheath; leaf solitary, well-developed; raceme short, 4–8-flowered; floral bracts ample, a third or half the length of the ovary; flowers bright purple; sepals obtuse, the dorsal sepal ovate, concave; the laterals almost semi-orbicular; petals narrow, curved, obtuse; lip 6-lobed with apical 4 lobes fimbriate-dentate on the margins; basal lobes auriculate; spur narrowly cylindrical, then clearly inflated in the apical third, 32–35 mm long. Antsiranana, Fianarantsoa. In shade, amongst wet rocks; 2000–2200 m. Fl. April.

9.50

9.50

9.50a

9.50. **C. lowiana** Rchb.f.   (p. 91, top and bottom left)

(syn. *Cynosorchis purpurascens* Hook.f. non Thouars)

Epiphyte, terrestrial or lithophyte; leaf solitary, linear-lanceolate; inflorescence 1–few-flowered; flowers large; lip rose-purple with a darker spot on the disk; lip 3-lobed, 2.2 × 2.7 cm; spur up to 4.5 cm long. Antsiranana, Fianarantsoa, Toamasina. Humid, evergreen forest in moss; shaded and humid rocks; 200–1200 m. Fl. August–May.

9.50a. C. × mirabile Hermans & P.J.Cribb, the natural hybrid of *C. lowiana* and *C. gibbosa* (p. 96), is intermediate between its parents.   (p. 91, bottom right)

9.51. **C. orchioides** Schltr.

Terrestrial or lithophyte up to 35 cm tall; leaf basal, narrowly lanceolate, up to 20 cm long; inflorescence densely 20–25-flowered; flowers pink; sepals 4.5 mm long; lip 4-lobed, 10 mm long, side lobes slightly larger than midlobes; spur subclavate, 4.5 mm long. Mahajanga. Seasonally dry, deciduous forest or woodland; rocky woods and outcrops. Fl. November–January.

9.52. **C. sambiranoensis** Schltr.

Terrestrial, 10–25 cm tall; leaf solitary, suberect, lanceolate, 6–15 × 1–2.4 cm; inflorescence 5–20-flowered; lip 10–12 mm, 4-lobed in basal half; side lobes divergent, obovate-cuneiform; midlobes elliptic-obovate; spur 4–4.5 mm long, clavate; related to *C. orchioides* and to *C. purpurascens* (p. 102) in habit, but the flowers are smaller and the shape of the spur is different. Antsiranana, Mahajanga. On rocks; on hills; in wet grassland; 300–500 m. Fl. October–January.

III. Section Cynorkis (syn. Sect. *Gibbosorchis* H.Perrier)

Flowers generally large and brightly coloured; base of the petals and lateral sepals free or more or less fused to the base; lip often 4-lobed (3-lobed with the middle lobe bifid). Rostellum entire for a certain length in front of the anther, then divided into 3 lobes (2 arms and a middle lobe), with the supplementary lower middle lobe; arms at least 1 mm long; middle lobe obliquely ascending at the front.

Key to groups of species in Section *Cynorkis*

1. Lip entire, 3- or 5-lobed, the lower lobes reduced in the latter case to small teeth 2
1. Lip 4-lobed, actually 3-lobed but the midlobe deeply bifid 4
2. Leaf solitary **9.53–9.57**
2. Leaves 2 or more 3
3. Inflorescence glandular hairy **9.58–9.62**
3. Inflorescence glabrous **9.63–9.64**
4. Inflorescence glandular hairy **9.65–9.68**
4. Inflorescence glabrous **9.69–9.75**

9.53. **C. boinana** Schltr.

Terrestrial or lithophyte; leaf ligulate, 8–14 × 2–2.8 cm; inflorescence 3–10-flowered; flowers mauve or pinkish-violet, c. 2.5 cm long; sepals 7–8 mm long; lip 3-lobed above, 13–17 × 13–17 mm; spur 5–6 mm long, dilated at the tip. Antsiranana, Mahajanga. Seasonally dry, deciduous forest or woodland; rocky limestone outcrops; grassland; sea level–500 m. Fl. September–February.

9.54. **C. catatii** Bosser

Terrestrial up to 35 cm tall; leaf lanceolate, acute, 5–6 × 2–2.5 cm; inflorescence many-flowered, glandular; flowers rose-purple; dorsal sepal 4 mm long; lateral sepals 7–7.5 mm long; lip obscurely 5-lobed, 6–8 mm long, the 2 basal lobes minutely triangular, the middle lobes rounded, the midlobe ovate, acute; spur filiform 13–14 mm long. Toamasina. Humid, evergreen forest in humus. Fl. September–October.

9.55. **C. latipetala** H.Perrier

Terrestrial, 2–12 cm tall; leaf ovate or oblong, 1–3 × 1–2 cm; Inflorescence 2-flowered; flowers purple-red, tinted green; sepals 8 mm long; lip 3-lobed, 15 × 15 mm; midlobe larger than side lobes, denticulate in front; spur 10 mm long, inflated towards the tip. Antsiranana, Toamasina. Marshes in peaty soil, mossy and lichen-rich forest, and rocky outcrops; 1200–2050 m. Fl. unknown.

9.56. **C. petiolata** H.Perrier

Terrestrial, 25–40 cm tall; leaf oblong-lanceolate, 5–9 × 2–3.5 cm, petiolate; inflorescence 3–7-flowered; flowers white with pink spots on the petals and a sulphur-yellow lip with 3 dark red spots on each side; sepals 11 mm long; lip 3-lobed, 11 × 8 mm, the midlobe obovate and larger than the side lobes; spur 4.5 mm long, truncate. Antsiranana. Lichen-rich forest; c. 1500 m. Fl. December.

9.57. **C. sylvatica** Bosser

Terrestrial, 12–25 cm tall; leaf ovate-oblong, acute, 1.5–6 × 0.8–1.8 cm; inflorescence 6–20-flowered, glabrous; flowers glabrous; sepals 3–4 mm long; lip oblong, 3-lobed, 3.5–4 mm long, the side lobed short, truncate; spur 6 mm long; the flowers resemble those of *C. tristis* (p. 112) but are distinguished by their 3-lobed, oblong lip and especially the rostellum. Antsiranana. Lichen-rich forest; 2000–2500 m. Fl. unknown.

9.58. **C. elata** Rolfe   *(opposite)*

Small terrestrial; leaves 2, prostrate, oblong-lanceolate, 4–9 × 1.3–3.8 cm, marbled; inflorescence 3–20-flowered, glandular; flowers pink or white and pink; sepals 6–6.5 mm long; lip 5-lobed, 8 mm long; spur 13 mm long, a little contracted in the middle and very slightly expanded toward the tip. Fianarantsoa, Toamasina, Toliara. Coastal forest in sand and in humus; sea level–50 m. Fl. April–May.

9.59. **C. hispidula** Ridl.

(syn. *Cynosorchis hispidula* (Ridl.) T.Durand & Schinz)

Small lithophyte, up to 10 cm tall; leaves 4, lanceolate, acute, up to 3 cm long; inflorescence glandular, 7–10-flowered; flowers with yellow-green sepals and petals spotted with maroon-brown and a white lip; sepals 3.5–4 mm long; lip spathulate, 4 mm long; spur less than 5 mm long. Antananarivo, Fianarantsoa. Shaded and humid banks and granite and gneiss rocks; c. 2000 m. Fl. February–March.

9.60. **C. humbertii** Bosser

Terrestrial 30–40 cm tall; leaves 2–3, linear or linear-lanceolate, 4–8 × 0.6–0.7 cm; raceme laxly 20–25-flowered; flowers rose-purple, glandular; sepals 3–5 mm; lip entire, subpandurate, 5–6 mm long; spur 2 mm long. Close to *C. brachyceras* (p. 104) but distinguished by the rostellum and lip structure. Mahajanga. Lichen-rich, evergreen forest; 1600–2000 m. Fl. unknown.

9.61. **C. laeta** Schltr.

Terrestrial or lithophyte similar to *C. lindleyana* (p. 106), 14–22 cm tall; leaves 3–5, linear or linear-lanceolate, 5–9 × 0.8–1.5 cm; inflorescence densely many-flowered; glandular; flowers purple; sepals 5–7 mm long; lip 10 mm, 3-lobed; midlobe ligulate; spur 9 mm long, cylindrical-obtuse, expanded into a club at apex. Antananarivo, Fianarantsoa. Shaded, humid, granite and gneiss rocks; 1000–2000 m. Fl. February–May.

9.61a. **C. laeta** Schltr. var. **angavoensis** H.Perrier ex Hermans

Terrestrial; differs from the typical variety by its 7–8 radical leaves that are only 4–8 mm broad. Shaded, humid, granite and gneiss rocks; 1000–2000 m. Antananarivo, Fianarantsoa. Fl. March–May.

9.62. **C. villosa** Rolfe ex Hook.f. (p. 97)

Epiphyte or terrestrial; leaves 3–5, elliptic-oblong, 10–20 cm long; inflorescence glandular hairy; flowers rose-purple, not opening much, very hairy; lip 3-lobed, pale; spur less than half the length of the ovary. Antananarivo, Antsiranana, Toamasina, Toliara. In ravines; in humid, low- and medium-elevation forest; 1000–1200 m. Fl. March–June.

9.63. **C. spatulata** H.Perrier ex Hermans
(syn. *Cynosorchis spatulata* H.Perrier)

Terrestrial or lithophyte, 15–20 cm tall; leaves 2, radical, spreading on the soil, ovate or ovate-lanceolate, shortly acute , 4–9 × 2–4.5 cm; inflorescence 2–6-flowered; flowers large, pink, spotted with purple; sepals 10–15 mm long; lip entire, narrow, the base rectangular, expanded and a little angular in the lower quarter, indented in the middle and again a little broadened at the rounded spathulate apex, 15 × 2.5–3 mm; spur cylindrical at the base, a little dilated towards tip, 25–30 mm long. Toliara. On sandstone rocks and wet sand, not far from the sea; grassland; sea level–100 m. Fl. June.

9.58

9.64. C. souegesii Bosser & Veyret

Rhizomatous terrestrial, lacking tubers; leaves 2–4, basal, lanceolate, acute, 7–9 × 0.8–1.2 cm; inflorescence many-flowered; flowers pink; sepals 4.5–6 mm long; lip oblong-ovate, 3.5–4 mm long; spur slightly dilated above, 3–4 mm long; close to *C. rosellata* (p. 107) which has a longer spur and a different rostellum. Toliara. Marshes; up to 200 m. Fl. April.

9.65. C. confusa H.Perrier

Terrestrial, 10–30 cm tall; leaf shortly petiolate, lanceolate, 2–4 × 0.8–2 cm; raceme 1–3 cm long, subdensely 1–7-flowered, glandular hairy; sepals 6–7 mm long; lip 10 mm long, cureate at base, 3-lobed in apical half; side lobes small, 1 mm long; midlobe 6 × 1 mm, acute; spur 2.2–2.3 mm long, cylindrical, truncate; close to *C. petiolata* (p. 93) but differs in leaf and petal shape, with the lip in the shape of a halberd. Antsiranana. Mossy forest; shaded and humid rocks; marshland in peaty and marshy soil; 1400–2000 m. Fl. March–April.

9.66. C. gaesiformis H.Perrier

Plant 25–30 cm tall; leaf oblong, 8–15 × 3–4 cm; raceme subcorymbiform, 4 × 3.3 cm, sparsely glandular-hairy; dorsal sepal 4.7 mm long; lateral sepals 11 mm long; lip 10 mm long, dilated at base, javelin-shaped. Close to *C. spatulata* (p. 95) but differs in having a single radical leaf and the lip in the shape of a halberd. Toamasina. Humid, evergreen forest; 600–800 m. Fl. January.

9.67. C. gibbosa Ridl. *(p. 99, top and bottom left)*

Large lithophyte or terrestrial plant, 25–50 cm tall; leaf solitary, with the young growths carrying dark spots; inflorescence 10–40-flowered, glandular; flowers large, orange-red to raspberry red with a red mark at base of lip; sepals 12–15 mm long; lip 3-lobed, up to 29 mm long; spur 2–2.5 cm long. Antananarivo, Fianarantsoa, Toamasina, Toliara. Edges of forest; shaded, wet rocks in open but sheltered situations; generally in peaty or humus-rich soil in areas of seepage; on granite rock outcrop; 600–1800 m. Fl. August–May.

Cynorkis gibbosa Ridl. forma **aurea** Hermans has the same mottled foliage but the flowers are pale yellow. *(p. 97, bottom right)*

9.62

9.68. **C. calanthoides** Kraenzl.   *(p. 101, top left)*
(syn. *Cynorkis uncinata* Perrier ex Hermans)

Epiphyte, 10–25 cm tall; leaf radical, large, ovate-oblong, acute, obtuse at the base, almost as long as the inflorescence, 7.5–22 × 2.8–8.3 cm; inflorescence glabrous, dense, subglobose, 7–15-flowered; flowers pink or mauve with a white disc to the lip, rather large, with a few glandular hairs on the ovary and the outside of the sepals; sepals 10 mm long; lip 4-lobed, 15 mm long; side lobes subrectangular, subcrenulate at the apex; midlobe divided into 2 obovate-cuneiform lobules, crenulate at the apex; spur subcylindrical, 2.5 cm long. Antananarivo, Toamasina, Toliara. On *Pandanus*; 600–1000 m. Fl. February–March.

9.69. **C. angustipetala** Ridl.   *(p. 101, top right)*
(syn. *Cynosorchis angustipetala* (Ridley) T.Durand & Schinz)

Terrestrial, 25–45 cm tall; leaves 2, ligulate or linear-lanceolate, 10–30 × 1.5–2 cm; inflorescence many-flowered; flowers large, green with a rose-purple lip; sepals 10–11 mm long; lip 2–3 cm long, with 4 almost equal lobes, the front edge crenulate; spur 2.2–3 cm long, narrowed towards the tip. Antananarivo, Fianarantsoa, Toamasina. Common in grassland; forest and woodland margins; marshland in peaty soil; rocky outcrops; 900–2000 m. Fl. November–June.

9.69a. **C. angustipetala** var. **amabilis** (Schltr.) H.Perrier
(syn. *Cynosorchis amabilis* Schltr.)

Flowers with a light violet-red lip, the disc darker; spur reddish, up to 4 cm long. Fianarantsoa. Marshland; grassland; rocky outcrops; 1200–1800 m. Fl. January–February.

9.69b. **C. angustipetala** var. **bella** (Schltr.) H.Perrier
(syn. *Cynosorchis bella* Schltr.)

Flowers larger; lip 3–3.5 cm long. Fianarantsoa, Toliara. Near forest and on rocky outcrops; 1000–2000 m. Fl. February–March.

9.69c. **C. angustipetala** var. **moramangensis** H.Perrier ex Hermans
Differs from the typical variety in having the midlobe of the rostellum longer than the arms and not apiculate. It has many of the characteristics of *C. angustipetala* var. *oxypetala* (p. 100), but the anther is not apiculate and the flowers are shortly glandular on their exterior surface. Antananarivo, Toamasina. Grassland and rocky outcrops. Fl. January–February, July.

9.67

9.67

9.67

9.69d. C. angustipetala var. oligadenia (Schltr.) H.Perrier
(syn. *Cynosorchis oligadenia* Schltr.)

Leaves 2, radical; anther without apicule; middle lobe of the rostellum with the front margin entire. Antananarivo, Toamasina. Grassland and rocky outcrops. Fl. February.

9.69e. C. angustipetala var. oxypetala (Schltr.) H.Perrier
(syn. *Cynosorchis oxypetala* Schltr.)

Spur 20 mm long; anther apiculate, the apicule furrowed; middle lobe of the rostellum a little longer than the arms. Antananarivo, Fianarantsoa. Grassland and rocky outcrops; 1500–1900 m. Fl. February, June.

9.69f. C. angustipetala var. speciosa (Ridl.) H.Perrier
(syn. *Cynosorchis speciosa* Ridl.)

Leaves 2; spur shorter than the ovary; anther apiculate; rostellum with the middle lobe rectangular, shorter than the arms. Antananarivo. Grassland and rocky outcrops. Fl. February-April.

9.69g. C. angustipetala Ridl. var. tananarivensis H.Perrier ex Hermans
Differs from the species type in having flowers with a few small glands and an anther that has an obtuse apicule. It is similar to *C. angustipetala* var. *moramangensis* but the flowers have a few glands, the anther has an obtuse not furrowed apicule, and the midlobe of the rostellum is bi-sinuate at the front. Antananarivo. Grassland and rocky outcrops. Fl. February–April.

9.70. C. fastigiata Thouars   *(opposite, bottom left)*
(syn. *Orchis obcordata* Willem.; *Cynorkis obcordata* (Willem.) Schltr.; *C. isocynis* Thouars; *Orchis fastigiata* (Thouars) Spreng.; *O. mauritiana* Sieber ex Lindl.; *Habenaria cynosorchidacea* C.Schweinf.)

Terrestrial, occasionally epiphyte, 10–30 cm tall; leaves 1–2; inflorescence few-flowered; flowers c. 3 cm long, creamy white variably toned reddish-purple; sepals 5–6 mm long; lip 10–15 mm long, unequally 4-lobed; spur cylindrical-narrow, 18–30 mm long. Antananarivo, Antsiranana, Fianarantsoa, Toamasina, Toliara; also in the Comoros, Mascarenes and Seychelles. On edges of woods; riverine forest; littoral forest; evergreen forest, on base of trees; amongst rocks; and in grassland; sea level–2000 m. Fl. November–June.
LOCAL NAME *Sofinampory.*

9.68

9.69

9.70

9.71

9.70a. C. fastigiata var. ambatensis H.Perrier

Differs from the other varieties in its small flowers, 4 mm long sepals and petals, 5 mm long lip, 9 mm long spur, glabrous ovary and entire middle lobe of the rostellum. Antsiranana. Dry slopes, in the shade of scrub. Fl. January.

9.71. C. flexuosa Lindl.   *(p. 101, bottom right)*

(syn. *Gymnadenia lyallii* Steud.; *Cynorkis fallax* Schltr.; *Cynosorchis flexuosa* subsp. *fallax* (Schltr.) H.Perrier)

Terrestrial, 15–30 cm tall; leaf 1, rarely 2, linear-ligulate; inflorescence few-flowered; flowers with green sepals and petals and a bright yellow lip, generally with two small reddish spots at base; sepals 8–9 mm long; lip 4-lobed, 18 mm long; spur 16–18 mm long. Antananarivo, Antsiranana, Fianarantsoa, Mahajanga, Toamasina; also in the Comoros. Grassland; rocky outcrops; sea level–1700 m. Fl. October–August.

9.71a. C. flexuosa var. bifoliata Schltr.

(syn. *Cynosorchis flexuosa* var. *ambongensis* H.Perrier)

Terrestrial; differs from other varieties by the greater number of radical leaves (2–3), and the larger, white or pink lip, with the spot on the disc a rich violet-red; spur 15 mm. Mahajanga. Marshes in peaty soil. Fl. January.

9.72. C. gigas Schltr.   *(opposite, top left and right)*

Lithophyte or terrestrial, 20–30 cm tall; leaf lanceolate-elliptic, 14–18 × 3–4 cm; flower large, white spotted with purple on the sepals and with a large dark purple spot at the base of the lip; sepals 4 cm long, glandular; lip 4-lobed, 4 × 2.3 cm; spur 10 cm long, cylindrical, pendent; somewhat similar to *C. uniflora* (p. 104) but larger with a white lip and a 10 cm long spur. Antananarivo, Fianarantsoa, Toliara. Rocky outcrops; 800–2000 m. Fl. February–April.

9.73. C. purpurascens Thouars   *(opposite, bottom left)*

(syn. *Orchis purpurascens* (Thouars) Spreng.; *Cynorkis purpurascens* var. *praecox* (Schltr.) Schltr.; *C. praecox* Schltr.)

Epiphyte or terrestrial; leaf ovate-lanceolate, 20–40 × 4–10 cm, often developing after flowering; flowers mauve; sepals 10 mm long; lip 18–25 mm long, the base narrow and hollowed into a gutter, 4-lobed; spur up to 4 cm long. Antananarivo, Antsiranana, Mahajanga, Toamasina; also in the Mascarenes and possibly the Comoros. Shaded and humid rocks; humid forest, moss and lichen-covered trees; sea level–1500 m. Fl. September–May. LOCAL NAME *Katsakandrango, Fitsotsoka, Tsimpelany.*

9.72

9.72

9.73

9.74

9.74. C. uniflora Lindl.

(p. 103, bottom right)

(syn. *Cynosorchis grandiflora* Ridl.)

Terrestrial 10–20 cm tall; leaf lanceolate, 7.5–20 cm long, green with brownish-red spots below especially towards the base; flowers 1–2, large; sepals 2.2–2.6 cm long; lip c. 4 × 2.5 cm, 4-lobed, purple, with 2 paler spots on the disk; spur 3.5–4 cm long. Antananarivo, Fianarantsoa. Grassland; rocky outcrops, often on inselbergs; 1200–1450 m. Fl. December–March.

9.75. C. violacea Schltr.

Terrestrial up to 40 cm tall; leaves 4, basal, linear-lanceolate, 8–15 × 0.5 cm; inflorescence few–many-flowered; flowers rich pink-purple; sepals 6 mm long; lip 10–12 mm long, 4-lobed, the lateral lobes larger than the middle ones, erose; spur cylindrical, 26 mm long. Antsiranana, Fianarantsoa, Mahajanga. Rocky outcrops and in sand; sea level–1200 m. Fl. February–March.

IV. Section Imerinorchis

Lateral sepals and petals more or less fused at the base of the lip; rostellum divided into 2 arms up to the base of the anther filament, the middle lobe folded back down to the base, reduced to a small protrusion, or more or less developed in a blade narrowly placed against the filament, between the 2 anther chambers.

Key to groups of species in Section Imerinorchis

1. Leaves 3–7 at base or along stem 2
1. Leaves 1–2 3
2. Ovary and rachis glandular . . . . **9.76. C. brachyceras**
2. Ovary and rachis glabrous or very sparsely glandular at base **9.77–9.81**
3. Spur cylindrical, much longer than the ovary, at least 1.5 times as long **9.82–9.83**
3. Spur shorter than to slightly longer than the ovary **9.84–9.96**

9.76. C. brachyceras Schltr.

Terrestrial, 30–35 cm tall; leaves (2)3–4, narrowly linear, acute, 8–10 × 0.4–0.8 cm; flowers small, white spotted with rose-pink, the ovary and exterior of the dorsal sepal reddish, glandular; sepals 5–6 mm long; lip oblong-pandurate, 3-lobed in basal half, 6 × 2.5 mm; spur conical, 1.5–2 mm long. Antananarivo, Fianarantsoa. Marshland in peaty and marshy soil; c. 1500 m. Fl. November.

9·79

9.77. C. lindleyana Hermans

(syn. *Bicornella gracilis* Lindl.; *Cynorkis gracilis* (Lindl.) Schltr. non (Blume) Kraenzl.)

Slender terrestrial, 25–35 cm tall; leaves 3–5, linear, acuminate; flowers purple-pink; lip entire, a little pandurate; spur almost cylindrical, wide at the base and a little contracted towards the middle, 7–9 mm long. Antananarivo, Fianarantsoa, Toamasina. Grassland; marshes in peaty soil; dry river-beds; 1000–2500 m. Fl. September–April.

9.78. C. minuticalcar Toill.-Gen. & Bosser

Terrestrial 5–8 cm tall; leaves basal, grass-like, 2.5–5 × 0.15–0.35 cm; rachis up to 15-flowered, glabrous; flowers white flushed with rose-purple; sepals 4–5 mm long; lip flabellate, 4–5 × 3.5–4 cm, minutely 3-lobed at apex; spur saccate, 0.5 mm long; close to *C. brachyceras* (p. 104) but the lip shape is different and it has a much shorter spur and column wings. Antananarivo. Grassland; marshes; 2000–2500 m. Fl. February.

9.79. C. purpurea (Thouars) Kraenzl.   (p. 105)

(syn. *Habenaria purpurea* Thouars)

Terrestrial or lithophyte, 20–35 cm tall; leaves 4–5, lanceolate, 7.5–13 × 1–1.5 cm; inflorescence laxly 9–13-flowered; flowers purplish; sepals 6–6.5 mm long; lip 6 mm long, angular at the base, 3-lobed in the upper quarter; spur 4 mm long, hook-shaped. Antananarivo, Fianarantsoa; also in the Mascarenes. Wet rocks bordering streams; river margins; 800–2000 m. Fl. January–June.

9.80. C. guttata Hermans & P.J.Cribb   (p. 109, top left and right)

Terrestrial herb 15–35 cm tall. Leaf solitary, basal, strap-shaped, 10–50 × 3–8 cm. Inflorescence up to 35 cm long, produced from the new developing shoot, densely 10–35-flowered, the peduncle green suffused with brown-red; floral bracts 18–22 mm long, green with a reddish tip,. Flowers creamy white suffused with purple, the lip with a carmine-red disc with carmine-purple spots above, the spur green suffused with dark red; pedicellate ovary 38–42 mm long, slightly glandular hairy below. Sepals 7–10 × 2–5 mm. Lip 4-lobed, 18 × 20–23 mm; side lobes 8–9 × 8–9 mm; apical lobes 9–11 × 7–9 mm, crenate-serrate on the margins; spur parallel to the ovary, 26–32 mm long, slender, slightly swollen towards the apex. Antananarivo, Fianarantsoa, Toliara. On rocks and banks in montane and riverine forest; 800–1800 m. Fl. Nov.

9.81. C. rosellata (Thouars) Bosser

(syn. *Satyrium rosellatum* Thouars; *Habenaria mascarenensis* Spreng.; *Habenaria imerinensis* Ridl.; *Cynorkis imerinensis* (Ridl.) Kraenzl.)

Terrestrial or lithophyte, 15–40 cm tall, leaves 4–6, lanceolate, acute, 4.5–7 × 1–1.5 cm; raceme laxly 3–12-flowered; flowers small, lilac, spotted with red; sepals 3–4 mm long; lip lanceolate, 4–6 × 1.3 mm; spur 6–9 mm long, cylindrical. Antsiranana, Fianarantsoa, Toamasina, Toliara; also in the Mascarenes. Mossy forest; river margins; shaded and humid rocks; 500–1500 m. Fl. August–November, March.

9.82. C. betsileensis Kraenzl.

Flowers pale yellow, marked with brown stripes; sepals 4–5 mm long; lip linear, c. 10 × 2 mm, a little expanded towards the apex; spur slender, filiform, 15 mm long. Fianarantsoa. Fl. unknown.

9.83. C. tenuicalcar Schltr.

Terrestrial, 13–25 cm tall; leaf solitary, oblanceolate, 4–10 × 0.6–1.4 cm; inflorescence 3–10-flowered; flowers reddish violet; sepals 4 mm long; lip oblong, 5–8 × 2.5 mm; spur filiform, 2.7 cm long; related to *C. lancilabia* (p. 110) but the lip is oblong and has a small callus at the base and the spur is different. Antsiranana. Lichen-rich forest; 1700–2000 m. Fl. October–January.

9.83a. C. tenuicalcar subsp. andasibeensis H.Perrier ex Hermans

Terrestrial; flowers pale red; lateral sepals 7–9-veined; petals 3-veined; lip with a lateral tooth on each side towards the base; ovary and the floral bracts with a few glands. Antananarivo, Toamasina. Mossy forest; c. 1000 m. Fl. unknown.

9.83b. C. tenuicalcar subsp. onivensis H.Perrier ex Hermans

Epiphyte similar in habit to the typical subspecies; flowers pale red; spike, floral bracts, ovary and the outside of the perianth covered by scattered hairs; lateral sepals 5–6-nerved; lip ovate-lanceolate or oblanceolate, wider in the middle or above, more-or-less elongate (but always narrowed towards the base) and lacking a callus at the base. Antananarivo, Toamasina. Lichen-rich forest on moss- and lichen-covered trees; c. 1300 m. Fl. February.

9.84. C. ambondrombensis Boiteau

Slender terrestrial with a small leaf; lip lozenge-shaped, the front margin finely fimbriate, 18 × 15 mm; spur cylindrical, 3.5 cm long. Fianarantsoa. Lichen-rich forest on wet rock; 1800 m. Fl. unknown.

9.85. C. disperidoides Bosser

Terrestrial or lithophyte, 35–42 cm tall; leaf lanceolate, acuminate, 8–9 × 1 cm, petiolate; inflorescence glandular, 5–6 cm long; flowers rose, spotted with purple; sepals 5–8 mm long; lip filiform, 5–6 mm long; spur clavate, 8–10 mm long. Similar in appearance to *C. papilio* (p. 112) but glandular hairy, in general appearance reminiscent of *Disperis*. Fianarantsoa. Mossy forest; on wet rocks; marshland; 1500–2000 m. Fl. April.

9.86. C. ericophila H.Perrier ex Hermans

Plant slender, 30–40 cm tall; leaf lanceolate, acute, 6–7 × 1.1–1.6 cm, narrowing below into a long petiole; inflorescence 3–6-flowered; flowers pink, papillose; sepals 6–10 mm long; lip oblong, 8 × 4 mm, with an obtuse tooth on each side in the lower third; spur 6 mm long, a little expanded towards the tip. Antsiranana. Montane, ericaceous scrub; c. 2400 m. Fl. April.

9.87. C. hologlossa Schltr.

Terrestrial, 12–22 cm tall; leaf solitary, lanceolate or elliptic-lanceolate; flowers pink or lilac, 16 mm long; sepals 5–6.5 mm long; lip entire, ligulate, 7–8 × 3–4 mm; spur narrowly cylindrical, obtuse, 6.5 mm long. Antsiranana. Lichen-rich, montane forest; c. 2000 m. Fl. January.

9.87a. C. hologlossa var. angustilabia H.Perrier ex Hermans

Differs from the typical variety in being taller, and in having shorter floral bracts which are about a third the length of the ovary, a shorter dorsal sepal, a shorter lip that is slightly expanded in the lower quarter, and a shorter spur. Antsiranana. Montane, ericaceous scrub; c. 2600 m. Fl. April.

9.80

9.80

9.92

9.99

9.87b. C. hologlossa var. gneissicola Bosser

Terrestrial or lithophyte; differs from the typical form in the shape of the rostellum. Antsiranana. Lichen-rich forest; on granite and gneiss; 2000–2200 m. Fl. March.

9.88. C. lancilabia Schltr.

Terrestrial, 22–26 cm tall; leaf solitary; flowers violet; sepals 7 mm long; lip entire, lanceolate, acute, 7 × 2.3 mm; spur incurved in a complete curl, c. 8 mm long. Antsiranana. Forest; c. 2000 m. Fl. January.

9.89. C. marojejyensis Bosser

Terrestrial up to 30 cm tall; inflorescence 4–5-flowered; flowers glabrous; sepals 4–5 mm long; lip rhombic, entire, obtuse, 8 × 4 mm; spur slightly incurved, globular at apex, 1 cm long; column and the rostellum with the midlobe developed into a flattened blade. Antsiranana (Marojejy). Marshy hollows in peaty soil; 2000–2100 m. Fl. March–April.

9.90. C. mellitula Toill.-Gen. & Bosser

Epiphyte or terrestrial; lip pale yellow, spotted violet; spur green, flattened and bilobed at the tip. Antananarivo, Fianarantsoa. Humid, mossy, evergreen forest; 1900–2000 m. Fl. January.

9.91. C. muscicola Bosser

Terrestrial up to 50 cm tall; inflorescence axis glandular, 18–20 cm long, laxly flowered; flowers glabrous, sepals 4 mm long; lip ligulate, obtuse, 7 × 3 mm; spur cylindrical, 8–9 mm long, narrow at the base but enlarged and globulose at the tip. Fianarantsoa, Toliara. Mossy forest in humus; 1500–2000 m. Fl. unknown.

9.92. C. nutans (Ridl.) H.Perrier　　　　　　*(below and p. 109, bottom left)*
(syn. *Habenaria nutans* Ridl.; *Cynosorchis nutans* var. *campenoni* H.Perrier)
Terrestrial or epiphyte, up to 30 cm tall; leaves 1–2, broadly lanceolate, acute, 10–12 × 2–2.5 cm; inflorescence many-flowered; flowers white and rose-purple; sepals 7–13 mm, the laterals twice as long as the dorsal sepal; lip narrow and entire, ligulate, 8 × 1.3 mm; spur c. 12 mm long. Antananarivo, Antsiranana, Toamasina, Toliara. Humid, highland, evergreen forest, in moss; 600–1400 m. Fl. January–October.

9.92

9.93. C. papilio Bosser

Terrestrial 20–40 cm tall; leaf solitary, linear-lanceolate, 4–8 × 0.5 cm; raceme 4–6-flowered; flowers rose-purple; dorsal sepal 6 mm long; lateral sepals erect, falcate, 9–9.5 mm long; petals, falcate, lanceolate, acute, with undulate margins, spotted with purple; lip linear, acute, 6–7 × 0.6–0.7 mm, curled at tip; spur 6 mm long, slightly apically dilated. Anatanarivo, Fianarantsoa, Toamasina. Montane *Philippia* scrub on quartz soil; 1100–1200 m. Fl. November, March–June.

9.94. C. stenoglossa Kraenzl.

Plant 13–25 cm tall; leaves 1–2, very broad; flowers up to 10, medium-sized, violet or purple; sepals 4.5–9 mm long; lip with a narrow blade, 5–8 × 2.5–2.8 mm; spur 10–14 mm long, cylindrical. Antananarivo, Antsiranana, Fianarantsoa, Toamasina. Humid, evergreen forest; 600–1500 m. Fl. August–December.

9.94a. C. stenoglossa var. pallens H.Perrier ex Hermans

Epiphyte or terrestrial; differs from the typical variety by the white or lilac-tinted flowers, the lip with 9 nerves, the 10–11 mm long spur being shorter than the 14–17 mm long ovary, the narrower viscidia, more acute at both ends, and the stigmatic processes that are fused together at the base and apex, with an elliptic gap in the middle. Antsiranana. Humid, medium-elevation forest; c. 1200 m. Fl. unknown.

9.95. C. tristis Bosser

Terrestrial 10–25 cm tall; leaf elliptic or lanceolate, acute, 2–3.5 × 0.7–1.2 cm; raceme 3–10-flowered, glabrous; flowers glabrous; sepals 4–5 mm long; lip oblong or narrowly lanceolate, subacute, 4–5 × 1.5–2 mm, with a small callus at the base; spur clavate, 5–6 mm long; rostellum short. Mahajanga. Mossy forest. Fl. January.

9.96. C. zaratananae Schltr.

Epiphyte 11–17 cm tall; leaf oblong-elliptic, 6–12 × 1.5–3 cm; inflorescence glandular, 5–10-flowered; flowers whitich green; sepals 4.5–6.5 mm long; lip ligulate with a small triangular tooth at each side near base, 7 × 3 mm; spur 14–15 mm long; related to *C. stenoglossa*, but with smaller flowers and a longer pointed spur. Antsiranana. Mossy forest; 1700–2200 m. Fl. January.

V. Section Monadeniorchis

Lip entire, folded into a narrow keel, very concave below. Rostellum very elongated, 5–6 mm from the anther to the front extremity, attenuate-folded into a keel in lower half and entire at the front, with a small median hollowed keel on top, truncate at the front and extended by a small rostrum incurved into a gutter above the viscidium; caudicles and canals united by a common viscidium, placed below the small rostrum which ends the rostellum.

9.97. C. monadenia H.Perrier

Terrestrial 20–35 cm tall; leaves 2, prostrate, ovate-suborbicular, 8–13 × 7–11 cm; inflorescence many-flowered, glandular; sepals 4–7 mm long; lip entire, lanceolate, obtuse, 7 × 2 mm; spur 6 mm long. Fianarantsoa. Quartz rock, rocky outcrops; 1600–1800 m. Fl. February.

VI. Section Lemuranthe

Rostellum with short arms, adnate to the anther chambers and not longer than them, the middle lobe an obtuse blade narrowly placed against the filament, between the 2 anther chambers; canals and caudicles minute, scarcely 0.1 mm long.

9.98. C. gymnochiloides (Schltr.) H.Perrier

(syn. *Habenaria gymnochiloides* Schltr.; *Lemuranthe gymnochiloides* (Schltr.) Schltr.)

Lithophyte or terrestrial, 15–25 cm tall; leaves 2–5, oblanceolate-elliptic, 4–6.5 × 0.8–2 cm; raceme short, densely-flowered, glandular-hairy; flowers glabrous, pink or violet; sepals 3 mm long; lip entire, narrowly ligulate and obtuse, 4 mm long; spur cylindrical, obtuse, 4 mm long. Antananarivo, Fianarantsoa. Rocky outcrops; 1700–2200 m. Fl. February–March.

9.99. C. subtilis Bosser   *(p. 109, bottom right)*

Terrestrial 6–15 cm tall; leaves 3–5, basal, oblong, acute, 2–4.5 × 0.3–0.5 cm; inflorescence up to 10-flowered; flowers pink, sparsely hairy; sepals 2.5 mm long; lip ovate, 2.5 mm long; spur 2.5 mm long, hooked at summit. Toamasina (Maroantsetra). In humid coastal forest; up to 10 m. Fl. October–November.

10. PHYSOCERAS

A genus of ten species endemic to Madagascar. Small terrestrials or epiphytes growing from small tubers. Stem, erect, slender, covered at the base with black or brown sheaths, with a sessile leaf in the middle. Inflorescence terminal, 2–few-flowered. Dorsal sepal forming a hood with the petals over the column. Lip 3-lobed, the midlobe more or less bilobulate, spurred at base; spur often somewhat dilated in middle or at apex, straight or recurved. Column with two anther loculi and canals.

Key to species of *Physoceras*

1. Spur curved into a hook at apex 2
1. Spur diverse, not hooked at tip 3
2. Leaf cordate, 4.5–10 × 2.4.6 cm .... **10.1. P. bellum**
2. Leaf lanceolate or elliptic-lanceolate, 10–13 × 2.7–3.5 cm .... **10.2. P. epiphyticum**
3. Spur bifurcate at tip .... **10.5. P. bifurcum**
3. Spur not bifurcate at tip 4
4. Inflorescence 1–2-flowered; flowers white; rostellar arms absent; sepals and petals 20 mm long .... **10.4. P. betsomangense**
4. Inflorescence usually 3- or more-flowered; flowers pink to violet; rostellar arms present; sepals and petals 11 mm or less long 5
5. Leaf obovate; spur 3 mm long, slightly inflated at tip .... **10.7. P. mesophyllum**
5. Leaf elliptic, ovate or lanceolate; spur more than 7 mm long, variously shaped 6
6. Lip side lobes lacerate; sepals 12 mm or more long .... **10.6. P. lageniferum**
6. Lip side lobes entire; sepals 10 mm long or less 7
7. Spur 10 mm long; flowers pink .... **10.3. P. australe**
7. Spur 8 mm long or less; flowers mauve to purple 8
8. Leaf 10–11 cm long; spur vesicular at tip .... **10.8. P. perrieri**
8. Leaf 5.5 cm long or less; spur clavate but not vesicular at tip 9
9. Leaves cordate at base; flowers orange-purple; ovary glabrous ...... **10.9. P. rotundifolium**
9. Leaves ovate; flowers violet-red; ovary glandular .... **10.10. P. violaceum**

10.1. P. bellum Schltr.

Lithophyte or terrestrial; leaf 4.5–10 × 2–4.6 cm, cordate at the base; flowers 1.5 cm long, white or lilac; dorsal sepal 7.5 mm long; lip angular at the base, deeply 3-lobed, 1.5 cm long, 1.75 cm broad; spur 7 mm long. Antsiranana. Lichen-rich, montane forest; 1500–2000 m. Fl. October–January.

10.2. P. epiphyticum Schltr.

Epiphyte differing from *P. bellum* (p. 114) by the narrower lanceolate leaves, 10–13 × 2.7–3.5 cm, larger lilac flowers, dorsal sepal 1 cm long, deeply 3-lobed lip 1.3 cm long, 2.5 cm broad, 8 mm long spur and distinct anther. Antsiranana. Mossy, montane forest; river margins on moss- and lichen-covered trees; 1600–1700 m. Fl. April, October.

10.3. P. australe Boiteau

Epiphyte or terrestrial plant up to 15 cm tall; flowers c. 10 mm long, pink; lip very narrowly 4-lobed; spur c. 1 mm long. Fianarantsoa. Mossy, montane forest; 1800 m. Fl. unknown.

10.4. P. betsomangense Bosser

Terrestrial distinguished by its narrowly ovate leaf, 6.5 7.5 × 1.2–2 cm, 1–2-flowered inflorescence, and large white flowers with a bottle-shaped spur, conical at the tip; sepals 10–11 mm long; lip 4-lobed 25 × 13 mm. Antsiranana, Fianarantsoa. Montane, ericaceous scrub; 1000–1500 m. Fl. November.

10.5. P. bifurcum H.Perrier

Terrestrial; leaf linear-lanceolate, 5–12 cm long; flowers up to 10, rose-lilac with a rose-violet lip; sepals 12 mm long; lip 4-lobed, 20 × 26 mm; spur 6 mm long, terete, apex expanded, flattened and forked. Antsiranana (Marojejy). Peaty hollows at high elevation; 2000–2500 m. Fl. March–April.

10.6. P. lageniferum H.Perrier

Terrestrial or lithophyte, up to 60 cm tall; leaf narrowly lancolate, 5–16 x 1–4 cm; flowers 6–12; sepals and petals 12 mm long; lip 4-lobed, 16 × 22 mm, side lobes lacerate; spur 13 mm long, inflated in apical 2/3; rostellum distinct by the absence of arms; anther fused. Antsiranana (Mt Beondroka). Humid, mid-elevation forest and rocks; 1000–1500 m. Fl. unknown.

10.7. **P. mesophyllum** (Schltr.) Schltr.

(syn. *Cynosorchis mesophylla* Schltr.)

Terrestrial plant 12–16 cm tall; leaf 4.5–6 × 1–1.7 cm; flowers 3–6, mauve; sepals 6 mm long; lip clawed at base, 4-lobed, 9 mm long; spur obovoid-apiculate, 3 mm long, slightly inflated at the tip. Antsiranana. Mossy and lichen-rich forest; c. 2000 m. Fl. October, December.

10.8. **P. perrieri** Schltr.

Terrestrial plant 30–45 cm tall; leaf lanceolate or elliptic-lanceolate, acuminate, 10–11 × 2–3 cm; inflorescence 3–10-flowered; flowers purplish; ovary glabrous; sepals 7.5 mm long; lip deeply 3-lobed, 13 × 16 mm; spur 6 mm long, inflated-vesicular into a sphere at the tip. Antsiranana. Dry forest; humid, evergreen forest on plateau; 1000–1800 m. Fl. January.

10.9. **P. rotundifolium** H.Perrier

Epiphyte; leaf broadly ovate, acute, c. 3 × 2.5 cm, the base rounded-subtruncate; lip 3-lobed in front; spur obtuse, claviform, 8 mm long. Antsiranana. Humid, evergreen forest on plateau. Fl. July.

10.10. **P. violaceum** Schltr.   *(opposite, top left)*

Epiphyte; leaf elliptic, acuminate, 5.5 × 3.3 cm; flowers violet-red; ovary glandular; sepals 8 mm long; lip deeply 3-lobed above, 18 × 10 mm; spur 7.5 mm long, in a long cylindrical clavate, contracted towards the base. Antananarivo, Antsiranana, Toamasina. Humid, evergreen forest; c. 500 m. Fl. August.

11. DISA

A genus of about 130 species in tropical and South Africa, Madagascar and the Mascarenes. Four species in Madagascar. Terrestrial herbs growing from underground ellipsoidal to ovoid tubers. Stems leafy. Leaves often merging into the bracts. Flowers often brightly coloured. Dorsal sepal hooded or helmet-shaped, spurred on outer surface. Lateral sepals more or less spreading. Petals free from the dorsal sepal. Lip small, linear or linear-spathulate, pendent, lacking a spur or callus. Column short; anther with two parallel loculi; pollinia 2, sectile, each attached to its own viscidium; stigma cushion-like below the small 3-lobed rostellum.

10.10

11.1

11.2

11.2

Key to species of *Disa*

1. Flowers red; dorsal sepal erect, orange, red-spotted; lateral sepals spreading widely . . . . **11.4. D. incarnata**
1. Flowers greenish white, lilac or blueish mauve; lateral sepals not spreading widely 2
2. Flowers greenish white, almost hidden by the bracts . . . . **11.1. D. andringitrana**
2. Flowers lilac to bluish mauve, not hidden by the bracts 3
3. Flowers small, 7 mm long, bluish violet; spur cylindrical-subclavate, 5 mm long . . . . **11.2. D. buchenaviana**
3. Flowers 15–20 mm long, light mauve; spur conical-tapering, 8–15 mm long . . . **11.3. D. caffra**

11.1. **D. andringitrana** Schltr. *(p. 117, top right)*

Plant up to 60 cm tall; flowers 5–7 mm long, greenish-white, almost hidden by the floral bracts. Fianarantsoa, Toliara. Montane, ericaceous scrub; 2400–2600 m. Fl. January.

11.2. **D. buchenaviana** Kraenzl. *(p. 117, bottom left and right)*
(syn. *Disa rutenbergiana* Kraenzl.)

Plant 35–70 cm tall; flowers c. 7 mm long, bright blue-mauve, not hidden by the floral bracts. Antananarivo, Fianarantsoa. Wet grassland; montane ericaceous scrub; rocky outcrops; 1500–2400 m. Fl. November–February.
LOCAL NAME *Vary mangatsiaka*.

11.3. **D. caffra** Bolus *(opposite, top and bottom left)*
(syn. *Disa compta* Summerh.; *D. perrieri* Schltr.)

Plant 25–50 cm tall; flowers 15–20 mm long, light mauve; lip lanceolate-spathulate; spur conical. Antananarivo, Fianarantsoa; also in E and S Africa. Marshes and wet grassland; sea level–2000 m. Fl. October–February.

11.4. **D. incarnata** Lindl. *(opposite, bottom right)*
(syn. *Disa fallax* Kraenzl.)

Plant c. 50 cm tall; flowers 8–12 mm long, bright orange-red. Antananarivo, Fianarantsoa. Grassland, marshes or near streams; 1300–2000 m. Fl. November–March.

11.3

11.3   4

12. BROWNLEEA

A small genus of seven species in tropical and South Africa and Madagascar. Two species in Madagascar. Small terrestrials, less commonly epiphytes or lithophytes growing from ovoid or ellipsoid tubers. Stems leafy. Leaves along stem, lanceolate. Inflorescence terminal, few–many-flowered. Flowers resupinate, small. Dorsal sepal hooded, spurred at back; spur short, often hooked. Lateral sepals spreading. Petals adnate to dorsal sepal. Lip linear, often curved upwards. Column erect; stigmas very short; pollinia 2, sectile.

Key to species of *Brownleea*

1. Plant glabrous on basal sheath; flowers mauve or bluish; spur 13–26 mm long ....**12.1. B. coerulea**
1. Plant hispid on basal sheath; flowers white, tinged with green and brown; spur 3–5 mm long ....**2.2. B. parviflora**

12.1. B. coerulea Harv. ex Lindl.   *(opposite, top left)*
(syn. *B. madagascarica* Ridl.)

Plant terrestrial, epiphytic or lithophytic; inflorescence subimbricate; flowers c. 20 mm long, mauve or bluish; dorsal sepal galeate; lip very small, linear, erect in front of the stigma, c. 1 mm long; spur slender, cylindrical, 13–26 mm long. Antananarivo, Fianarantsoa; also in South Africa. In forest and woodland remnants; humid, evergreen forest on plateau; mossy forest; and rocky outcrops; 1500–2500 m. Fl. February–June.

12.2. B. parviflora Harv. ex Lindl.
(syn. *B. perrieri* Schltr.)

Plant terrestrial, tall; basal sheath black-hispid; inflorescence densely 20–60-flowered; flowers small, white with a slight green or brown tone; lip minute, 1 mm long; spur 3–5 mm long. Antananarivo, Fianarantsoa; also in E and W Africa. In grassland, marshes in peaty soil and on rocky outcrops; 1000–2500 m. Fl. February–April.

13. SATYRIUM

A large genus of about 100 species, mostly in tropical and South Africa but extending to tropical Asia and China. Five species in Madagascar. Terrestrials growing from subterranean ovoid tubers. Stems erect, leafy, some species have separate sterile and fertile stems. Leaves 2–several, prostate to suberect, lanceolate to circular. Inflorescences terminal, densely many-flowered; bracts leafy, often reflexed. Flowers non-resupinate, white, yellow, pink, purple or red. Sepals and petals subsimilar, small, ligulate to obovate, united at base. Lip uppermost in flower, hooded, two-spurred on back; spurs short to elongate. Column within hood, stalked; anther-loculae saccate and pendent; pollinia 2, sectile.

12.1

13.2

13.5

Key to species of *Satyrium*

13.1. S. rostratum Lindl.   (*opposite*)

(syn. *Satyrium gigas* Ridl.)

Plant up 100 cm tall; largest leaf up to 60 cm long; raceme 5 cm in diameter; flowers pink; spurs 35–45 mm long. Antananarivo, Fianarantsoa, Mahajanga. Grassland, rocky outcrops, and forest and woodland margins; 1200–2000 m. Fl. December–April.

LOCAL NAME *Rasamoala*.

13.2. S. amoenum (Thouars) A.Rich.   (*p. 121, bottom left*)

(syn. *Satyrium gracile* Lindl.)

Plant up to 45 cm tall; leaves 2–3, basal, elliptic, largest up to 20 cm long; flowers white to pink, sometimes speckled with dark pink, hood 7–9 mm long; spurs 20–35 mm long. Antananarivo, Antsiranana, Fianarantsoa, Mahajanga, Toliara; also in the Comoros and Mascarenes. In dry meadows, grassland, *Philippia* scrub and rocky outcrops; 1000–2200 m. Fl. October–June.

13.2a. S. amoenum var. tsaratananae H.Perrier ex Hermans

Differs from the typical variety in having 4 mm long sepals and petals and 12 mm long spurs; sepals not apiculate, dorsal sepal is 3-veined, the petals 1-veined. Antsiranana. Montane, ericaceous scrub; c. 2400 m. Fl. April.

13.3. S. baronii Schltr.

Leaves suberect in lower half of stem; inflorescence cylindrical, densely many-flowered; flowers white to purple-pink, c. 6 mm long; sepals and petals spathulate, fused for more than half their length; spurs 7–11 mm long, as long as the ovary. Antananarivo. Grassland and amongst rocky outcrops; 1100–2200 m. Fl. March–May.

13.1

13.4. **S. perrieri** Schltr.

Leaves 2, basal; inflorescence long and cylindrical, 13–25 × 2 cm, narrow and densely flowered; flowers rose-pink; hood 5–7 mm long; spurs up to 12 mm long. Very close to *S. amoenum* (p. 122) but flowers smaller and spurs shorter. Antananarivo, Antsiranana, Fianarantsoa, Toliara. Grassland, montane ericaceous scrub and rocky outcrops; 1000–1600 m. Fl. March–April.

13.5. **S. trinerve** Lindl.   *(p. 121, right)*
(syn. *Satyrium atherstonei* Rchb.f.)

Inflorescence densely flowered, pyramidal to cylindrical; bracts white, much longer than flowers; flowers c. 5 mm long, white, the segments fused for over almost 2 mm; spurs obtuse, less than 2.2 mm long. Antananarivo, Antsiranana, Fianarantsoa, Mahajanga, Toamasina, Toliara; also in the Comoros and tropical E Africa. Grassland or marshland, usually in moist localities; 500–2000 m. Fl. December–June.

14. DISPERIS

A genus of about 75 species, mostly in tropical and southern Africa and Madagascar but extending east into tropical Asia and the Malay Archipelago. In Madagascar, 21 species. Small erect terrestrials growing from spherical to elliptic tubers. Stems glabrous, leafy. Leaves 1–4, sessile or shortly stalked, sometimes patterned on upper surface and purple beneath. Inflorescence 1–several-flowered. Dorsal sepal hooded, sometimes spurred; lateral sepals spreading, often saccate or shortly spurred. Petals hyaline, adnate to dorsal sepal. Lip complex, hidden within the dorsal sepal. Column complex, anther loculi at the end of long curled stalks; pollinia 2, sectile.

Key to species of *Disperis*

1. Hood conical, pointed like a dunce's cap .... **14.1. D. anthoceros** var. **humbertii**
1. Hood not conical 2
2. Leaves two, opposite, sessile, more or less in the middle of the stem 3
2. Leaves 1–2, alternate 7
3. Lateral sepals joined for basal half to a third 4
3. Lateral sepals free to the base 5
4. Hood glabrous .... **14.4. D. oppositifolia**
4. Hood ciliate at apex and on inner face .... **14.6. D. ciliata**
5. Lateral sepals 6–10 mm long, lacking a spur or pouch; hood broader than long .... **14.5. D. latigaleata**
5. Lateral sepals up to 6 mm long, spurred or pouched; hood longer than broad 6
6. Lip apex 5-lobed; lateral sepals 3–4 mm long .... **14.2. D. trilineata**
6. Lip apex trilobed; lateral sepals 4–6 mm long .... **14.3. D. similis**
7. Hood distinctly broader than long 8
7. Hood distinctly longer than broad or more or less orbicular 13

8. Lateral sepals joined in basal half 9
8. Lateral sepals free to the base or joined for the basal quarter 11
9. Dorsal sepal shorter than the lateral sepals; lateral sepals obscurely pouched in middle; lip papillose ....**14.12. D. masoalensis**
9. Dorsal sepal longer or equalling lateral sepals; lateral sepals pouched; lip pubescent or tomentose 10
10. Sepals of equal length; lip midlobe appendage sessile ....**14.8. D. hildebrandtii**
10. Dorsal sepal 12–13 mm long; lateral sepals 8–10 mm long; lip midlobe stalked ....**14.10. D. lanceana**
11. Lip with a simple ovate appendage covered in dense papillae ....**14.7. D. discifera**
11. Lip appendage variously lobed, glabrous with a terminal papillose or hairy lobule 12
12. Hood shallow, obovate, 12–15 mm long, 16–20 mm broad; lip terminal lobe reniform, 2–3 mm long, not hairy ....**14. 14. D. perrieri**
12. Hood very concave, 8 mm long, 15 mm broad; lip terminal lobe tongue-shaped, 3.5–5 mm long, hairy ....**14.9. D. ankarensis**
13. Lowermost leaf with a petiole forming a sheathing base 14
13. Leaves sessile, not sheathing at the base 16
14. Lateral sepals joined for the basal quarter ....**14.11. D. humblotii**
14. Lateral sepals free to the base 15
15. Lateral sepals 8–11 mm long; lip terminal lobe reniform, 1.8–3 mm long, papillose ....**14.15. D. saxicola**
15. Lateral sepals 13–14 mm long; lip terminal lobe tongue-shaped, glabrous, 5.5. mm long ....**14.19. D. bathiei**
16. Lateral sepals up to 5 mm long; lip appendage 3-lobed, side lobes conical, inrolled, midlobe narrowly triangular, flat ....**14.16. D. majungensis**
16. Lateral sepals 6 mm long; lip appendage not as above 17
17. Leaves distinctly cordate at the base ....**14.18. D. tripetaloides**
17. Leaves rounded at the base 18
18. Base of leaves clasping the stem ....**14.21. D. bosseri**
18. Base of leaves not clasping the stem 19
19. Laterals sepals fused in basal half.............................**14.13. D. falcatipetala**
19. Lateral sepals free to base 20
20. Leaves narrowly ovate or ovate, 5.8–6.2 x 1.8–3.6 cm; lip apical part 7 mm long, oblong, deeply notched at the base, dilated into a densely tomentose subrhombic lobe ....**14.17. D. erucifera**
20. Leaves lanceolate, 1.8–3.6 x 0.7–1.5 cm; lip apical part 5.5 mm long, with two lateral rounded papillose lobes and with an apical ovate papillose lobe ....
....**14.20. D. lanceolata**

14.1. D. anthoceros Rchb.f. var. **humbertii** (H.Perrier) la Croix
(syn. *D. humbertii* H.Perrier)

Flowers white, with an erect, conical, 7–9 mm long spur on hood. Antsiranana. In ravine with bamboos or on basalt and sandstone; 950–1250 m. Fl. March. The typical variety is widespread in tropical Africa.

14.2. **D. trilineata** Schltr.

Leaves dull green on the upper side with 3 main rose-pink veins, the lower surface red-purple; flowers dull pink or violet, rather small; hood 3–4 mm long; lateral sepals 3–4 mm long, free more or less to the base, obliquely elliptic, each with a conical spur around the middle; lip bearing at the top a 5-lobed appendage, the side lobes short and farinose-tomentose, the midlobe larger, lanceolate and glabrous. Antsiranana; also in the Comores. In humus in moist forest; 200–400 m. Fl. January–March.

14.3. **D. similis** Schltr.   *(opposite, top left)*

Similar to *D. trilineata* but lip with a 3-lobed appendage at the apex, the 2 side lobes diverging and curved, pubescent, the midlobe triangular, shorter than the side lobes, glabrous, obtuse at the apex. Antananarivo, Antsiranana, Fianarantsoa. In humus in mossy forest; 1500–2000 m. Fl. December–January.

14.4. **D. oppositifolia** Sm.   *(opposite, top right)*

Flowers off-white to pale pink with red, magenta or mauve lines on the upper part of the petals, glabrous; lateral sepals clearly joined at the base for almost half of their length; lip clawed, with a 3-lobed appendage, slightly pubescent, the midlobe short, triangular, the side lobes linear, curving back, 3–4 times as long as the midlobe. Antananarivo, Antsiranana, Fianarantsoa, Toamasina, Tolanaro; also in the Comores, Mauritius and Réunion. In humid forests, sometimes in moss on the base of tree trunks, on rocks by streams, in riparian forest with *Pandanus* sp., and sometimes in pine plantations; up to 1200 m. Fl. September–February.

14.5. **D. latigaleata** H. Perrier

Similar to following species but lateral sepals free to base or a little united, 6–10 × 5–7 mm, lacking a spur or pouch; lip with a claw c. 4 mm long, the basal half joined to the column, bearing a 3-lobed appendage in front; midlobe pendent, pubescent on top except for the subspathulate apex which is glabrous; side lobes slightly shorter, cylindrical, slightly thickened and rounded the tips, slightly curved, pubescent on both surfaces. Antananarivo, Fianarantsoa. In shade, terrestrial or on moss-covered boulders, on river sides, in forest on laterite and gneiss; epiphytic; 600–2000 m. Fl. September–December.

14.3

14.4

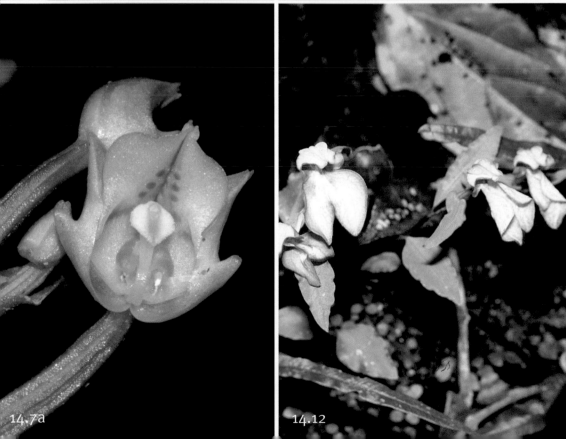

14.7a

14.12

14.6. **D. ciliata** Bosser

Similar to *D. latigaleata* (p. 126) but distinguished by its ciliate petals and by the terminal appendage of its lip in which the midlobe is relatively short and covered by a short papillate pubescence; in *D. latigaleata*, that lobe is lengthened into a linear glabrous extension that is slightly dilated at the tip. Antsiranano, Mahajanga. Under trees in humid montane forest on laterite of gneiss, transition to lichen-rich forest; 1700–1900 m. Fl. January–February.

14.7. **D. discifera** H.Perrier

Flowers clear rose-pink with two lines of green or brown spots on the basal part of the hood; hood 8–12 mm long; lateral sepals 11–13 × 4.5–7 mm, scarcely joined for a short distance at the base, obliquely obovate, each with a sac-like spur c. 1.5 mm long; lip with a narrow claw c. 3.8 mm long, adnate to the column at the base then bent over and bearing at the tip an ovate appendage, with a dense papillose yellow pubescence above, glabrous below, with a glabrous apiculus at the tip. Antananarivo, Antsiranana, Fianarantsoa, Mahajanga, Toamasina. In moist forest, in shade and open areas; 800–2250 m. Fl. April–July.

14.7a. **D. discifera** H.Perrier var. **bourbonica** (H.Perrier) Bosser

(p. 127, bottom)

Differs from the species type in its flowers being much smaller with a concave hood, 6–7 × 7–8 mm, rounded or subquadrate at apex, and in its 6.5–7.5(–8) mm long lateral sepals. Antananarivo, Antsiranana, Fianarantsoa; also in the Mascarenes. Under trees in humid forests and ericaceous scrub; in the Ankaratra massif it also grows in pine (*Pinus patula*) plantations; 1200–2350 m. Fl. February–August.

14.8. **D. hildebrandtii** Rchb.f.

Leaves 1–3 in upper half of stem, 1.5–4 × 0.5–1.8 cm, lanceolate, acute, clasping the stem at the base; flowers white, lined or striped with pale rose-pink or rose-purple; hood 8–10 × 1 mm; lateral sepals 8–10 × 3–6 mm, joined in the basal half, each with a deep spur near the base; lip joined to the column at base, narrowly cylindrical above and then enlarged, bearing at the top a 3-lobed appendage, the 2 lateral lobes short, thick, erect but slightly curved, densely covered with a short pubescence of papillose hairs, the midlobe 5 mm long, pendent, arched, tongue-shaped, with a short pubescence of small papillose hairs on the sides and beneath, and with longer dense calvate-vesicular hairs on the upper surface at the tip. Antananarivo, Antsiranana, Fianarantsoa, Mahajanga, Toamasina, Tolagnaro; also in Comoros. Quite frequent in the understorey of humid to semi-dry woodland and forest on laterite soils derived from gneiss or from basalt, 90–1400 m. Fl. September–February.

14.9. **D. ankarensis** H.Perrier

Leaves 2, in the upper half of the stem, 3.5–5 × 0.5–1 cm, narrowly lanceolate or linear-lanceolate, forming a funnel-shaped sheath at the base that clasps the stem; flowers with green sepals, veined with pink, petals pink; hood 8 × 15 mm; lateral sepals 7 × 3 mm, joined near the base, falcate, the tips curling in, each with a sac-like spur in the basal half; lip erect, joined to the column for most of its length; appendage 3-lobed; midlobe fleshy, ligulate, carrying long hairs on the upper surface and short papillae beneath; side lobes linear-oblong, glabrous or with short papillae on the upper surface, twisted at the tips. The flower shows affinities with *D. hildebrandtii* (p. 128) but it is distinguished readily by the nature of the hairs of the terminal appendage of the lip. Antsiranana. In semi-dry forest, growing in the humus amongst calcareous rocks; c. 300 m. Fl. January.

14.10. **D. lanceana** H.Perrier

This species resembles *D. hildebrandtii* in its habit but is readily distinguished by the terminal appendage of its lip, which is stalked not sessile, and by the indumentum of that appendage, which lacks the vesicular hairs; flowers pink, relatively large; hood 13 mm long; lateral sepals 8–10 mm long, joined for about half their length. Lip with a narrow claw with 3 tomentose appendages at the apex. Antsiranana, Toamasina. Habitat and flowering time not known.

14.11. **D. humblotii** Rchb.f.

(syn. *Disperis comorensis* Schltr.)

Leaves 2–3, set around the middle of the stem, narrowly ovate to ovate, acute, 2–4 × 1–2 cm; hood 6–7 mm long; lateral sepals narrowly ovate, 9–10 × 2.5–3 mm, slightly falcate, obtuse at the apex, adnate in the basal quarter, bearing a conical obtuse, 0.6–0.7 mm long spur in the basal third; lip linear, 2 mm long, united at the base to the column; terminal part 3-lobed; the two lateral lobes slightly divergent, oblong, outer face papillose-pubescent; midlobe stalked, fleshy, saddle-shaped, 1.5–2.8 mm long, obtuse at the tip, rounded and gibbous at the base, upper surface shortly papillose-pubescent. The flower is similar to that of *D. lanceana*, but the terminal appendage of the lip is larger and of a different form. Antananarivo, Antsiranana, Fianarantsoa, Toamasina; also in the Comoros. Under trees in humid forest at intermediate elevations; 700–1400 m. Fl. August–October.

14.12. **D. masoalensis** P.J.Cribb & la Croix   *(p. 127, bottom right)*

Leaves 2, narrowly ovate, rounded at base, 25–32 × 11–12 mm, green; flowers pale pink inside, whitish tinged with pink on the outside; hood c. 10 mm long; lateral sepals joined for about half their length; lip fleshy, 3-lobed, 7 × 2.5 mm, papillose; basal lobes incurved, clavate; midlobe tongue-shaped, deflexed, concave, blunt. Antsiranana, Toamasina. Growing in moss on top of boulders in river-bed; sea level–25 m. Fl. October.

14.13. D. falcatipetala P.J.Cribb & la Croix

Leaves 3, suberect to spreading, ovate, obtuse, up to 1.8 × 0.8 cm, green with purple markings; flowers clustered, white with purple dots; hood 4 mm long; lateral sepals deflexed, obliquely spathulate, united in basal half to form a narrow claw, 9.5–10 × 3.5 mm; lip 3-lobed, 3.5 mm long and wide when spread; side lobes fleshy, arcuate, truncate, villose in apical part; midlobe spathulate with an undulate margin. Antsiranana. On a rotten branch on the ground in ridge-top montane forest; 1700–1730 m. Fl. April.

14.14. D. perrieri Schltr.

Leaves 2–4, irregularly spaced, 2.8–5 × 0.9–1.2 cm, lanceolate, acute, clasping the stem at base, dull green above with a paler median nerve and the underside is red; flowers mauve or rose-coloured with a green hood and olive or brownish radiating lines on the inner face; hood 10 mm long with 3 keels on the back; lateral sepals 8.5–10 × 3.5–4 mm, free almost to base, each with a small, sac-like spur; lip glabrous, 4.5 mm long, adnate to the column for 2 mm and edged there with a crenate membrane; free part narrow then gradually dilating to form a wide V, prolonged by a linear 1-nerved stipe, bearing a terminal fleshy reniform appendage, with a large hollow in the middle of a papillose lobe. Antsiranana. In humus and amongst lichens in damp woodland on gneiss; 1000–2250 m. Fl. March–May.

14.15. D. saxicola Schltr.

Leaves 3, the lowest just above the middle of the stem, ovate, acute, 2.4–6 × 1–2.8 cm, upper leaves amplexicaul; flowers purplish; hood 8 mm long; lateral sepals free to the base, 8–11 × 5 mm; lip joined to the column for 2 mm, the margin more or less dilated on each side to form 2 lobes, the free part enlarged into a heart-shaped lamina 3 mm high and wide, papillose pubescent towards the top. Antananarivo, Fianarantsoa, Mahajanga. In humus in dense forest, and among shaded rocks, 400–1000 m. Fl. June.

14.16. D. majungensis Schltr.

Leaves 2, ovate, acute, 2–4.3 × 1.2–2 cm, dark green above, red below; flowers violet; hood 6 mm long; lateral sepals free more or less to base, more or less 5 mm long, deeply saccate, with an obtuse spur 1.5 mm long; lip 4.5 mm long, the claw 3 mm long, the basal 2 mm fused to the column; appendage 3-lobed, like a trident, the side lobes erect, densely shortly papillose-pubescent; midlobe triangular, flat, often bent over to the front, densely shortly papillose-pubescent. Mahajanga. In humus, in rocky wood on calcareous soil; sea level–100 m. Fl. January–February.

14.17

14.17. **D. erucifera** H.Perrier   *(page 131)*

Leaves 2 sessile, ovate to lanceolate, acute or apiculate, 5.8–6.2 × 1.8–3.6 cm, green above with 3 violet veins, violet below; flowers violet or deep mauve-pink, relatively large; hood 11 mm long; lateral sepals 7–10 × 3–4 mm, more or less semicircular, acute, free almost to base, each with a sac-like spur; lip 4–7 mm long, more or less hollowed at the base, narrowed above, then dilated into a round or densely tomentose, subrhombic lobe. Antsiranana. Calcareous rocks and limestone crevices in shade, in leaf debris and humus; 30–350 m. Fl. November–March.

14.18. **D. tripetaloides** (Thou.) Lindl.

Leaves 2–3, ovate, subacute, 1.5–5 × 0.8–2.5 cm, dark green with 5 pale nerves above, dark violet below; flowers white or rose-coloured, the nerves on the back of the sepals green or brownish; the terminal appendage of the lip yellow-hairy; hood 8–9 mm long; lateral sepals free more or less to base, with a sac-like spur in the middle; lip joined to the column at the base, the lower part triangular–cordate, covered with short, dense papillae, extended above into a glabrous stalk bearing at its tip a fleshy appendage, 1–1.5 mm long, excavated on the upper surface and bearing the same short papillose puberulence. Antsiranana, Mahajanga, Toliara; also in Seychelles, Mauritius, Réunion and Rodrigues. Rich soil and medium shade in the understorey of semi-dry forests on tsingy or rarely in humid forest; sea level–700 m. Fl. November–March, June.

14.19. **D. bathiei** Bosser & la Croix

Leaves 3, in upper half of stem, ovate, up to 3.3 × 1.8 cm. It resembles *D. lanceana* (p. 129) in the shape of the flowers and the morphology of the lip, but the appendage is completely glabrous and is borne on a glabrous stalk. Antsiranana, Mahajanga. In humid montane forest; 1000–2000 m. Fl. May.

14.20. **D. lanceolata** Bosser & la Croix   *(p. 137, top left)*

This recalls *D. humblotii* (p. 129) and *D. saxicola* (p. 130) in its habit but is distinguished by its much smaller and narrower, sessile leaves, 1.8–3.6 × 0.7–1.5 cm, and by the terminal appendage of its lip which is a flat elliptic or suborbicular plate, having at its base a rounded protuberance. Antananarivo, Antsiranana, Mahajanga, Toamasina. In humus in shade in moist forest, edge of forest in shade of small trees in deep leaf litter; 1200–2000 m. Fl. March–May.

14.21. **D. bosseri** la Croix & P.J.Cribb

Leaves 3, suberect, more or less evenly spaced along the stem, ovate to lanceolate, acute, 1–2 cm long; flowers pale pink; hood c. 6 mm long; lateral sepals free to base, divergent, with a prominent, obtuse spur 2–3.5 mm long, c. 1.5 mm wide; lip ascending, obtriangular, 6–7 mm long, attached to the base of the column for 1–1.5 mm, rounded or subtruncate at the base which is adorned with a small median subcircular, papillose lobule, obtuse and subtriangular at the tip; the upper face having in the basal half 2 slender, erect wings with a thick and convex median lobule between them, margins finely papillose towards the summit. Antananarivo. Humid places, mist forest on steep slopes; 2400 m. Fl. February.

15. CHEIROSTYLIS

A genus of about 22 species in Old World tropics and Australasia. One species in Madagascar and the Comoros. Small terrestrials with creeping fleshy rhizomes and erect fleshy, brittle glandular leafy stems. Leaves several, membraneous, often withered at flowering time, stalked. Inflorescence terminal, densely many-flowered, glandular. Flowers small, resupinate, glandular. Dorsal sepal forming a hood with the petals over the column. Lateral sepals joined for half their length and to the dorsal sepal. Lip saccate at base with 2 entire or lobed calli, transversely oblong at apex. Column with a dorsal anther; pollinia 2, sectile.

15.1. **C. nuda** (Thouars) Ormerod

(syn. *Goodyera nuda* Thouars; *Monochilus gymnochiloides* Ridl.; *Cheirostylis gymnochiloides* (Ridl.) Rchb.f.; *C. humblotii* Rchb.f.; *C. micrantha* Schltr.; *Gymnochilus recurvum* Blume; *Monochilus boryi* Rchb.f.; *Zeuxine gymnochiloides* (Ridl.) Schltr.; *Z. boryi* (Rchb.f.) Schltr.; *Z. sambiranoensis* Schltr.)

Plant small with erect leafy stem arising from fleshy rhizomes; leaves 4–5, withered at flowering time; flowers small, mostly whitish; sepals 3–3.5 mm long; lip saccate, with a 3–4-lobed callus on each side at base. Antananarivo, Antsiranana, Fianarantsoa, Toamasina, Toliara; also in the Mascarenes and Comoros, South Africa and Tanzania. In humid, medium-elevation forest; seasonally dry, deciduous forest and woodland; humid evergreen forest, in humus and detritus; up to 1700 m. Fl. August–March.

16. ZEUXINE

A genus of perhaps 60 species, in Old World tropics and subtropics as far north as Japan. A single species in Madagascar. Small terrestrial plants with a creeping, fleshy rhizome and erect leafy flowering stems. Stems relatively short. Leaves entire, ovate. Inflorescence terminal, racemose, many-flowered, often glandular or pubescent. Flowers small, resupinate. Sepals and petals connate, forming a hood over the column. Lip bipartite, the basal part saccate, the apical part transversely oblong. Column short, anther dorsal; pollinia sectile.

16.1. Z. madagascariensis Schltr.

Plant up to 60 cm tall; flowers downy on the outside, small; lip 10 mm long, inflated-gibbose at the base, with 2 rounded calli at the base. Antsiranana. In humid, high-elevation forest; 1500–1700 m. Fl. April.

17. GOODYERA

A genus of some 80 species; 3 in Madagascar. Plants with fleshy creeping rhizomes and erect fleshy leaf stems. Leaves ovate to lanceolate, often dark greenish black above, shortly stalked. Inflorescence terminal, often glandular. Flowers small, glandular, resupinate. Dorsal sepal forming a hood with the petals over the column and lip. Lateral sepals spreading. Lip deeply saccate at base, ovate-ligulate or transversely oblong in front; callus of keels or papillae at base of lip. Column short; anther dorsal; pollinia sectile, soft.

Key to species of _Goodyera_

1. Leaves rounded or obtuse at the base; pedicel ovary more than twice as long as the sepals 2
1. Leaves attenuate at the base; ovary much less than twice as long as the sepals 3
2. Rachis 15 cm long; sepals 5 cm long; petals elliptic-rhombic; lip hemispherical, with dense papillae inside .... **17.1. G. afzelii**
2. Rachis 10 cm long; sepals 4.5 mm long or less; petals oblong; lip verrucose on the outside, with 4 callosities within .... **17.2. G. perrieri**
3. Plant slender, 30 cm or less tall; leaves 3.5–7 × 0.7–1.2 cm; flowers 5–10; sepals and petals 1-nerved .... **17.3. G. flaccida**
3. Plant robust, 40–60 cm tall; leaves 6.2–10 × 1.8–2.1 cm; flowers 20, densely arranged; sepals 3-nerved .... **17.4. G. humicola**

17.1. G. afzelii Schltr.   (p. 137, top right)

Plant terrestrial, up to 40 cm tall; leaves 5, 5–7.5 × 1.7–2.7 cm, petiolate; flowers many; glandular; sepals 5 mm long; lip 4.5 × 4 mm, bipartite, the hypochile globose, deeply saccate, bearing numerous papillae on inner surface, the epichile recurved, ovate-ligulate, obtuse, 2 mm long. Antsiranana, Fianarantsoa, Toamasina; also Mozambique. Humid, mid-elevation forest; 700–1800 m. Fl. October.

17.2. **G. perrieri** (Schltr.) Schltr.

(syn. *Platylepis perrieri* Schltr.)

Plant up to 30 cm tall; leaves 5–6, oblong, 4–5 × 1.3–2 cm; bracts longer than the flowers; flowers 8–12, glandular pubescent; sepals 3.7–4.5 × 1.2–1.8 mm; petals linear-oblong, obtuse; lip 3.7–4 mm long, with 4 callosities in the basal part, the apical lobe longer than wide, acute. Antsiranana, Fianarantsoa, Toamasina. In montane and lower montane forest. Fl. unknown.

17.3. **G. flaccida** Schltr.

Epiphyte or terrestrial, 20–30 cm tall; leaves 5–6, narrowly lanceolate, 3–7 × 0.7–1.2 cm; flowers 5–10, orange; sepals 4–4.5 mm long; lip 4 × 3 mm, 3-lobed above the middle, white, the lateral lobes very small, the midlobe expanded at the tip into a transversal blade, 1 × 1.8 mm, the disc with 2 diverging keels within. Antsiranana, Toamasina. Humid, evergreen forest; 200–1000 m. Fl. September.

17.4. **G. humicola** (Schltr.) Schltr.   *(p. 137, bottom left)*

(syn. *Platylepis humicola* Schltr.)

Terrestrial plant 40–60 cm tall; leaves c. 8, elliptic-lanceolate, 6.2–10 × 1.8–2.2 cm, petiolate; flowers 20; sepals 3.5–4.5 mm long; lip very concave, with 2 verrucose keels within, the apical lobe transversely elliptic. Toamasina. Humid, evergreen forest, in humus; 500–1200 m. Fl. September–October.

18. PLATYLEPIS

A genus of about 10 species in tropical and South Africa, the Comoros and Madagascar. Three species in Madagascar. Terrestrials with a creeping fleshy rhizome and erect fleshy, leafy stems. Leaves several, ovate, stalked, somewhat fleshy. Inflorescence terminal, glandular, densely many-flowered. Dorsal sepal adnate with petals; lateral sepals deflexed. Lip adnate to column in basal part, saccate at base with included calli, reflexed at tip. Column elongate; anther dorsal; pollinia 2, sectile.

Key to species of *Platylepis*

1. Anther 3 mm long; petals 1-nerved; lip fleshy, 5-nerved .... **18.3. P. polyadenia**
1. Anther 2 mm or less long; petals 3-nerved; lip translucent, 7-nerved 2
2. Lip strongly bigibbous at the base, with 6 callosities within, the outer 2
   on each side spatulate; bracts more than 3 times as long as wide .... **18.1. P. bigibbosa**
2. Lip scarcely bigibbous at the base, with 2 obsolete callosities within ...... **18.2. P. occulta**

18.1. P. bigibbosa H.Perrier

Plant 25–40 cm tall; leaves 8–10 × 2.8–4.2 cm; inflorescence glandular; floral bracts more than 3 times longer than wide; lip strongly gibbose at the base, with 6 calli at the base of the lip, the front keels of the lip protruding and clearly ridged. Antsiranana, Mahajanga. Humid, highland forest; 600–800 m. Fl. unknown.

18.2. P. occulta (Thouars) Rchb.f.

(syn. *Goodyera occulta* Thouars; *Platylepis goodyeroides* A.Rich.; *P. densiflora* Rolfe)

Similar in habit to *P. bigibbosa* but the inflorescence is almost glabrous, the floral bracts 3 times longer than wide, and the lip and the rostellum distinct. Antsiranana, Toamasina; also in the Comoros, Mascarenes and Seychelles. Montane forest; 1000–1600 m. Fl. January, July.

18.3. P. polyadenia Rchb.f.   (*opposite, bottom right and p. 139, top*)

Plant 25–40 cm tall; leaves elliptic, 12–15 × 3–5 cm; inflorescence glandular; floral bracts more than 3 times longer than wide; sepals 8–9 mm long; lip 7 mm long, verrucose on the outside, bigibbous at the base, thickened and semicircular above, with 2 large calli at the base. Antananarivo, Mahajanga, Toamasina, Toliara; also in the Comoros. Seasonally dry, deciduous woods; humid, evergreen forest; in marshes; up to 1200 m. Fl. November–February.

19. CORYMBORKIS

A small cosmopolitan genus of seven species. A single species in Madagascar. Terrestrials with short woody rhizomes, stout bootlace-like roots and clustered, erect, rigid, leafy stems. Leaves pleated, broadly lanceolate. Inflorescences 1–4, lateral, axillary, branched, shorter than the leaves, several-flowered in succession. Flowers not opening widely, somewhat tubular, white, yellow or greenish white. Sepals and petals linear-oblanceolate. Lip spathulate, apical lobe broadly ovate to subcircular, apiculate. Column clavate, elongate; anther dorsal; pollinia 2, sectile.

19.1. C. corymbis Thouars   (*p. 139, bottom left*)

(syn. *Corymbis thouarsii* Rchb.f.; *Corymborkis disticha* Lindl.; *Corymborkis thouarsii* (Rchb.f.) Blume)

Plant up to 2 m tall, often forming large clumps; flowers white to greenish; lip 60–80 mm long, linear; column erect, terete, clavate towards the apex. Antsiranana, Fianarantsoa, Toamasina, Toliara; also in the Mascarenes, tropical and southern Africa. In humid, evergreen forest and lowland forest, often at base of trees in densely shady undergrowth, often forming large colonies; 100–1300 m. Fl. January–April.

14.20

17.1

17.4

18.3

20. NERVILIA

A genus of about 30 species widespread in tropical and South Africa, across Asia to the SW Pacific islands. Eight species in Madagascar. Terrestrial herbs growing from underground tubers. Leaf solitary, erect or prostrate, ovate, to reniform or fan-shaped; stalked. Inflorescence appearing before the leaf, erect, 1–several-flowered; inflorescence stalk long; rachis short. Flowers resupinate or erect. Sepals and petals linear-oblanceolate, similar. Lip obovate, flabellate or 3-lobed, with or without a spur. Column elongate; pollinia 2, granular.

Key to species of *Nervilia*

1. Inflorescence 2- or more-flowered; flowers nodding or spreading 2
1. Inflorescence 1-flowered; flower erect 4
2. Column 7–11 mm long .... **20.6. N. kotschyi**
2. Column 15–22 mm long 3
3. Leaf with a long stalk 10–25 cm long; column 15–17 mm long .... **20.7. N. bicarinata**
3. Leaf prostrate on the soil, the petiole very short; column 20–22 mm long .... **20.8. N. renschiana**
4. Leaf erect, sagittate-cordate .... **20.5. N. lilacea**
4. Leaf prostrate, cordate-reniform 5
5. Leaf less than 15 × 25 mm, often marked above with silver veins and purple below .... **20.4. N. petraea**
5. Leaf usually more than 30 × 60 mm, green 6
6. Leaf somewhat 5-angled, very sparsely pilose above; flowers violet-red; lip side lobes acute .... **20.1. N. affinis**
6. Leaf reniform, pubescent above; lip side lobes obscure 7
7. Leaf up to 8 × 10 cm; lip with a cillus of 3 lines of papillae .... **20.2. N. simplex**
7. Leaf up to 19 × 20 cm; lip covered with short club-shaped hairs .... **20.3. N. leguminosarum**

20.1. **N. affinis** Schltr.

(syn. *N. perrieri* Schltr.; *N. pilosa* Schltr. & H.Perrier)

Leaf reniform, slightly angled, up to 3 × 6 cm; lip c. 16 × 15 mm, 3-lobed, shortly pubescent on upper surface; side lobes subacute, shorter than ovate and erosely margined midlobe. Antananarivo, Antsiranana, Fianarantsoa, Toamasina; also in the Mascarenes. In rather dry to very wet, humid, evergreen forest; 700–1500 m. Fl. October–January.

20.2. **N. simplex** (Thouars) Schltr.   (*opposite, bottom right*)

(syn. *Bolborchis crociformis* Zoll. & Mor.; *Nervilia françoisii* H.Perrier ex François; *N. bathiei* Senghas; *N. crociformis* (Zoll.& Mor.) Seidenf.)

Leaf reniform, up to 8 × 10 cm, shortly pubescent on upper surface; lip 3-lobed, oblong-cuneate, 12–18 × 9–11 mm, inner surface with 3 lines of papillae. Antananarivo, Antsiranana, Mahajanga; also in the Mascarenes, tropical and southern Africa, S and SE Asia, Indonesia, Philippines, New Guinea, Australia. In grassland; humid evergreen forest; riverine forest; plantations; dry *Uapaca* forest; up to 2000 m. Fl. October–January.

18.3

19.1   20.2

20.3. **N. leguminosarum** Jum. & H.Perrier

Leaf prostrate, reniform, shortly pubescent, large, up to 19 × 20 cm; lip 3-lobed, covered by very short, clavate hairs; side lobes longer than semicircular midlobe. Mahajanga. In dry, sandy forest, amongst *Tamarindus*; up to 200 m. Fl. December–January.

20.4. **N. petraea** (Afzel. ex Sw.) Summerh.

(syn. *Arethusa simplex* Thouars; *Stellorkis aplostellis* Thouars; *Nervilia simplex* (Thouars) Schltr.; *Pogonia thouarsii* Perrier)

Leaf prostrate, cordate, 6–15 × 8–25 mm, often with silver veins on upper surface, purple below; lip obovate-cuneate, 3-lobed, the lateral lobes oblong-triangular, the midlobe suborbicular, with long papillae on surface, lacerate on margin. Antsiranana, Mahajanga; also in the Mascarenes and tropical Africa. In grassland; coastal forest; seasonally dry, deciduous woods; plantations; up to 1500 m. Fl. October–January.

20.5. **N. lilacea** Jum. & H.Perrier

Leaf erect, sagittate-cordate, 1.5 × 2 cm, long-stalked, glabrous; flowers pinkish lilac; lip 3-lobed, the side lobes acute, shorter than the lanceolate midlobe, front margin finely toothed. Antananarivo, Antsiranana. In humid, evergreen forest; 900–1200 m. Fl. November–December.

20.6. **N. kotschyi** (Rchb.f.) Schltr.

(syn. *Pogonia kotschyi* Rchb.f.; *Nervilia sakoae* Jum. & H.Perrier; *N. similis* Schltr.)

A variable species; leaf prostrate or shortly stalked, heart-shaped, up to 12 × 14 cm, pubescent on veins on upper surface; stalk short; flowers 1–2; lip elliptical, 10–19 × 7–12 mm, obscurely 3-lobed, bearing 2 parallel fleshy pubescent ridges running from base of the lip to base of midlobe. Antsiranana, Mahajanga; also in the Comoros and tropical Africa. In grassland; plantations; dry savannah; and semi-deciduous, western forest; up to 2300 m. Fl. September–January.

20.6a. **N. kotschyi** var. **purpurata** (Rchb.f. & Sond.) Börge Pett.

(syn. *Pogonia purpurata* Rchb.f. & Sond.; *Nervilia purpurata* (Rchb.f. & Sond.) Schltr.; *N. dalbergiae* Jum. & H.Perrier)

Leaf erect, ovate-elliptic, stalked, glabrous; flowers like those of var. *kotschyi*. Mahajanga; also in E and S Africa. In grassland and marshes, under Dalbergia; 475–2300 m. Fl. November–January.

20.8

22.1

22.1

20.7. N. bicarinata (Blume) Schltr.

(syn. *Pogonia bicarinata* Blume; *P. commersonii* Blume; *P. barklayana* Rchb.f.; *Nervilia viridiflava* (Blume) Schltr.; *N. barklayana* (Rchb.f.) Schltr.; *N. commersonii* (Blume) Schltr.)

Leaf erect, very large, heart-shaped, up to 22 × 26 mm, on a long stalk; lip 3-lobed, ovate, 20–31 × 17–25 mm, with 2 parallel fleshy pubescent ridges running from the base of the lip to base of midlobe; side lobes oblong in front. Antsiranana, Mahajanga, Toamasina, Toliara; also in the Comoros and Mascarenes, E, S and W Africa. In coastal forest; riverine forest often in *Syzygium* thicket; deciduous western forest; and highland forest; up to 1500 m. Fl. October–December.
LOCAL NAME *Agoagoala.*

20.8. N. renschiana (Rchb.f.) Schltr.   (p. 141, top)

(syn. *Pogonia renschiana* Rchb.f.; *Nervilia insolata* Jum. & H.Perrier)

Leaf prostrate, broadly heart-shaped, c. 15 × 16 cm; petiole short; flowers large; lip ovate, obscurely 3-lobed, with 2 parallel fleshy pubescent ridges from base to base of midlobe. Antananarivo, Antsiranana, Mahajanga, Toamasina; also in E and S Africa. In seasonally dry deciduous woods; riverine forest; rocky outcrops; up to 1760 m. Fl. September–December.

21. AUXOPUS

A small genus of 3 species in Africa and Madagascar; 1 species in Madagascar. Small, leafless saprophytes, growing from underground cylindrical tubers. Inflorescences erect, few-flowered, not branched. Flowers small, short-lived, yellowish brown. Sepals and petals slender, united into a tube in basal part, subsimilar. Lip petaloid, clawed, entire, lacking a spur or callus. Column elongate, clavate; pollinia 4, granular. Fruit stalks elongating after pollination.

21.1. A. madagascariensis Schltr.

Up to 12 cm tall; sepals 4–5 mm long, glabrous; lip 3 × 2.5 mm, shortly clawed, sagittate-biauriculate at the base, then broadly rhomboid, very obtuse and a little thickened at the apex. Mahajanga. Seasonally dry, deciduous forest and woodland and rocky outcrops on limestone. Fl. January.

22. DIDYMOPLEXIS

An Old World tropical genus of some 20 species. One species in Madagascar. Leafless saprophytes growing from elongated, fleshy, underground tubers. Stem erect, distantly few-flowered. Inflorescence terminal, unbranched. Flowers resupinate, short-lived, white, pale buff-coloured or pink. Sepals and petals subsimilar, fused in basal half, free parts elliptic, obtuse. Lip free, 3-lobed in apical part, callose, lacking a spur. Column clavate; pollinia 4, mealy. Capsule on a stalk that elongates greatly after pollination.

22.1. D. madagascariensis (Schltr. ex H.Perrier) Summerh.

(p. 141, bottom left and right)

(syn. *Gastrodia madagascariensis* Schltr. ex H.Perrier)

Terrestrial plant 20–40 cm. Antsiranana, Toamasina. Humid, evergreen forest; riverine forest, in humus; sea level–200 m. Fl. July, November.

23. LIPARIS

A large cosmopolitan genus of perhaps 200 species. Some 38 species in Madagascar. Pseudobulbs clustered or well-spaced on a rhizome, ovoid to stem-like, leafy at or near apex. Leaves 1–6, often pleated, lanceolate, ovate or cordate, shortly stalked. Inflorescence terminal, few–many-flowered, usually rather lax. Flowers resupinate. Sepals free, subsimilar, erect to reflexed. Petals often linear, reflexed, rarely narrowly elliptic. Lip strongly reflexed, entire or somewhat bilobed, rarely 3-lobed, apical margins entire or toothed, usually callose at base; callus bilobed or bifid, fleshy. Column elongate, often incurved; pollinia 4, waxy.

Key to species of *Liparis*

1. Pseudobulbs several, ovoid to conical; leaves persistent, coriaceous or rigid 2
1. Pseudobulbs 1–2, stem-like, elongate; leaves deciduous, membranous 13
2. Pseudobulbs small, short, less than 2–3 times as long as broad 3
2. Pseudobulbs stem-like, more than 10 times as long as broad 6
3. Pseudobulbs well-spaced on an elongate rhizome; leaves 1.8 cm or less long ....
   .... **23.2. L. bulbophylloides**
3. Pseudobulbs clustered on a short rhizome; leaves 2 cm or more long 4
3. Leaves (2–)3 or more towards top of an elongated pseudobulb 10
4. Leaf ovate, 3–4.5 × 0.8–1 cm; inflorescence 1–2-flowered; sepals 15–18 mm
   long .... **23.4. L. warpurii**
4. Leaves 1–2, oblanceolate to narrowly lanceolate; inflorescence 5- or more-
   flowered; sepals less than 7 mm long 5

5. Leaf 1, erect, oblanceolate, 2–5 × 0.4–0.5 cm; sepals and petals 1.5–2 mm
   long; lip lanceolate, ecallose .... **23.3. L. caespitosa**

5. Leaves 2–3, narrowly lanceolate, 11–14 × 0.9–2.5 cm; sepals and petals 5–7
   mm long; lip broadly ovate, emarginated .... **23.1. L. anthericoides**

6. Leaf solitary at apex of pseudobulb .... **23.5. L. clareae**

6. Leaves 2, subopposite at apex of pseudobulb 7

7. Plant 40–70 cm tall; pseudobulb 20 cm or more tall, angular .... **23.8. L. longicaulis**

7. Plant 35 cm or less tall; pseudobulbs 13 cm or less long 8

8. Sepals 12 mm or more long; lip 10 × 8 mm, broadly ovate-subrhomic,
   ecallose .... **23.10. L. zaratananae**

8. Sepals 10 mm or less long; lip 6 mm or less long and broad, callus present 9

9. Leaves linear-lanceolate, 12–16 × 0.8–0.9 cm; lip with two small basal calli ....
   .... **23.9. L. stenophylla**

9. Leaves ovate to elliptic-lanceolate, 1.5 cm or more broad 10

10. Leaves broadly ovate, 2.4–4.5 × 2.3 cm; sepals 8–10 mm long; lip callus
    conical .... **23.6. L. danguyana**

10. Leaves ovate to elliptic-lanceolate, 3.5–7.5 × 1.5–2.8 cm; sepals 6–8 mm
    long; lip callus an elongate horn .... **23.7. L. listeroides**

11. Stem 20–30 cm long, sheaths spotted; leaves oblong-lanceolate, less than
    1.8 cm broad .... **23.12. L. puncticulata**

11. Stem 20 cm or less long, unspotted; leaves elliptic to lanceolate, 2 cm or
    more broad 12

12. Inflorescence with 25 or more yellowish flowers; lip callus obscure .... **23.13. L. rivalis**

12. Inflorescence with 4–8 green flowers; lip callus 3-toothed .... **23.11. L. gracilipes**

13. Leaves more than 3 times as long as broad, if broader less than 2 cm long;
    leaf attenuate at base 14

13. Leaves less than 3 times as long as broad; leaf obtuse or rounded at base 22

14. Plants 10 cm or less tall 15

14. Plants 11 cm or more tall 19

15. Leaves 3 cm or less long, 9 mm or less broad 16

15. Leaves 3 cm or more long 18

16. Plant 6 cm or less tall; inflorescence 5 cm or less long; lip callus crescent-
    shaped .... **23.20. L. microcharis**

16. Plant 7–10 cm tall; inflorescence 7 cm or more tall; lip callus rectangular
    or obscure 17

17. Lip suborbicular with a large fleshy rectangular basal callus .... **23.21. L. parva**

17. Lip oblong, lacking a callus .... **23.22. L. xanthina**

18. Sepals 2.5 mm long; lip 2 mm long, the basal callus transverse with 2
    lateral higher keels .... **23.15. L. cladophylax**

18. Sepals 3.5 mm long; lip 3 mm long, the callus bilobed .... **23.17. L. dryadum**

19. Pseudobulbs ovoid, sometimes compressed 20

19. Pseudobulbs elongate, fusiform 21

20. Leaves 2, oblanceolate, 6–10 x 0.8–1.2 cm; lip oblong, shortly apiculate ....
    .... **23.18. L. longipetala**

20. Leaves 3, lanceolate, 5 x 1.2 cm; lip ovate-cordate, obtuse .... **23.19. L. lutea**

21. Flowers yellow; sepals 6.5 mm long; lip oblong, ecallose .... **23.16. L. densa**

21. Flowers green, turning reddish; sepals 4 mm long; lip obcuneiform, the
    basal callus bilobed .... **23.14. L. bicornis**

23.1. **L. anthericoides** H.Perrier   *(p. 147, top left)*

Epiphyte or terrestrial, 20–30 cm tall; pseudobulbs ovoid; leaves 2–3, narrowly lanceolate, 11–14 × 0.9–2.5 cm; inflorescence 10–20-flowered; flowers green; sepals 5–7 mm long; lip broadly ovate above, indented in the middle, 3 × 2.3 mm; anther with a straight, acute beak. Antananarivo, Antsiranana, Toamasina, Toliara. Mossy, montane forest; 900–1300 m. Fl. February–August.

23.2. **L. bulbophylloides** H.Perrier   *(opposite, top right)*

Epiphytic plant up to 10 cm tall; pseudobulbs bifoliate; leaves ovate, 1–1.8 × 0.8–1.5 cm; inflorescence 3–10-flowered; flowers yellow; sepals 5–6 mm long; lip dilated and suborbicular above, auriculate at base; anther with a wide beak. Toamasina, Toliara. Humid, evergreen forest. 500–1000 m. Fl. February.

23.3. **L. caespitosa** (Lam.) Lindl.
(syn. *Epidendrum cespitosum* Lam.)

Epiphyte or terrestrial plant less than 9 cm tall; pseudobulbs 1-leafed; leaf erect, oblong or oblanceolate, 1.8–5 × 0.4–0.5 cm; inflorescence 5–12-flowered; sepals 1.5–2 mm long; lip recurved, lanceolate, lacking a callus. Antananarivo, Fianarantsoa, Toamasina; also in the Comoros, the Mascarenes, E Africa, Sri Lanka, NE India to the Philippines, New Guinea, Solomon Is. and Fiji. Mossy, evergreen forest; 400–1400 m. Fl. February.

23.4. **L. warpurii** Rolfe   *(opposite, bottom left)*

Plant 6–9 cm tall; pseudobulbs unifoliate; leaf ovate, acute, 3–4.5 × 0.8–1 cm; inflorescence 1–2-flowered; sepals 15–18 mm long; lip obovate, denticulate, 12 × 10 mm, verrucose; callus bilobed. Related to *L. parva* (p. 153), but flowers bigger and lip crenulate. Fianarantsoa, exact locality unknown.

23.5. **L. clareae** Hermans   *(opposite, bottom right)*
(syn. *Liparis cardiophylla* H.Perrier)

Epiphyte or terrestrial, 8–16 cm tall; pseudobulbs cylindrical, 4–10 cm tall; leaf heart-shaped, 5.5–10 × 4–5.3 cm; inflorescence 6–10-flowered; flowers green, turning pale orange with age; sepals 7 mm long; lip 7 × 4 mm, auriculate at the base, the front margin 3-toothed in the middle, lacking a callus. Antananarivo, Fianarantsoa, Mahajanga, Toamasina. Mossy, lowland and montane forest, often near water; 100–1500 m. Fl. February–May.

23.5a. **L. clareae** var. **angustifolia** H.Perrier ex Hermans
(syn. *Liparis cardiophylla* var. *angustifolia* H.Perrier)

Epiphyte or terrestrial; differing from the typical variety in having narrower leaves, 3–5 flowers, 10–11 mm long sepals, petals with 2 veins, a larger lip, with a rectangular blade and a disk with a pronounced callus at the base, a taller 5.5 mm high column and a smaller anther, the beak being wider and more obtuse. Toamasina. Humid, mossy, evergreen forest; c. 700 m. Fl. February.

23.1

23.2

23.4

23.5

23.6. **L. danguyana** H.Perrier

Epiphyte or terrestrial, 15–25 cm tall; pseudobulbs stem-like, 8–12 cm long; leaves 2, broadly ovate, 2.4–4.5 × 2–3 cm, dull green; inflorescence 8–12 cm long; sepals 8–10 mm long; lip orbicular, apiculate, 6 × 5 mm, auriculate at base, with a small conical callus; related to *L. listeroides* but with broadly ovate-elliptic leaves, a rounded lip, apiculate at the front and an anther lacking a beak. Antananarivo. Mossy, montane forest; 1500–1800 m. Fl. March.

23.7. **L. listeroides** Schltr.   (p. 151, top left)

Epiphyte or terrestrial, 16–35 cm tall; pseudobulbs stem-like 8–13 cm tall; leaves 2, ovate or elliptic-lanceolate, 3.5–7.5 × 1.5–2.8 cm; flowers 5–20, yellow; sepals 6–8 mm long; lip obcordate, 5–6 × 5–6 mm, auriculate at base, calli elongate, horn-shaped, oblique at the front; anther with a short, obtuse beak. Antananarivo, Antsiranana, Fianarantsoa. Mossy, montane forest; 1250–1500 m. Fl. January–April.

23.8. **L. longicaulis** Ridl.   (p. 151, top right)

Epiphyte or terrestrial, 40–70 cm tall; pseudobulbs stem-like, 20–60 cm long, 5 mm in diameter at base, angular; leaves 2, ovate, 6–10 × 4.5–6 cm; flowers 5–12, yellow-green; sepals 12–15 mm long; lip broadly ovate-lanceolate, 12 × 7–9 mm, with a large transversal callus on base, indented at the front. Antananarivo, Antsiranana, Toamasina, Toliara. Montane forest, in deep shade; 700–1800 m. Fl. January–April.

23.9. **L. stenophylla** Schltr.

Epiphyte or terrestrial, 22–35 cm tall; pseudobulbs cylindrical, 6.5–11 cm long, 2-leaved; leaves linear, acute, 12–16 cm × 8–9 mm; flowers yellow-green turning pale orange with age; sepals 7.5 mm long; lip 5 × 2.5 mm, obovate-suborbicular, with 2 small calli above the base. Antsiranana. Humid, evergreen forest; 400–800 m. Fl. April–May.

23.10. **L. zaratananae** Schltr.   (opposite)

Epiphytic plant 15–20 cm tall; pseudobulbs cylindrical, 3–10 cm tall; leaves 2–3, broadly elliptic, acute, 7–9 × 3.5–4.5 cm, rounded at the base and petiolate; flowers 4–8, green to brownish yellow; sepals 12–14 mm long; lip broadly ovate, 10 × 8 mm, auriculate at base, lacking a callus; anther extended by a 1 mm long, narrow beak. Antsiranana. Mossy, montane forest; 1500–2000 m. Fl. April, October, December.

23.10

23.11. L. gracilipes Schltr.

Lithophyte or terrestrial, 16–20 cm tall; pseudobulbs stem-like, less than 13 cm long; leaves 2–3, asymmetric, elliptic or elliptic-lanceolate, 4–7.5 × 1.5–3.2 cm; inflorescence 4–8-flowered; flowers green; sepals 7.5 mm long; lip 6.5 × 7.5 mm, wider than long, the disk with a 3-toothed callus near the base. Antsiranana. Mossy, montane forest; river margins; and shaded and humid rocks; c. 1800 m. Fl. January.

23.12. L. puncticulata Ridl.

(opposite, bottom left)

Epiphyte or terrestrial, 20–40 cm tall; pseudobulbs stem-like, 20–30 cm long, 3–4-leaved; leaves oblong-lanceolate, 6–6.5 × 1.6–1.8 cm; sheaths finely punctuate; flowers 6–15, olive-green; sepals 10 mm long; lip obtriangular, 10 × 5 mm, auriculate at base, lacking a callus; anther with an obtuse beak. Antananarivo, Antsiranana, Fianarantsoa, Toamasina. Humid evergreen forest; 800–2000 m. Fl. January–May.

23.13. L. rivalis Schltr.

Terrestrial, 25–40 cm tall; pseudobulbs stem-like, up to 20 cm long; leaves 3–4, lanceolate, 10–12 × 2–4.3 cm; raceme with at least 25 flowers, flowers yellowish; sepals 7 mm long; lip broadly obovate, 5.5 × 4.5 mm, very obtuse at the front and at the margins a little crenulate; with a small shallow callus near the base. Antsiranana. Mossy, montane forest; river margins; 1200–1600 m. Fl. December–January.

23.14. L. bicornis Ridl.

Lithophyte or terrestrial, 20–40 cm tall; pseudobulbs elongate; leaves 3–4, lanceolate, acute, 6–16 × 2–3.5 cm; rachis 3 times shorter than the peduncle; flowers 15–50, greenish turning reddish with age; sepals 4 mm long; lip obcuneiform, 4 × 4 mm, with a small 2-horned callus at the base. Antananarivo. Marshes; rocky outcrops; c. 1400 m. Fl. January–March.

23.15. L. cladophylax Schltr.

Epiphyte, 6–8 cm tall; pseudobulbs less than 1 cm tall, often with 3 leaves; leaves ligulate, subacute, 3–6 × 0.6–1.2 cm; inflorescence as long as the leaves; flowers 15–20, yellow-orange; sepals 2.5 mm long; lip suborbicular, 2 × 2 mm, with a transverse callus at the base and with 2 higher lateral keels; anther clearly truncate at the front. Antsiranana. Humid, evergreen forest; c. 800 m. Fl. December.

23.7

23.8

23.12

23.17

23.16. L. densa Schltr.

Lithophyte or terrestrial, 11–16 cm tall; pseudobulbs oblong, compressed-ancipitous, 3–4 cm long; leaves 2–5, oblong-lanceolate, 4–8 × 1.2–2.5 cm; rachis longer than the peduncle; flowers 10–20, yellow; sepals 6.5 mm long; lip oblong, 6 × 3 mm, without a callus. Fianarantsoa, Toliara. Shaded and humid rocks; 1500–2400 m. Fl. January.

23.17. L. dryadum Schltr. (p. 151, bottom right)

Epiphyte or lithophyte, 2–8 cm tall; pseudobulbs compressed-ovoid, 5–9 mm tall and broad; leaves 2, ligulate, 10–35 × 3–8 mm; inflorescence more or less as long as the leaves, few–many-flowered; flowers yellow-brown to pale orange; sepals 3.5 mm long, 3-nerved; lip suborbicular, emarginate, 3 mm long, the callus near the base bilobed. Antsiranana. Lichen-rich forest and forest remnants; 1200–1600 m. Fl. January.

23.18. L. longipetala Ridl.

Terrestrial, 12–18 cm tall; pseudobulbs ovoid, small; leaves 2, oblanceolate, acute, 6–10 × 0.8–1.2 cm; inflorescence 6–10-flowered; sepals 4–5 mm long; petals much longer than the dorsal sepal; lip oblong, very shortly apiculate, 3 mm long, the disk with a small obtuse callus; anther with an acute beak. Fianarantsoa. Humid, evergreen forest; c. 1500 m. Fl. March.

23.19. L. lutea Ridl.

Terrestrial, 15–20 cm tall; leaves 3, lanceolate, 5 × 1.2 cm; inflorescence 6–8-flowered; flowers small, yellow; sepals 3–4 mm long; petals much longer than dorsal sepal; lip ovate-cordate, obtuse. Fianarantsoa. Marshes; c. 1500 m. Fl. unknown.

23.20. L. microcharis Schltr.

Small lithophytic or terrestrial plant, 4–6 cm tall; pseudobulbs broadly ovoid, compressed, 3–4.5 × 2–3 mm; leaves 2, narrowly lanceolate, 8–18 × 3–6 mm; inflorescence up to 5 cm long; flowers 3–6, yellow; sepals 3–4 mm long; lip broadly suborbicular, 3 × 3.5 mm, callus crescent-shaped. Fianarantsoa. On humid rock piles; c. 2000 m. Fl. February.

23.21. **L. parva** (Kuntze) Ridl. *(p. 155)*

Small epiphyte, 8–10 cm tall; leaves 2, 2–3 × 0.3–0.5 cm, with narrow sheath-like bases; inflorescence twice as long as the leaves; flowers olive-coloured; lateral sepals lanceolate; petals 1-nerved; lip suborbicular, with a large subrectangular, erect, fleshy callus, near the base; anther obtuse at the front. Antananarivo, Fianarantsoa, Toamasina. Mossy, montane forest on moss- and lichen-covered trees; c. 1500 m. Fl. February.

23.22. **L. xanthina** Ridl.

Epiphyte, 7–10 cm tall; pseudobulbs ovoid-conical, about 1 cm across; leaves 2–3, elliptic-lanceolate, 2–4 × 0.5–0.9 cm; inflorescence more than twice as long as the leaves; flowers 10–15, yellow; sepals 3 mm long; lip oblong and obtuse. Fianarantsoa. Humid evergreen forest; c. 1500 m. Fl. unknown.

23.23. **L. ambohimangana** Hermans *(p. 157, top right)*
(syn. *Liparis monophylla* H.Perrier)

Terrestrial plant, 8–10 cm tall, with slender stolons; pseudobulb 5–10 mm tall; leaf lanceolate, acute, 2–9 × 0.4–1.5 cm; flowers 1–7, greenish or pale yellow; sepals 4–6 mm long; lip broadly ovate or suborbicular, 4–5 × 3.5–5 mm, with a small basal callus. Antananarivo. In shady areas by rocks, on clay; 1200–1600 m. Fl. March–May.

23.24. **L. andringitrana** Schltr.

Lithophyte, 8–15 cm tall; pseudobulbs compressed ovoid, 2.5 cm tall; leaves 2–3, elliptic, acute, 9–14 × 3–4.5 cm; inflorescence as long as the leaves; flowers 7–12, greenish; sepals 5–7 mm long; lip obovate-cuneate, 4.5–5 × 3.5–4 mm, with a transverse callus. Fianarantsoa. Shaded and humid rocks and rocky outcrops; c. 1800 m. Fl. February.

23.25. **L. bathiei** Schltr.

Terrestrial, 6–12 cm tall; pseudobulbs short and sub-globose, 7–10 mm tall; leaves 2, oblong or elliptic, acute, 5–7 × 1.3–3 cm; flowers 10–20, greenish yellow; sepals 4–5 mm long; lip 4–6 × 3.5 mm, orbicular or obovate, angular at the front; callus obsolete; anther with a narrow beak, 0.6 mm long; column 2 mm long. Antananarivo. In shade of *Acacia*; 1500–1600 m. Fl. February–March.

23.26. **L. flavescens** (Thouars) Lindl. (p. 157, top left)
(syn. *Malaxis flavescens* Thouars)

Leaves 3–4, plicate, suberect, 4–6.5 × 1.2–3.5 cm; inflorescence longer than the leaves; lip rounded, crenulate and cuspidate; anther truncate. Antananarivo, Fianarantsoa; also in the Comoros, Mascarenes and Seychelles. In forest. Fl. June.

23.27. **L. henrici** Schltr.
(syn. *Liparis latilabris* Schltr.; *L. verecunda* Schltr.)

Lithophyte or terrestrial, 11–23 cm tall; pseudobulbs elongate, 2–4 cm long; leaves 2–3, elliptic or ovate-elliptic, apiculate, 2.5–9 × 1.5–4.5 cm; flowers 3–10, greenish with a yellowish lip; sepals 6–8 mm long; lip orbicular or obovate, 5 × 4 mm, rounded at the front; callus obsolete; anther with a short triangular beak; column 3–4 mm long. Antananarivo, Fianarantsoa, Toliara. Shaded and humid rocks in forest; 800–1800 m. Fl. January–March.

23.28. **L. imerinensis** Schltr. (p. 157, middle left)

Epiphyte, lithophyte or terrestrial, 15–30 cm tall; pseudobulbs elongate, 4–6 cm long; leaves 2–3, ovate-elliptic, shortly acuminate, 4.5–13 × 2.2–4.5 cm; flowers 5–16; sepals 7–8 mm long; lip ovate-rhombic, acute at the tip, 7 × 5.5 mm, with a basal callus. Antananarivo, Fianarantsoa, Toliara. Mossy, montane forest; 700–1500 m. Fl. December–March.
LOCAL NAME Famany.

23.29. **L. nephrocardia** Schltr.

Epiphyte or terrestrial, 7–25 cm tall; pseudobulbs conical, 1–2.5 cm tall; leaves 2–3, elliptic or elliptic-lanceolate, 6–12 × 1.5–3.2 cm; inflorescence as long as the leaves; flowers 4–12, yellow-brown; sepals 7–8 mm long; lip expanded into a wide, subcordate blade and rounded in front, 5–6 × 5–5.5 mm; callus semicircular or V-shaped. Antsiranana, Toamasina. Humid, mossy, evergreen, montane forest; 600–1600 m. Fl. December–March.

23.30. **L. perrieri** Schltr.

Lithophyte 30–40 cm tall; pseudobulbs narrow; leaves 3–4, elliptic, acute, 10–25 × 4–8 cm; peduncle as long as the leaves; raceme carrying 15–45 small green flowers with a red-purple lip; sepals 4–5 mm long; lip obcordate, 3.5 × 4 mm, with 2 small calli at the base. Antsiranana, Fianarantsoa, Mahajanga, Toliara. Humid, evergreen forest and shaded and humid rock; 400–1200 m. Fl. September, December–March.

23.30a. L. perrieri var. trinervia H.Perrier ex Hermans

(syn. *Liparis perrieri* var. *trinervia* H.Perrier)

Epiphyte or terrestrial; differing from the typical variety in having lateral sepals with 5 veins, a 3-veined lip that is shortly apiculate in the middle of the front indentation, the middle vein being unbranched, and in having a distinct, very pronounced, conical callus, and an anther with a subrectangular front lobe. Toamasina. In the Savoka, humid, evergreen forest; c. 700 m. Fl. February.

23.31. L. rectangularis H.Perrier

Epiphyte, 7–15 cm tall; pseudobulbs rounded, 1.5 cm long; leaves 2–3, elliptic, shortly acute, 2.1–9 × 1.3–2.6 cm; flowers 3–10, yellow-orange; sepals 4–6 mm long; lip 3-nerved, 5 × 4 mm, abruptly expanded above the middle in a transversely subrectangular blade, subcordate at the lower edge, the disk with 2 small, obsolete swelling at the base. Toamasina. Mossy, montane forest; c. 1000 m. Fl. February.

23.32. L. salassia (Pers.) Summerh. *(opposite, bottom left)*

(syn. *Epipactis salassia* Pers.; *Malaxis purpurascens* Thouars; *Liparis purpurascens* (Thouars) Lindl.)

Reddish or violet epiphyte or terrestrial, 6–15 cm tall; pseudobulbs stem-like, 5–10 cm long; leaves 2–3, ovate, acute, 2–5 × 1–4 cm; raceme short, subcorymbose; flowers 7–12, white with purple streaks or purple; sepals 5–6.5 mm long; lip suborbicular, emarginate and denticulate. Antsiranana, Antananarivo; also in the Mascarenes. Humid, evergreen forest; 1000–1500 m. Fl. December–April, September.

23.33. L. trulliformis Schltr.

Small epiphyte 10–13 cm tall; pseudobulbs short, up to 2 cm long; leaves 2, elliptic, acuminate, 4–6 × 1.3–2.3 cm; flowers 5–10, yellowish; sepals 7–8 mm long; lip 6 × 3 mm, trowel-shaped and thick, with a basal callus. Antsiranana. Mossy, montane forest on moss and lichen-covered trees; c. 1800 m. Fl. January.

23.34. L. jumelleana Schltr.

Epiphyte or terrestrial, 13–16 cm tall; pseudobulbs at an angle to one another, elongate; leaves 2–3, elliptic, 5–9 × 2.7–4 cm; flowers 5–8, yellowish; sepals 11 mm long; lip much wider than long, 7–8 × 10–11 mm, shortly acuminate at the front; callus very thick and very obtuse. Antsiranana, Fianarantsoa. Humid, evergreen forest; c. 800 m. Fl. December.

23.26

23.23

23.28

23.32

23.36

23.35. L. ochracea Ridl.   *(opposite)*

(syn. *Liparis connata* Ridl.; *L. hildebrandtiana* Schltr.)

A very variable epiphyte or terrestrial, 15–30 cm tall; pseudobulbs elongate, up to 7 cm long; leaves petiolate, elliptic or ovate-elliptic, 4.5–14 × 2.8–7 cm; flowers 6–12, yellow-green; sepals 12–15 mm long; lip oblong, 12–15 mm long, the disk with 2 very obvious calli. Antananarivo, Antsiranana, Fianarantsoa, Toamasina. Humid, lichen-rich, evergreen forest; 700–2000 m. Fl. February–March.

23.36. L. ornithorrhynchos Ridl.   *(p. 157, bottom right)*

Terrestrial or epiphyte up to 30 cm tall; leaves 2–3, ovate, acute; flowers 3–12, greenish; sepals 10–11 mm long; lip broadly cordate, 10–11 mm long, broadest in basal half, lacking a callus, the base narrow; anther with a long beak. Fianarantsoa, Antananarivo. Humid areas near evergreen forest, at the base of or low down on trees. Fl. January–February.

23.37. L. panduriformis H.Perrier

Lithophyte 25–30 cm tall; pseudobulbs at an angle to each other; leaves 4–5, broadly ovate, acute, 7.5–14 × 4.8–6 cm; flowers 15–20; sepals 12–15 mm long; lip broadly 3-lobed, pandurate, 10 × 6 mm, the lateral lobes small, angular-obtuse and recurved towards the base, the front edge crenulate-dentate; callus large, bilobed and thick. Antsiranana. Shaded and humid rocks; 500–800 m. Fl. February–May.

23.38. L. sambiranoensis Schltr.

Epiphyte or terrestrial, 17–30 cm tall; pseudobulbs narrow and elongate, 8–10 cm long; leaves 3–4, elliptic, acuminate, 11–18 × 4.5–5.8 cm; flowers many, greenish yellow; sepals and petals 11–14 mm long; lip 5-nerved, 10 × 5 mm, obovate-suborbicular, abruptly widened above the middle, the front part broadly obovate-suborbicular; callus large, broadened and bilobed at the front. Antsiranana; also in the Comoros. Coastal forest; humid, evergreen forest; 200–500 m. Fl. October, December–February.

24. MALAXIS

A large cosmopolitan genus of perhaps 250 species. Four species in Madagascar. Terrestrial or less commonly lithophytic or epiphytic plants with creeping fleshy rhizomes and erect leafy stems. Leaves pleated, fleshy to thin-textured, often shortly stalked and sheathing at base. Inflorescence terminal, densely to laxly few–many-flowered. Flowers often non-resupinate, small, rather flat, green, orange or purple. Sepals and petals subsimilar spreading. Lip entire to 3-lobed, deflexed, flat, callose or not, spurless; calli shallow hollow or low mounds. Column very short; pollinia 4, waxy.

23.35

Key to species of *Malaxis*

1. Lip strongly 3-lobed .... **24.1. M. madagascariensis**
1. Lip entire or only obscurely lobed 2
2. Inflorescence 50- or more-flowered .... **24.2. M. physuroides**
2. Inflorescence with 20 or fewer flowers 3
3. Lip transversely semicircular, broader than long, bicallose; sepals
   3–3.5 mm long .... **24.3. M. atrorubra**
3. Lip orbicular, with 2 dot-like calli at the base; sepals 2 mm long .... **24.4. M. françoisii**

24.1. **M. madagascariensis** (Klinge) Summerh.

(syn. *Microstylis madagascariensis* Klinge)

Plant 20–30 cm tall; pseudobulbs conical; leaves 4–5, ovate-lanceolate, 6–9 × 2–4 cm; flower stalk 4-angled; flowers violet; lip clearly 3-lobed, the lobes obtuse, the midlobe a little bigger. Madagascar, exact locality unknown. Fl. unknown.

24.2. **M. physuroides** (Schltr.) Summerh.

(syn. *Microstylis physuroides* Schltr.)

Epiphyte or lithophyte, 15–20 cm tall; pseudobulbs 6–8 cm long; leaves 3–4, ovate, acute, 3–7.5 × 2.4–4 cm; raceme densely more than 50-flowered; flowers red; sepals 2 mm long; lip suborbicular, 2 mm long, papillose, with a single callus. Antsiranana, Toamasina. Humid, evergreen forest; shaded and humid rocks; 400–700 m. Fl. February–April.

24.3. **M. atrorubra** (H.Perrier) Summerh.

(syn. *Microstylis atro-ruber* H.Perrier)

Epiphyte or terrestrial, 10–12 cm tall; pseudobulb 5–8 cm long; leaves 3–4, ovate, acute, 1.5–5 × 1.5–3.5 cm; raceme 6–8-flowered; flowers dark purple; sepals 3–3.5 mm long; lip transversely semicircular, 2 × 2.6 mm, carrying 2 calli. Antananarivo. Humid, evergreen forest on plateau. c. 1500 m. Fl. October.

24.4. **M. françoisii** (H.Perrier) Summerh.

(syn. *Microstylis françoisii* H.Perrier)

Epiphyte or terrestrial, up to 12 cm tall; pseudobulb 5 cm long; leaves 3–4, ovate, acute; 2.5–3.5 × 1.4–1.6 cm; raceme with at least 15 flowers; sepals 2.5 mm long; lip broadly orbicular, 2 mm across, carrying 2 small dot-like calli near the base. Antananarivo. Humid, evergreen forest on plateau; c. 1200 m. Fl. February.

25. OBERONIA

A genus of about 80 species, predominantly in tropical Asia, the Malay archipelago, the Philippines, New Guinea, N and E Australia and the SW Pacific islands. One species in Madagascar, the Mascarenes, Comores and tropical Africa. A small twig or branch epiphyte with wiry basal roots. Stem short. Leaves distichous, imbricate, bilaterally flattened and iris-like. Inflorescence terminal, densely many-flowered in whorls. Flowers minute, yellow, translucent. Sepals and petals subsimilar. Lip flat. Column very short; pollinia 4, waxy.

25.1. **Oberonia disticha** (Lam.) Schltr.   *(p. 163, top left)*

(syn. *Epidendrum distichum* Lam.; *Oberonia brevifolia* Lindl.; *O. equitans* (G. Forst.) Schltr.)

Stems 2–15 cm long, often clustered; leaves 2–5 cm long; flowers tiny, c. 1 mm in diameter, yellow. Antananarivo, Antsiranana, Fianarantsoa, Toamasina, Toliara; also in the Comoros, Mascarenes, E and S Africa. In humid, evergreen and ridge-top forest; up to 2000 m. Fl. September–May.

LOCAL NAMES *Fontsilahinjanahary, Ahipisabato*.

26. CALANTHE

A large genus of perhaps 200 species, cosmopolitan but mostly in Old World tropics and subtropics but extending to Japan, Korea and into the Pacific as far east as Tahiti. Terrestrial or, less commonly, epiphytic plants with a short, stout rhizome and stout roots. Pseudobulbs small, often obscure, several-leaved. Leaves suberect to arching, pleated, lanceolate to oblanceolate or elliptic. Inflorescence erect, racemose, few–many-flowered. Flowers white, pink or rose-purple, turning bluish when bruised. Sepals and petals subsimilar, spreading. Lip fused at base to column, 3-lobed, often obscurely so, spurred at base; side lobes small, spreading; midlobe often bilobed; callus usually of 3 verrucose ridges at base of lip; spur elongate, usually cylindrical, curved or not. Column short, fleshy; pollinia 8, clavate.

Key to species of *Calanthe*

1. Flowers white with a red callus . . . . **26.5. C. millotae**
1. Flowers pink, purple or violet with a white or yellow callus 2
2. Lip entire . . . . **26.1. C. humbertii**
2. Lip 3-lobed, sometimes obscurely so 3
3. Plant less than 20 cm tall; leaves with strongly undulate margins . . . .
. . . . **26.2. C. madagascariensis**
3. Plant 30 cm or more tall; leaves with straight margins 4
4. Plant with a creeping habit; lip side lobes obscure . . . . **26.3. C. repens**
4. Plant with a clustered growth; habit; lip distinctly 3-lobed . . . . **26.4. C. sylvatica**

26.1. C. humbertii H.Perrier

Flowers violet; lip entire, 12 × 10 mm, the disc with a tripartite callus; spur cylindrical up to 3 cm long; ovary more-or-less pilose-glandulose. Antsiranana. In lichen-rich forest on granite, gneiss and limestone; 1700–2000 m. Fl. March.

26.2. C. madagascariensis Watson ex Rolfe   *(opposite, top right)*
(syn. *Calanthe warpuri* Watson ex Rolfe)

Plant small, less than 20 cm tall; leaves 6–10 × 3–4 cm, undulate; flowers small; lip 3-lobed, side lobes narrow, 5–6-times longer than wide. Antananarivo, Antsiranana, Fianarantsoa, Toliara. In shady places, in moist clay in humid, evergreen forest on plateau; river margins; on *Asplenium nidus*; on rocks in humus; 1000–1500 m. Fl. February–September.

26.3. C. repens Schltr.   *(opposite, bottom left)*

Plant 30–50 cm tall; leaves 10–13 × 3–4.4 cm; inflorescence 20–27 cm tall; lip entire, slightly expanded-rounded on each side towards the base, contracted in the middle, then a little widened at the apex. Antsiranana, Fianarantsoa. In mossy, evergreen forest; 1200–1800 m. Fl. February–March, October.

26.4. C. sylvatica (Thouars) Lindl.   *(opposite, bottom right)*
(syn. *Centrosia auberti* A.Rich.; *Calanthe violacea* Rolfe; *C. durani* Ursch & Genoud; *C. perrieri* Ursch & Genoud)

Plant large, 40–80 cm tall; leaves 14–30 × 4–9 cm, generally wider and longer than the other species; flowers large; lip 3-lobed, the side lobes 2 times longer than wide, bearing a basal callus of 3 verrucose ridges. Antananarivo, Antsiranana, Fianarantsoa, Toamasina, Toliara; also in the Comoros, Mascarenes, Seychelles, tropical and South Africa. In humid, evergreen forest from sea level to plateau; 300–2000 m. Fl. December–July.

26.5. C. millotae Ursch & Genoud ex Bosser

Inflorescence short; spur shorter than the ovary; flower pure white, including the lip; lip with 4 more-or-less equal lobes and a red callus. Antsiranana. In humid, eastern forests; up to 500 m. Fl. March–April.

25.1

26.2

26.3

26.4

27. PHAIUS

A large genus of perhaps 200 species, in Old World tropics and subtropics but extending to China and into the Pacific as far east as Samoa. One species in Madagascar. Terrestrial or, less commonly, epiphytic plants with a short to long, stout rhizome and stout roots. Pseudobulbs present in Madagascan species, small, often obscure, several-leaved. Leaves suberect to arching, pleated, lanceolate to oblanceolate or elliptic. Inflorescence erect, racemose, few–many-flowered. Flowers with yellow, yellow-green, white, pink or rose-purple sepals and a purple lip, turning bluish when bruised. Sepals and petals subsimilar, spreading. Lip free from column, 3-lobed, often obscurely so, spurred at base; side lobes erect; midlobe with a ridged callus; spur short, usually cylindrical, curved or not. Column elongate, slightly incurved; pollinia 8, clavate.

27.1. P. pulchellus Kraenzl. *(p. 167, top left)*

Plant 40–80 cm tall; leaves 3–4, 12–30 × 2–2.5 cm; flowers purple; sepals 2.5–2.8 cm long; lip 3-lobed, 2–2.8 × 2 cm, with the side lobes longitudinally fused to and surrounding the column, with a small spur, 3 mm long. Antananarivo, Antsiranana, Toamasina, Toliara; also in Mauritius. In humid, evergreen forest; 700–1700 m. Fl. December–March.

LOCAL NAME *Tadidiala*.

27.1a. P. pulchellus var. **ambrensis** Bosser

Sepals and petals white, more than 3 cm long, the callus ridges warty. Antsiranana. In humid, evergreen forest on plateau; 1000–1600 m. Fl. unknown.

27.1b. P. pulchellus var. **andrambovatensis** Bosser

Leaves with very undulating edges; sepals and petals pinkish-purple, reflexed at the tip. Fianarantsoa. In humid, mid-elevation forest. Fl. October–November.

27.1c. P. pulchellus var. **sandrangatensis** Bosser   *(p. 167, top right)*
Sepals and petals pale green; lip completely dark purple. Fianarantsoa, Toliara, Toamasina. In humid, evergreen forest; 500–1000 m. Fl. February.

28. GASTRORCHIS

A genus of eight species, endemic to Madagascar. Terrestrial or, less commonly, epiphytic plants with a short, stout rhizome and stout roots. Pseudobulbs small, often obscure, several-leaved. Leaves suberect to arching, pleated, lanceolate to oblanceolate or elliptic. Inflorescence erect, racemose, few- to many-flowered. Flowers white, pink or rose-purple, lip usually marked with pink or purple and with a yellow callus, turning bluish when bruised. Sepals and petals subsimilar, spreading. Lip free from column, deeply concave, 3-lobed, as broad or broader than long, often obscurely so, saccate at base; side lobes large, rounded, erect; midlobe often smaller than side lobes; callus usually of 2 short pubescent ridges at base of lip. Column elongate, clavate; pollinia 8, clavate. Differs from *Phaius* in its lip and callus shape and lack of a spur.

Key to the species of *Gastrorchis*

1. Sepals and petals pink to rose-purple 2
1. Sepals and petals white or pale yellow 4
2. Lip side lobes larger than the midlobe, yellow spotted with purple ...... **28.1. G. françoisii**
2. Lip side lobes smaller than the midlobe, purple or brownish 3
3. Leaves greyish green with undulate margins; lip midlobe white, blotched
   with purple; callus of 3 raised ridges and verrucose in front .... **28.2. G. peyrotii**
3. Leaves green with straight margins; lip midlobe with a white base and broad
   purple margin; lip callus of 2 ridges .... **28.3. G. humblotii**
4. Sepals and petals pale yellow, up to 6 cm long .... **28.4. G. lutea**
4. Sepals and petals white, usually less than 5 cm long 5
5. Plant epiphytic; rhizome elongate, creeping .... **28.6. G. simulans**
5. Plant terrestrial; rhizome short, the pseudobulbs usually clustered 6
6. Lip midlobe larger than the side lobes .... **28.3b. G. humblotii** var. **schlechteri**
6. Lip midlobe smaller than the side lobes 7
7. Lip side lobes spreading widely, yellow, heavily blotched with purple; callus
   with several verrucose ridges in front .... **28.8. G. tuberculosa**
7. Lip side lobes suberect, white, heavily blotched with purple; callus lacking
   verrucose ridges in front, covered with clavate hairs 8
8. Lip midlobe transversely rectangular-reniform, retuse, spotted purple on
   side lobes and with a purple midlobe .... **28.5. G. pulchra**
8. Lip midlobe ovate-triangular, acute, marked with pink on side lobes and with
   a white midlobe .... **28.7. G. steinhardtiana**

28.1. G. françoisii Schltr.   *(opposite, bottom left)*
(syn. *Phaius françoisii* (Schltr.) Summerh.)

Terrestrial 45–60 cm tall; leaves elliptic-lanceolate, up to 62 × 7 cm; inflorescence 10–25-flowered, 60–90 cm tall; flowers pinkish purple; sepals 18 mm long; lip 3-lobed, 18 × 30 mm; side lobes erect, obliquely semi-orbicular; midlobe broadly flabellate, emarginate; callus V-shaped with a lower central ridge, sparsely hairy, with 3 verruculose lines in front; flowers somewhat similar to *G. humblotii* but a paler pink colour and the shape of the lip is very different; lip side lobes yellow, spotted with pink; midlobe not clawed, emarginated, pink. Antananarivo, Antsiranana, Fianarantsoa, Toamasina. In humid, evergreen forest on plateau; 1200–1800 m. Fl. January–March.

28.2. G. peyrotii (Bosser) Senghas   *(opposite, bottom right)*
(syn. *Phaius peyrotii* Bosser)

Terrestrial; pseudobulbs discoidal, small; leaves relatively small, linear-oblong, acute, 20–45 × 2–4.5 cm, undulate and crispate at the edges and greyish-green in colour; flowers 5–12; sepals 22–30 mm long; lip 3-lobed, 22–30 × 12–18 mm; midlobe orbicular with an undulate margin; callus of 3 glabrous ridges, raised highest at base; column 12–14 mm long. Antananarivo, Toamasina. In humid, evergreen forest on plateau; 500–1500 m. Fl. December–January.

28.3. G. humblotii (Rchb.f.) Schltr.   *(p. 169, top left)*
(syn. *Phaius humblotii* Rchb.f.; *Gastrorchis schlechteri* var. *milotii* Ursch & Genoud)

Plant 45–80 cm tall; leaves 2–4, lanceolate, 25–40 × 6–10 cm; inflorescence 12–15-flowered; flowers pink with a darker lip; sepals and petals reflexed, 2.5–3.2 cm long; lip obscurely 3-lobed, 3 × 3 cm, callus V-shaped at the base of the lip, yellow; lip side lobes orange, heavily marked with pink; midlobe shortly, broadly clawed, rose-purple, emarginate. Antananarivo, Antsiranana, Fianarantsoa, Toamasina. In humid, evergreen forest on plateau and on the margins of sphagnum bogs; 850–1500 m. Fl. October–March.

28.3a. G. humblotii var. rubra (Bosser) Bosser & P.J.Cribb
(syn. *Phaius humblotii* var. *ruber* Bosser)

Differs from the typical variety in its deep violet-red sepals and petals and in the colour and shape of the lip. Antananarivo. In humid, mossy, evergreen forest. Fl. January–March.

27.1   27.1C

28.1   28.2

28.3b. G. humblotii var. **schlechteri** (H.Perrier) Senghas ex Bosser & P.J.Cribb

(syn. *Gastrorchis schlechteri* François; *Phaius humblotii* var. *schlechteri* (H.Perrier) Bosser; *P. schlechteri* (H.Perrier) Summerh.)

Differs from the typical variety in sepals and petals, which are white instead of pink; floral bracts white instead of green. Antsiranana. In mossy forest; 1200–2000m. Fl. January–April.

28.4. G. lutea (Ursch & Genoud ex Bosser) Senghas   *(opposite, top right)*

(syn. *Phaius luteus* Ursch & Genoud ex Bosser; *P. geffrayi* Bosser; *Gastrorchis luteus* Ursch & Genoud; *G. geffrayi* (Bosser) Senghas)

Terrestrial; pseudobulbs ovoid 2–2,5 cm long; leaves narrowly lanceolate, acute, 2–45 × 4–6 cm; flowers 5–10, with fleshy, yellowish-green to off-white petals and sepals, yellowing with age, and up to 6 cm long floral bracts; sepals 2.5–3 cm; lip 2.5–3 cm long, with spreading, purple side lobes and a white mucronate midlobe; callus V-shaped, with yellow hairs. Antsiranana, Toamasina. In humid, evergreen forest; marshes; 500–1100 m. Fl. October.

Note. *P. geffrayi* is a peloric form with a petaloid lip.

28.5. G. pulchra Humbert & H.Perrier   *(opposite, bottom left)*

(syn. *Phaius pulcher* (Humbert & H.Perrier) Summerh.)

Flower with pure white tepals; lip pinkish-red, trapezoid and widely oblong with undulating margins and with a central raised mound covered in yellow hairs. Antsiranana, Toamasina, Toliara. In humid, mossy, evergreen forest, in shade and near rocks; 400–2000 m. Fl. November–December.

28.6. G. simulans (Rolfe) Schltr.   *(opposite, bottom right)*

(syn. *Phaius simulans* Rolfe; *P. fragrans* Grignan)

Epiphyte; close to *G. tuberculosa* (p. 170), the main difference being that it is an epiphyte with a relatively slender, elongated rhizome and has smaller flowers. Toamasina. In montane forest. Fl. September–February.

28.7. G. steinhardtiana Senghas   *(p. 171, top left)*

(syn. *Gastrorchis pulchra* var. *perrieri* (Bosser) Bosser & P.J.Cribb; *Phaius pulcher* var. *perrieri* Bosser; *G. tuberculosa* var. *perrieri* Ursch & Genoud)

Plant tall; pseudobulbs 2.5 × 1.5 cm, 3–5 cm apart on rhizome; leaves 3–4, 30– 60 × 1.5–2.5 cm; inflorescence up to 40 cm tall; flowers 7–15, 6 cm across; sepals and petals white; sepals 2.5–3.5 × 1.5 cm; lip obscurely 3-lobed in apical half, 24–25 × 30–32 mm, white, speckled with pink; side lobes semi-elliptic, undulate; midlobe ovate, acute, undulate on margins; callus slightly raised, bearing yellow clavate hairs. Antsiranana, Toamasina. In humid, mossy, evergreen forest; marshes; 500–1420 m. Fl. October–November.

28.3

28.4

28.5

28.6

28.6

28.8. G. tuberculosa (Thouars) Schltr. *(opposite, top right)*

(syn. *Limodorum tuberculosum* Thouars; *Phaius tuberculosus* (Thouars) Blume; *P. warpuri* Weathers; *Gastrorchis humbertii* Ursch & Genoud)

Terrestrial; leaves 5–6, narrowly lanceolate, 30–60 × 2–3 cm; flowers 6–8; sepals white, 32–44 mm long; lip obscurely 3-lobed, the side lobes yellow covered with red spots, the midlobe white with lilac- or violet-spotted margins, the disc bearing 3 yellow, verrucose ridges. Antananarivo, Toamasina. In humid, evergreen, lowland or montane forest; up to 1500 m. Fl. September–February.

29. AGROSTOPHYLLUM

A genus of c. 100 species, extending from the Seychelles to the islands of the Pacific; 1 species in Madagascar. Stem long, leaves 2-ranked, thin, with overlapping sheaths. Inflorescences in terminal heads, usually of many small flowers. Lip sac-shaped at the base.

29.1. A. occidentale Schltr.

(syn. *Agrostophyllum seychellarum* Rolfe)

A tall plant forming dense clumps, with flattened stems completely covered in rigid sheaths and inflorescences in terminal heads of small flowers. Antsiranana; also in the Seychelles. Under large trees; 300–800 m. Fl. December–March.

30. POLYSTACHYA

A large pantropical genus of over 200 species. About 21 species in Madagascar. Small to large epiphytes, lithophytes or rarely terrestrials. Stems usually pseudobulbous, clustered or less commonly superposed. Leaves 1–several, lanceolate to oblanceolate. Inflorescence apical, usually many-flowered. Flowers non-resupinate, small to medium-sized. Dorsal sepal free, lateral sepals oblique, forming a prominent chin with the column-foot. Petals small, usually oblanceolate. Lip entire or 3-lobed, usually uppermost in the flower, spurless, with or without a callus, sometimes with floury hairs on upper surface. Column short, with a prominent foot; pollinia 4, waxy, attached by a short stalk to a viscidium.

Key to species of *Polystachya*

1. Leaf solitary, terminal on stem .... **30.1. P. cultriformis**
1. Leaves 2 or more 2
2. Stems superposed, each new one arising from above the middle of the
   older stem 3
2. Stems clustered, not superposed 4
3. Inflorescence branching; flowers tiny with 1–2 mm long sepals .... **30.2. P. fusiformis**
3. Inflorescence simple; flowers with 4 mm long sepals .... **30.3. P. oreocharis**

28.7

28.8

30.1

30.2

4. Flowers green, greenish yellow, yellow, orange or yellow-orange 5
4. Flowers white, pink or green, sometimes with a yellow lip 10
5. Inflorescence glabrous 6
5. Inflorescence minutely puberulent; lip with a basal horn-like callus; lip
   margin undulate . . . . **30.7. P. cornigera**
6. Pseudobulbs elongate-cylindrical, 3.5 cm or more long 7
6. Pseudobulbs ovoid-conical, 2 cm or less long 9
7. Lip glabrous . . . . **30.8. P. henrici**
7. Lip farinaceous (with flour-like hairs) 8
8. Flowers orange; sepals 5–7 mm long . . . . **30.6. P. clareae**
8. Flowers greenish; sepals 3.5–4 mm long . . . . **30.9. P. virescens**
9. Plant up to 20 cm tall; leaves oblong-ligulate, 5–8 cm long; sepals 5 mm
   long . . . . **30.4. P. monophylla**
9. Plant 25–40 cm tall; leaves linear-lanceolate, 12–17 cm long; sepals 3 mm
   long . . . . **30.5. P. aurantiaca**
10. Sepals 8 mm or more long 11
10. Sepals 8 mm or less long 13
11. Flower-stalk covered by ancipitous sheaths; flowers red purple; lip entire,
    subrectangular, covered by golden hairs . . . . **30.12. P. tsaratananae**
11. Flower stalk lacking ancipitous sheaths; flowers white to mauve with a
    rose-pink lip with a yellow indumentum; lip 3-lobed, covered with a white
    or yellow farina 12
12. Inflorescence simple; lowermost bracts shorter than the flowers; lip longer
    than broad . . . . **30.10. P. heckeliana**
12. Inflorescence branched; lowermost bracts longer than the flowers; lip
    broader than long . . . . **30.11. P. rhodochila**
13. Inflorescence glabrous 14
13. Inflorescence hairy; ovary hairy or bristly 16
14. Plant often terrestrial; sepals rose-pink, 6–8 mm long, acute; inflorescence
    simple; leaves usually 2, 8–13 × 0.8–1.3 cm . . . . **30.15. P. rosea**
14. Plants often epiphytic or lithophytic; sepals purple, yellow or green, 3–6 mm
    long, apiculate; inflorescence usually branched; leaves 10–25 × 1.8–2.5 cm 15
15. Peduncle bearing ancipitous sheaths; branches not second . . . . **30.13. P. anceps**
15. Peduncle lacking ancipitous sheaths; branches second . . . . **30.14. P. concreta**
16. Pseudobulbs cylindrical or fusiform, 2 cm or more long 17
16. Pseudobulbs ovoid to conical, 1 cm or less long 20
17. Inflorescence branched; leaves 1–2, elliptic, 2.5–3.5 × 0.7–1.5 cm; lip
    apiculate . . . . **30.17. P. pergibbosa**
17. Inflorescence simple; leaves linear or ligulate, usually 3.5 or more cm long;
    lip not apiculate 18
18. Inflorescence 2–4 times as long as the leaves . . . . **30.18. P. perrieri**
18. Inflorescence 2 times as long as the leaves or less 19
19. Leaves linear-lanceolate, 3.5–6 cm long; inflorescence twice as long as the
    leaves, 16–24-flowered lip longer than broad . . . . **30.20. P. tsinjoarivensis**
19. Leaves ligulate, 18–20 cm long; inflorescence as long as the leaves, 6–12-
    flowered; lip broader than long . . . . **30.21. P. waterlotii**
20. Flowers white with a yellow lip; sepals 4–5 mm long; lip papillose, the callus
    large, obtuse . . . . **30.16. P. humbertii**
20. Flowers pink; sepals 2–2.5 mm long; lip not papillate, the callus horn-like . . . .
    . . . . **30.19. P. rosellata**

30.1. P. cultriformis (Thouars) Spreng. *(p. 171, bottom left)*

(syn. *Dendrobium cultriforme* Thouars; *Polystachya cultrata* Lindl.; *P. cultriformis* var. *humblotii* Rchb.f.)

Epiphyte, lithophyte or terrestrial; pseudobulbs cylindrical; leaf 1, oblanceolate-oblong, 12–25 × 2.5–3.5 cm; inflorescence branched, equal length to or longer than the leaf; flowers white, rarely pink or yellow; sepals 7–8 mm long; lip 3-lobed in middle, with a keel towards the base. Antananarivo, Antsiranana, Fianarantsoa, Toamasina, Toliara, Mahajanga; also in Comoros, Mascarenes, Seychelles and tropical and southern Africa. Coastal forest, seasonally dry deciduous woods, humid evergreen forest; up to 2500 m. Fl. throughout the year.

LOCAL NAME *Sonjomboae.*

30.2. P. fusiformis (Thouars) Lindl. *(p. 171, bottom right)*

(syn. *Dendrobium fusiforme* Thouars; *Polystachya minutiflora* Ridl.; *P. multiflora* Ridl.; *P. composita* Kraenzl.)

Epiphyte or lithophyte; a common and widespread species; stems cylindrical-fusiform, superposed and pendulous, 10–25 cm long; leaves 3–5, narrowly oblong-elliptic, 4–8 × 0.6–0.8 cm; inflorescence branched, pubescent or hirsute; flowers tiny, glabrous; sepals 1–2 mm long; lip 3-lobed; lobes similar in size. Antananarivo, Antsiranana, Fianarantsoa, Toamasina, Toliara; also in the Mascarenes, Seychelles and tropical and southern Africa. Dry forest; humid, evergreen forest on plateau; 600–2000 m. Fl. December–April.

30.3. P. oreocharis Schltr. *(p. 175, top left)*

Epiphyte, 9–25 cm tall; pseudobulbs sympodial and ascending, very rarely branched; leaves 6–10, distichous, linear-lanceolate, obtuse, 1.8–3.3 × 0.8–0.9 cm; inflorescence simple, 8–20-flowered, longer than the leaves, puberulent above; flowers reddish or greenish, the ovary hairy; sepals 4 mm long; lip 3-lobed in middle, 4 × 2 mm, with an incurved basal horn-like callus. Antsiranana, Fianarantsoa, Antananarivo, Toamasina, Toliara. Lichen-rich montane forest; 950–2000 m. Fl. January–April.

30.4. P. monophylla Schltr. *(p. 175, top right)*

Plant 15–20 cm tall; pseudobulbs cylindrical, 2.5 cm tall; leaves (1–)2, oblong-ligulate, 5–8 × 1.2–1.5 cm; inflorescence twice as long as the leaves, densely 7–12-flowered; flowers c. 11 mm long, yellow to reddish-orange, glabrous; sepals 5 mm long; lip 4-lobed, 5.5 × 4 mm, with 3 fleshy crests. Antananarivo, Antsiranana. On quartzite rocks; 1800–2300 m. Fl. February.

30.5. P. aurantiaca Schltr. *(opposite, bottom left)*

Epiphyte, 25–40 cm tall; pseudobulbs up to 6 cm long; leaves 3, linear-lanceolate, 12–17 × 1.2–1.7 cm; inflorescence longer than the leaves, branched; bracts very short; flowers small, yellow-orange, glabrous; sepals 3 mm long; mentum short; lip 3-lobed, 3.5 × 2.5 mm, the midlobe apiculate, the basal callus subglobose; ovary glabrous. Antananarivo, Mahajanga. On shaded rocks; semi-deciduous, low-elevation forest, remnant gallery forest; 150–700 m. Fl. February–April.

30.6. P. clareae Hermans *(opposite, bottom right)*

Epiphyte or terrestrial, 18–30 cm tall; pseudobulbs conical-cylindrical, 5–7 cm tall; leaves 2–4, linear-ligulate, 10–19 × 1.2–2.6 cm; inflorescences simple or up to 3-branched; flowers clustered, large, bright orange-yellow, glabrous; sepals 5–7 mm long; petals spathulate; lip 3-lobed in middle, 6 × 5 mm, with a central callus, farinaceous. Antsiranana, Antananarivo, Toamasina. In mid-elevation, humid, evergreen forest; 700–1550 m. Fl. February–March.

30.7. P. cornigera Schltr. *(p. 177, top left)*

Epiphyte, 5–11 cm tall; pseudobulbs ovoid, 6–10 × 3–6 mm; leaves 2–4, linear, 3–11 × 0.35–0.5 cm; inflorescence 4–15-flowered; flowers greenish yellow, glabrous; sepals 3–4 mm long; lip 3-lobed, 3 × 2.5 mm, with a distinct horn at the base, margins slightly undulate; ovary glabrous. Antananarivo, Antsiranana, Fianarantsoa, Toamasina. Lichen-rich montane forest; 1000–2000 m. Fl. January–March.

30.7a. P. cornigera var. **integrilabia** Senghas

Differs from the typical variety by its entire lip. Fianarantsoa. Humid, evergreen forest on plateau; c. 1200 m. Fl. October.

30.8. P. henrici Schltr.

Epiphyte or lithophyte, up to 20 cm tall; pseudobulbs oblong-cylindrical, 3.5–4 cm tall; leaves 2, ligulate, 5–10 × 0.8–1.8 cm; inflorescence up to 10-flowered, glabrous; flowers yellowish, glabrous; sepals 5 mm long; lip 3-lobed, 6 × 3.2 mm, the side lobes very small, obtuse. Fianarantsoa. Grassland and xerophytic scrub; rocky outcrops; 1600–1700 m. Fl. February.

30.3

30.4

30.5

30.6

30.9. P. virescens Ridl.

Epiphyte, 15–30 cm tall; pseudobulbs narrow; leaves 2, ovate-lanceolate to oblanceolate, 5–12 × 1.2–1.6 cm; inflorescence usually less than twice as long as the leaves, simple or branched; flowers greenish; sepals 3.5–4 mm long; mentum 4 mm tall; lip 3-lobed, 5 × 4.5 mm, farinose and with a basal callus; very close to *P. concreta* (p. 178). Antananarivo, Antsiranana, Fianarantsoa, Mahajanga, Toamasina. Humid, evergreen forest on plateau; 700–1800 m. Fl. October–April.

30.10. P. heckeliana Schltr.

Terrestrial; pseudobulbs 8 cm long; leaves 3–4, ligulate, up to 25 × 1.8–2.4 cm; inflorescence densely 8–15-flowered, not branched; flowers large, whitish pink or mauve with a sulphur yellow lip, slightly warty; sepals 10–11 mm long; lip 3-lobed below the middle, 10 × 6 mm. Antsiranana. Lichen-rich forest; dry forest; 1000–1700 m. Fl. May.

30.11. P. rhodochila Schltr.   *(opposite, top right)*

Terrestrial or rarely epiphytic, 30–50 cm tall; pseudobulbs stem-like; leaves 3–4, linear-ligulate, 17–25 × 1.2–2 cm; inflorescence few-branched; lowermost bracts much longer than flowers; flowers white with a rose-pink lip, slightly warty; sepals 11–15 mm long; lip 9 × 10 mm, 3-lobed below the middle, covered with yellow hairs. Antananarivo, Antsiranana, Mahajanga. Lichen-rich forest; 1400–2000 m. Fl. October–March.

30.12. P. tsaratananae H.Perrier   *(opposite, bottom left)*

Lithophyte, 15–25 cm tall; pseudobulbs 1.5–4 cm tall; leaves 4–5, linear, 10–20 × 1–2 cm, dark green or purplish-green; peduncle covered by ancipitous sheaths; flowers 10–20, large, purple-red; sepals 8–9 mm long; lip entire, subrectangular, 9 × 5 mm, covered by golden hairs. Antsiranana. On shaded, humid rocks; 1500–2000 m. Fl. April–May.

30.13. P. anceps Ridl.   *(opposite, bottom right)*

(syn. *Polystachya mauritiana* var. *anceps* (Ridl.) H.Perrier)

Epiphyte or lithophyte, up to 40 cm tall; pseudobulbs elongate; leaves 3–4, oblong or oblanceolate, 10–22 × 1.8–2 cm; inflorescence branched; sheaths of the scape large, ancipitous and reaching the base of the panicle; flowers yellow and purple or red; sepals 3–6 mm long; lip 3-lobed below the middle, 4–6 mm long. Antananarivo, Fianarantsoa, Toamasina; also in the Comoros. Humid, highland forest; 800–1700 m. Fl. December–March.

30.7

30.11

30.12

30.13

30.14. **P. concreta** (Jacq.) Garay & H.R.Sweet   *(opposite top left)*
(syn. *Epidendrum concretum* Jacq.; *Dendrobium polystachyum* Thouars; *Polystachya tessellata* Lindl.; *P. minuta* Frappier; *P. zanguebarica* Rolfe)

Epiphyte or lithophyte; a very variable and widespread species; plant with 3–many broad, oblanceolate leaves; inflorescence usually branched, the branches secund; flowers small, green, purplish or yellowish; sepals 3–4 mm long; lip oblong, 3-lobed, descending between the triangular petals and fused with them on both sides, farinose. Antananarivo, Antsiranana, Fianarantsoa, Mahajanga, Toamasina, Toliara; also in the Comoros, Mascarenes, Seychelles, and the tropical Americas, Africa and Asia. Seasonally dry, deciduous woods; humid evergreen forest; shaded, moist rocks; up to 1400 m. Fl. December–August.

30.15. **P. rosea** Ridl.   *(opposite, top right)*
(syn. *Polystachya bicolor* Rolfe; *P. hildebrandtii* Kraenzl.)

Lithophyte or terrestrial, 15–20 cm tall; pseudobulbs stem-like; leaves 2–3, linear-lanceolate, 8–13 × 0.8–1.3 cm; inflorescence simple, rarely branched; flowers rose-pink, glabrous; sepals 6–8 mm long; mentum 4–5 mm long; lip long-clawed, 3-lobed in apical half, with a linear basal callus. Antananarivo, Fianarantsoa, Mahajanga, Antsiranana, Toamasina, Toliara; also in the Seychelles. Shaded, humid, rocky outcrops; lichen-rich forests; 800–1500 m. Fl. January–May.

30.16. **P. humbertii** H.Perrier   *(opposite, bottom left)*
Epiphyte, 6–8 cm tall; pseudobulbs tiny, 1 cm tall; leaves broad, elliptic-oblong, 1.5–2.2 × 0.6–0.8 cm; inflorescence 6–7-flowered, the stalk with brown hairs; flowers c. 10 mm long, white with a yellow lip, spotted with purple in the centre, hairy on ovary; sepals 4–5 mm long; lip 3-lobed, papillose, with a large obtuse callus at its base. Fianarantsoa, Toliara, Antananarivo, Toamasina. High-elevation ericaceous scrub and humid forest; 1000–2000 m. Fl. November–December.

30.17. **P. pergibbosa** H.Perrier
Epiphyte; pseudobulbs small, stem-like; leaves 1–2, elliptic, obtuse, 2.5–3.5 × 0.7–1.5 cm; inflorescence few-branched; flowers with hairy ribs to ovary; sepals 3–5 mm long; lip 4 × 2.8 mm, 3-lobed, the side lobes suborbicular, the midlobe subrectangular, apiculate. Antsiranana. Humid, mossy, evergreen forest on plateau; 500–2000 m. Fl. January–March.

30.14

30.15

30.16

30.21

30.18. P. perrieri Schltr.

Epiphyte or lithophyte, 20–30 cm tall; pseudobulbs cylindrical, 3–4 cm tall; leaves 2–3, linear-ligulate, 8–15 × 0.6–1 cm; inflorescence 2–4 times as long as the leaves, simple; bracts short; flowers rose-purple, the ovary finely hairy; sepals 5.5–6 mm long; lip 3-lobed in basal half, 7 × 5 mm, the callus obovoid, covered in hairs. Antananarivo. On rocky outcrops; c. 1900 m. Fl. February.

30.19. P. rosellata Ridl.

Epiphyte, up to 10 cm tall; pseudobulbs tiny, obscure; leaves linear-ligulate, 2.5–5 × 0.4–0.6 cm; inflorescence few-flowered, the stalk hairy; flowers small, c. 6 mm long, pink; sepals 2–2.5 mm long; lip 3-lobed, 3 × 2 mm, with a horn-like basal callus; ovary bristly. Antananarivo, Antsiranana. High-elevation, ericaceous scrub and humid lichen-rich forest; 1000–1800 m. Fl. January–March.

30.20. P. tsinjoarivensis H.Perrier   (opposite)

Terrestrial, 20–25 cm tall; pseudobulbs 2–3 cm tall; leaves 3–4, linear-lanceolate, 3.5–6 × 0.8–1 cm; inflorescence twice as long as the leaves, 16–24-flowered, bristly; bracts as long as the ovary; flowers c. 8 mm long; sepals 4.5–5 mm long; lip 3-lobed, 4.8 × 2.5 mm, with a papillate flattened callus at the base; ovary bristly; similar to *P. rosellata*. Antananarivo, Fianarantsoa. Humid, evergreen forest on plateau; 1300–2000 m. Fl. February.

30.21. P. waterlotii Guillaumin   (p. 179, bottom right)

Plant 15–20 cm tall; pseudobulbs fusiform, 3–5 cm long; leaves 2–5, ligulate, 18–20 × 0.8–2 cm; inflorescence equalling the leaves, 6–12-flowered; bracts 1–2 mm long; flowers with a slightly hairy ovary; sepals 4–6 mm long; lip 3-lobed, wider than long, 5 × 5.5 mm; ovary slightly hairy. Antananarivo; also in the Comoros. High-elevation, ericaceous scrub and humid forest. Fl. June.

31. IMERINAEA

A monotypic genus endemic to Madagascar. Erect terrestrial plants with a short rhizome, and dense simple, hairy roots; pseudobulbs obscure, cylindrical, unifoliate. Leaf erect, narrowly ligulate-lanceolate, acuminate, somewhat petiolate at base. Inflorescence terminal, slender, erect, slightly longer than the leaf, racemose, few-flowered. Flowers non-resupinate. Sepals and petals subsimilar. Lip obovate, obtuse, parallel to column, apical margins undulate, papillose-pubescent at the base, warty-rugulose at the tip. Column slender, minutely papillose-puberulent at the base, lacking a foot; pollinia 4, superposed.

30.20

31.1. **Imerinaea madagascarica** Schltr.   (*p. 185, top left*)
(syn. *Phaius gibbosulus* H.Perrier, *nom. nud.*)

Plant 23–28 cm tall; leaf 10–17 × 1.5–1.8 cm; flowers 2–4, yellow with red-purple tips; sepals 10 mm long; lip uppermost, rolled, 9 × 7 mm, with a distinct callus. Antananarivo, Antsiranana, Fianarantsoa, Toamasina. In mossy and lichen-rich montane forest; on shaded and humid rocks; 1000–1500 m. Fl. December–May.

32. BULBOPHYLLUM

A very large cosmopolitan genus of perhaps 1500 species; 188 or more mostly endemic species in Madagascar. Epiphytic or less commonly lithophytic plants with short to elongate creeping, sometimes branching rhizomes; pseudobulbs proximate or well-spaced on rhizome, 1–2-leaved; leaves linear, oblong, elliptic or lanceolate, often leathery; inflorescence erect or, less commonly, arcuate or pendent, 1–many-flowered; flowers relatively small, often fleshy; sepals usually glabrous, sometimes ciliate; lateral sepals sometimes united into a synsepal; petals usually smaller than sepals; lip articulated by a hinge to apex of column-foot, often very fleshy, simple or 3-lobed, often recurved; callus ridges sometimes present, glabrous, papillose or pubescent, rarely with ligulate caduceus appendages in apical half. Column short, often with short to elongate stelids at apex on either side of the anther, with a distinct foot, forming a short mentum with the base of the lateral sepals.

Bulbophyllums are very difficult to identify and name. A ×10 hand-lens is necessary for the identification of most species. The sections used here follow Perrier's, in *Flore de Madagascar* (1941), but have been modified by the work of Fischer *et al.* (in press). The following key will take users to the appropriate section.

Key to sections of *Bulbophyllum*

1. Inflorescence a daisy-like head (umbel) .... **I. Sect. Cirrhopetalum**
1. Inflorescence 1-flowered or elongate; flowers densely to laxly arranged in a spike or head in a round head or elongate 2
2. Pseudobulbs densely clustered, bilaterally flattened, unifoliate; inflorescence few-flowered, sessile; flowers glabrous .... **III. Sect. Polyradices**
2. Pseudobulbs not as above; inflorescence not as above 3
3. Flower-stalk bristle-like or very short, 1–few-flowered, or 6–8-flowered but then the inflorescence very short, not passing the base of the leaves; flowers with a ciliate lip 4
3. Flower-stalk slender or fleshy, more than 5-flowered, or rarely with 1–3 flowers but then always 10 cm or more long, rigid, with a fleshy rachis; flowers fleshy, 8–9 mm long 7
4. Pseudobulbs 2-leaved .... **II. Sect. Lichenophylax**
4. Pseudobulbs 1-leafed 5
5. Lip ciliate; peduncle short; flower stalk shorter than the leaf .... **IX. Sect. Pantoblepharon**
5. Lip not ciliate; flower-stalk bristle-like; inflorescence longer than the leaf 6

6. Stelids at apex of column very short; free part of the column-foot curved up . . . .
. . . . **XV. Sect. Lepiophylax** (in part)
6. Stelids finely pointed, as long as the column; free part of column-foot shorter
than the rest of the column . . . . **IV. Sect. Elasmotopus** (in part)
7. Lateral sepals united to form a boat-shaped synsepal with a two-toothed
apex, with two lateral keels or wings; pseudobulbs always 2-leaved . . . . **V. Sect. Ploiarium**
7. Lateral sepals free or only connate at the base to the column-foot, never
keeled 8
8. Anther large with a large conical appendage in front; flowers large, 15–26
mm long, and subtended by broad bracts . . . . **VI. Sect. Alcistachys**
8. Anther not prolonged in front by a large appendage and flowers smaller 9
9. Lip manifestly ciliate 10
9. Lip not ciliate 12
10. Column with a marginal tooth or lobule on each side of the stigma below the
stelids; column slender . . . . **VII. Sect. Kainochilus**
10. Column lacking marginal teeth or lobule below the stelids 11
11. Pseudobulbs 1-leafed . . . . **VIII. Sect. Trichopus**
11. Pseudobulbs 2-leaved . . . . **X. Sect. Lupulina** (in part)
12. Pseudobulbs 1-leafed 13
12. Pseudobulbs 2-leaved 16
13. Pseudobulbs yellow, the sheaths decaying into white, cottony fibres;
peduncle very short; rachis many-flowered, ascending and dense; flowers
thin-textured, white or yellowish; rachis not fleshy . . . . **XI. Sect. Loxosepalum**
13. Pseudobulbs and sheaths not as above; peduncle often as long as the
straight or slightly recurved, often fleshy rachis; flowers fleshy 14
14. Stelids with a fleshy extension at the base, the latter differing from the
stelids in consistency, form and colour; inflorescence dense or subcapitate,
6–12-flowered . . . . **XII. Sect. Lyperocephalum**
14. Stelids simple or toothed, but, in the latter case, tooth not differing from the
stelid; inflorescence longer, bent or straight 15
15. Stelids simple and 1 mm or more long; peduncle short, 3.5 cm or more long;
rachis bent; dorsal sepal large and concave, very much the largest segment
of the flower . . . . **X. Sect. Lupulina** (in part)
15. Stelids simple or toothed on the lower margin; inflorescence 10 cm or more
in length; rachis bent or straight; dorsal sepal not the largest segment of the
flower . . . . **XIII. Sect. Lyperostachys**
16. Rachis more fleshy than the top of the peduncle; flowers fleshy; sheaths
of the young pseudobulbs often very fleshy, hard and long; rachis recurved . . . .
. . . . **XIV. Sect. Pachychlamys**
16. Rachis not fleshy, the same diameter as the top of the peduncle; flowers
very thin-textured; sheaths thin-textured 17
17. Stelids short, more or less as long as the anther 18
17. Stelids more than 1 mm long, subulate, as long as the column 19
18. Peduncle bristle-like; inflorescence 3–4 times as long as the leaves . . **XV. Sect. Lepiophylax**
18. Peduncle fleshier; inflorescence less than twice as long as the leaves . . **XVI. Sect. Lemuraea**
19. Rachis very sinuous; lip 2 two small pointed side lobes at the base . . . . . **XVII. Sect. Bifalcula**
19. Rachis not or only slightly sinuous; lip lacking basal lobes, cup-shaped at
apex . . . . **IV. Sect. Elasmatopus**

I. Section Cirrhopetalum

Medium-sized plants with elongated stout rhizomes; pseudobulbs ovoid, well-spaced on rhizome, 1-leafed; flowers borne in a false umbel, like a daisy; lateral sepals joined in upper two-thirds, very much longer than the other floral segments.

32.1. B. longiflorum Thouars   (*opposite, top right*)

(syn. *Epidendrum umbellatum* Forst.; *Cirrhopetalum thouarsii* Lindl.; *C. umbellatum* (Forst.) Frapp. ex Cordem.; *C. longiflorum* (Thouars) Schltr.)

Epiphyte; pseudobulbs 1-leafed; inflorescence subumbellate; flowers purple; lateral sepals c. 27 × 6 mm. Antananarivo, Antsiranana, Fianarantsoa, Mahajanga, Toamasina, Toliara; also in the Comoros, Mascarenes, Seychelles, E Africa, SE Asia and SW Pacific Islands. Mossy humid, evergreen forest; coastal forest; sea level–1200 m. Fl. October–April.

LOCAL NAME *Solofa*.

II. Section Lichenophylax

Tiny plants with branching rhizomes growing on rocks or trunks; pseudobulbs tiny, ovoid or oblong, approximate; inflorescences setiform 1-flowered; flowers with longly acuminate sepals; lip fleshy; stelids shortly triangular, sometimes didenticulate.

Key to species of Section Lichenophylax

1. Sepals 6 mm or more long 2
1. Sepals 5.5 mm or less long 7
2. Lip and petals ciliate 3
2. Lip and petals glabrous 4
3. Pseudobulbs obliquely obovoid-cylindrical; lip lacking horns and winged crests .... **32.5. B. lakatoense**
3. Pseudobulbs discoid; lip with 2 lateral horns and winged crests edging the depression .... **32.7. B. mangenotii**
4. Sepals 9–13 mm long, including the elongate tip 5
4. Sepals 6–7.5 mm long, including the tip 6
5. Leaves ovate-elliptic, 3 × 2 mm; sepals 13 mm long, including the aristate tip .... **32.4. B. hapalanthos**
5. Leaves linear-lanceolate, 6–23 mm long; sepals 9–10 mm long .... **32.2. B. afzelii**
6. Leaves narrowly elliptic; petals 3-nerved; lip with 2 keels diverging at the summit .... **32.3. B. forsythianum**
6. Leaves linear-lanceolate; petals 1-nerved; lip oblong, papillose ..... **32.6. B. lichenophylax**
7. Sepals long-aristate; lip callus warty .... **32.8. B. cataractarum**
7. Sepals shortly aristate or acuminate or not acuminate 8
8. Sepals 3 mm or less long; flowers white or pinkish white 9
8. Sepals 4 mm or more long; flowers red or reddish 10

31.1

32.1

32.2

32.4

32.2. **B. afzelii** Schltr.   *(p. 185, bottom left)*

Epiphyte; pseudobulbs ovoid, 3–4 mm tall, well-spaced on a branching rhizome; leaves 2, linear-lanceolate, 10–23 × 1–2.2 mm; inflorescence 1-flowered; flowers small; sepals 9–10 mm long; lip fleshy, curved, elliptic, with 2 ridges on top, the stelids triangular, acute, short and erect. Antsiranana. Humid, evergreen forest; 800–1400 m. Fl. September.

32.2a. **B. afzelii** var. **microdoron** (Schltr.) Bosser

(syn. *Bulbophyllum microdoron* Schltr.; *B. lichenophyllax* var. *microdoron* (Schltr.) H.Perrier)

Epiphyte or lithophyte; leaves shorter than typical variety, lanceolate, acute, subacuminate, up to 5 mm long, contracted at the base into a very short petiole; petals obovate, rounded at the tip and shortly mucronate. Antananarivo, Antsiranana, Toamasina. On wet rocks in humid, evergreen forest on plateau; c. 1000 m. Fl. July–October.

32.3. **B. forsythianum** Kraenzl.

Small lithophyte, less than 3 cm tall; pseudobulbs oblong, 1–3 mm tall; leaves 2, narrowly elliptic, 4–8 × 0.8–2 mm; inflorescence 1-flowered; sepals 6–7.5 mm long, hyaline; lip 3 × 0.8 mm, very curved, shortly angular at the apex, with 2 callosities on top joining together to form a central ridge. Fianarantsoa, Toamasina, Toliara. Shaded and humid rocks in river-bed, carpeting rocks; 300–1400 m. Fl. October–February.

32.4. **B. hapalanthos** Garay   *(p. 185, bottom right)*

Creeping branching epiphyte or lithophyte; pseudobulbs obovoid, 3 mm long; leaves 2, fleshy, ovate-elliptic, acute, 3 × 2 mm; inflorescence 1-flowered, 5 cm tall; sepals 9.5–10 × 2–3 mm, long aristate; petals linear-oblong, acute, 3 × 1 mm; lip fleshy, arcuate, 2.5 × 2 mm. Similar in habit and column structure to *B. lakatoense* (p. 187) but differs in the completely glabrous flowers and in the shape of the petals and lip. Madagascar. Fl. unknown.

32.5. B. lakatoense Bosser
(p. 189, top)

Creeping branching epiphyte or lithophyte; pseudobulbs obovoid, 2.5–3.5 mm long; leaves 2, linear-lanceolate, acute, 3–6 × 0.5–0.8 mm; inflorescence 1-flowered, 2–3 cm tall; sepals 13 × 1.2 mm, long acuminate; petals oblong, rounded, ciliate; lip fleshy, arcuate, 1.5 × 1 mm, ciliate. Similar to *B. lichenophylax* but distinguished by the lip which is ciliate at the edges. Toamasina. Lichen-rich forest; c. 1000 m. Fl. June.

32.6. B. lichenophylax Schltr.
(syn. *Bulbophyllum quinquecornutum* H.Perrier)

Small epiphyte; pseudobulbs tiny; leaves 2, linear-lanceolate, 6–8 mm long; inflorescence 1–more-flowered, the stalk bristle-like; flower dull red; sepals 6 mm long; lip oblong, without a callus; stelids narrow and acute. Antananarivo, Antsiranana, Toamasina. Mossy and lichen-rich forest; 800–2000 m. Fl. January–March, August.

32.7. B. mangenotii Bosser
(p. 189, bottom left)

Plant and flower tiny; pseudobulbs flattened-ovoid, 2.5–4 × 2–2.5 mm, yellow-green, flushed red; leaves 2, lanceolate, acute, 4–6 × 1.5–2 mm; inflorescence 4–5.5 cm tall, 1-flowered; flower greenish yellow with red venation on sepals and a dark red lip; sepals 11 mm long; petals obovate, ciliate; lip recurved, oblong, 3.5 mm long, ciliate, with a central depression on the upper surface that branches into 2 lateral horns, 2 winged crests edge this depression. Antananarivo, Fianarantsoa. Humid, mossy, evergreen forest on plateau; 1200–1300 m. Fl. August.

32.8. B. cataractarum Schltr.
(syn. *Bulbophyllum forsythianum sensu* Schltr.)

Lithophyte less than 3 cm high; pseudobulbs suborbicular, 1–1.5 mm across; leaves 2, linear-lanceolate, 7–9 × 1–1.5 mm; inflorescence 1-flowered; peduncle thread-like, 2–2.5 cm long; sepals 4.5 mm long; lip curved, ovate, cordate at the base, 2.5 × 1.5 mm, carrying a large, rounded, verrucose callus. Antsiranana. On very wet rocks in rivers; c. 1200 m. Fl. April.

32.9. B. debile Bosser
(p. 189, bottom right)

Creeping epiphyte; pseudobulbs obliquely obovoid-cylindrical, 1.5–2 mm long; leaves lanceolate, acute, 1.5–3 × 0.4–0.6 mm; inflorescence 1-flowered, 1.5–2 cm long; flowers white with a yellow lip, glabrous; sepals 2–2.4 mm long; lip linguiform, 1.5 × 0.5 mm, papillose in basal half. Close to *B. percorniculatum* (p. 188) but differs by the non-acuminate sepals and the shape of the lip. Toamasina. Mossy and lichen-rich forest; 900–1000 m. Fl. February.

32.10. **B. neglectum** Bosser

Plant tiny; pseudobulbs flattened-ovoid, 2–3 mm in diameter; leaves 2, lanceolate or linear-lanceolate, acute, 4–6.5 × 1–1.2 mm; inflorescence 1.5–1.7 cm tall, 1-flowered; flower red; sepals 5–5.5 mm long; petals ovate; lip strongly recurved, obscurely 3-lobed, glabrous, the base with 2 keels; stelids bilobed. Toamasina. Humid, evergreen forest, plant partly immersed in moss and liverworts on trunks; 800–1000 m. Fl. unknown.

32.11. **B. percorniculatum** H.Perrier

Plant small; pseudobulbs disc-like, 1–2 mm across; leaves 2, 3.5–5 × 0.6–1 mm; inflorescence 1-flowered; floral bracts extended at the tip; sepals shortly acuminate, 4–4.2 mm long; lip oblong-ligulate, obtuse, porrect, somewhat concave, 3.5 × 1.5 mm. Toliara. Humid, evergreen forest; c. 600 m. Fl. March.

32.12. **B. perpusillum** H.Wendl. & Kraenzl.   *(p. 191, top left)*

Plant very small; rhizome filiform; pseudobulbs subglobose, 2–4 mm across; leaves 2, lanceolate, 2–3 × 0.5 mm; inflorescence 1-flowered; flowers small, white-pink; sepals 3 mm long; lip ovate, obtuse, with a velvety callus. Toamasina. Fl. June (in cultivation).

III. Section Polyradices

Dwarf plants with densely clustered and bilaterally flattened, unifoliate pseudobulbs; inflorescence 1–2-flowered, sessile; the sepals, petals and lip glabrous but the sepals finely papillose on the margin and the lip slightly papillose beneath.

32.13. **B. petrae** G.A.Fischer, Sieder & P.J.Cribb

Roots numerous; rhizome creeping, subulate to ensiform; pseudobulbs 4–5 mm apart, orbicular, 8–12 × 8–12 mm, 1-leafed; leaf elliptic to obovate, 12–16 mm long × 5–8 mm wide, acute; inflorescence a short raceme, 13–18 mm long; rachis zigzag, not thickened, 2–3 mm long; flowers white with dark purple dotted sepals, with dark purple specks at the tips of the petals, with dark purple spots suffusing the lip; sepals ovate 7–8 mm long; petals ovate-spathulate, 1.8 × 2.8 mm, lower margins ciliate; lip recurved, ovate, finely papillose beneath; column with tooth-like wings; stelidia triangular. Toamasina. In wet montane forests, 1000–1400 m. Fl. Feb.

32.5

32.7

32.9

IV. Section Elasmotopus (including Sect. Hymenosepalum)

Medium-sized to small epiphytes; pseudobulbs 2-leaved, rarely 1-leafed; peduncle slender, rachis often recurved, many-flowered or rarely 1-flowered; sepals free, 3–7-nerved; petals 3-nerved; lip glabrous or rarely papillose, never ciliate; stelids elongate and acicular, longer than the anther.

Key to groups of species in Section Elasmotopus

1. Tiny plants; pseudobulbs 1-leafed; inflorescence 1–2-flowered; stalk setiform . . . . **32.14. B. analamazoatrae**
1. Pseudobulbs 2-leaved; inflorescence several-flowered; stalk not setiform 2
2. Pseudobulbs with 1 leaf . . . . **32.15. B. amphorimorphum**
2. Pseudobulbs 2-leaved **32.16–32.22**

32.14. B. analamazoatrae Schltr.   (opposite, top right)

Pseudobulbs distant from each other on rhizome, oblong, 6–8 × 3.5–4.5 mm; leaf ligulate to oblanceolate, 20–40 × 3–5 mm; inflorescence up to 5.5 cm long; flowers hyaline; sepals 5.5 mm long; lip subspherical, 1.5–2 mm long. Antsiranana, Toamasina. Hill forest; 400 m. Fl. January.

32.15. B. amphorimorphum H.Perrier   (p. 193)

Pseudobulbs 1-leafed; lip sigmoid-curved, 4 mm long, apex widened; stelids long. Antsiranana. Humid, highland forest; lichen-rich forest; c. 1400 m. Fl. December.

32.16. B. aubrevillei Bosser   (opposite, bottom left)

Pseudobulbs ovoid, conical, 10–25 × 3–7 mm; leaves divergent, ligulate, 3–9 × 0.5–0.9 cm; inflorescence 5–7 cm tall; 5–7-flowered; flowers red; sepals 6.5–7 mm long; lip red-purple, recurved, fleshy, ovate, 4 mm long, 2 mm wide, apex almost concave, base almost bilobed, 'V'-shaped; sepals spotted with purple; close to *B. amphorimorphum* which has a similar flower but has 1-leafed bulbs. Toamasina. Humid, mossy, evergreen forest; 500–1100 m. Fl. November–December.

32.17. B. françoisii H.Perrier   (opposite, bottom right)

Pseudobulbs fusiform, 20–40 × 4–5 mm; leaves oblanceolate, 5–7 × 1–1.3 cm; inflorescence as long as or longer than the leaves; 3–6-flowered; sepals 7–8 mm long; petals linear, 4 mm long; lip 5 mm long, oval-oblong, oscillating, set in the middle of the foot, the back part a little widened and ciliate and the front half a little narrowed and glabrous. Toamasina. Coastal and humid, evergreen forest; sea level–800 m. Fl. September–November, February.

32.12

32.14

32.16

32.17

32.17a. B. françoisii var. andrangense (H.Perrier) Bosser
(syn. *Bulbophyllum andrangense* H.Perrier)

The inflorescence has a very short stalk, much shorter than the leaves, the petals are oblong, less elongate, and the anther is also distinct. Fianarantsoa, Toamasina. In very shady, humid, evergreen forest on plateau; 1200–1400 m. Fl. October.

32.18. B. kieneri Bosser

Plant c. 12 cm tall; pseudobulbs cylindrical, canaliculated, 4.5–5 × 1–1.6 cm; leaves linear-oblong, 6–6.5 × 1–1.6 cm; inflorescence 3–4 cm long, slender, 4–5-flowered; flower fleshy, whitish spotted with red; sepals 3–4.5 mm long; lip fleshy, curved, whitish, 2.5 × 2 mm wide. Toamasina. Humid, lowland forest. Fl. unknown.

32.19. B. oxycalyx Schltr.
(syn. *Bulbophyllum rubescens* var. *meizobulbon* Schltr.)

Pseudobulbs ovoid, 10–15 × 5–8 mm; leaves 2, oblong-ligulate, 20–40 × 5–8 mm; inflorescence up to 18 cm long, 10–15-flowered; peduncle very slender, filiform, up to 11 cm long; flowers 7–7.5 mm long, red, striped or spotted; lip dark greenish-red, recurved, narrowly oblong-linguiform, with a small obsolete callus, and obscurely 2-ridged in the lower half. Antananarivo, Antsiranana. Lichen-rich, evergreen forest; c. 1500 m. Fl. March–February, October.

32.19a. B. oxycalyx var. rubescens (Schltr.) Bosser
(syn. *Bulbophyllum rubescens* Schltr.; *B. caeruleolineatum* H.Perrier; *B. loxodiphyllum* H.Perrier; *B. rostriferum* H.Perrier)

Petals are larger and more elongate than in the typical variety and do not have a red violet-red spot near the tip. Antsiranana. Lichen-rich forest; 2000–2500 m. Fl. November–December.

32.20. B. pandurella Schltr. *(p. 197, top left)*

Epiphyte, less than 5 cm tall; pseudobulbs ovoid, 3–5 × 3–3.5 mm; leaves 2, elliptic, 6–10 × 3.5–5 mm; inflorescence 1-flowered; flower yellow with red veins; sepals 5 mm long; lip somewhat pandurate; column with long aciculate stelids. Antsiranana. Lichen-rich, evergreen forest; c. 2000 m. Fl. January.

32.15

32.21. B. rauhii Toill.-Gen. & Bosser

Close to *B. oxycalyx* but can be distinguished by more slender, bare, rhizome, the linear narrow leaves, the smaller flowers and especially the size and shape of the lip. Antananarivo. Humid, evergreen forest on plateau; 1200–1300 m. Fl. February.

32.21a. B. rauhii var. andranobeense Bosser

Leaves shorter and wider than the typical variety; pseudobulbs smaller; flowers smaller; the dorsal sepal has 5 veins instead of 3. Toamasina. Humid, evergreen forest. Fl. March.

32.22. B. therezienii Bosser

Plant smallish; pseudobulbs flattened, 0.8–1 cm; leaves 2, oblong, 1–1.8 × 0.6–0.8 cm; inflorescence 10–12 cm long, laxly 8–12-flowered; flowers reddish, slightly fleshy; sepals 3.3–4 mm long; lip fleshy, arcuate, 2–2.5 mm across; stelids erect, triangular, acute, 0.2–0.3 mm long. Fianarantsoa. Mossy forest. Fl. February.

V. Section Ploiarium

Medium-sized to large plants; pseudobulbs 2-leaved; inflorescence many-flowered; lateral sepals united into an elliptic to semicircular synsepal, bifid or bidentate at apex, keeled on reverse, the keels extending onto the ovary; stelids short.

Key to groups of species in Section Ploiarium

1. Peduncle four times broader at apex than at the base and much broader than the rachis . . . . Group 1: 32.23. B. platypodum
1. Peduncle as broad at the apex as at the base 2
2. Rachis dilated and winged below each flower . . . . Group 2: 32.24. B. pleiopterum
2. Rachis dilated or not, but not winged below each flower 3
3. Keels on lateral sepals crenulate, lacerate or irregularly dentate . . . . . Group 3: 32.25–32.30
3. Keels on lateral sepals entire 4
4. Spike short, less than 4 mm long and densely flowered . . . . Group 4: 32.31–32.44
4. Spike longer; peduncle often more than twice as long as the spike 5
5. Spike lax, usually less than 30-flowered; bracts 2 mm or more apart . . . Group 5: 32.45–32.67
5. Spike dense or subdense, many-flowered; bracts less than 2 mm apart . . . .
   . . . . Group 6: 32.68–32.82

Ploiarium Group 1

Peduncle four times broader at apex than at the base and much broader than the rachis.

32.23. **B. platypodum** H.Perrier   *(p. 197, top right)*

Pseudobulbs ovoid, 10–15 × 8–10 mm, 2-leaved; leaves oblanceolate, 2.6–8.2 × 0.8–1.5 cm; inflorescence up to 32 cm tall; peduncle flattened, laxly flowered, many-flowered; flowers subdistichous; sepals 5 mm long; lip curved-geniculate, oblong, obtuse, bicarinate above the base and with a slightly concave speculum. Antananarivo, Fianarantsoa, Toamasina. Humid, evergreen forest; 400–1200 m. Fl. October.

Ploiarium Group 2

Peduncle as broad at the apex as at the base; rachis dilated and winged below each flower.

32.24. **B. pleiopterum** Schltr.   *(p. 197, bottom left)*

Pseudobulbs ovoid, 10–20 × 6–8 mm, with 5–7 verrucose angles, yellow; leaves 2, oblanceolate, 5–10 × 0.8–1.2 cm; inflorescence up to 34 cm tall; rachis with expanded 3 mm broad wings below each flower; flowers tiny; sepals 2 mm long. Antsiranana, Toamasina. Humid, lowland, evergreen forest; up to 300 m. Fl. October.

Ploiarium Group 3

Peduncle as broad at the apex as at the base; rachis dilated or not, but not winged below each flower; keels on lateral sepals crenulate, lacerate or irregularly dentate.

Key to species in Ploiarium Group 3

1. Inflorescence 1-flowered; pseudobulbs 5–7 mm long; leaves 4–6.5 mm long . . . . **31.27. B. insolitum**
1. Inflorescence many-flowered; pseudobulbs 20 mm long; leaves 9 cm or more long 2
2. Peduncle and rachis erect; sepals 4–8 mm long 3
2. Peduncle and rachis at an obtuse angle to each other; sepals 7.5–11 mm long 5
3. Rachis dark red, spongy with the flowers in depressions on its surface . . . . **31.25. B. coriophorum**
3. Rachis not dark red nor spongy, the flowers free on its surface 4
4. Leaves oblanceolate, 9–11 × 1–1.5 cm; lip elliptic-ligulate 2 × 1.5 mm with 3 longitudinal furrows on its surface . . . . **31.26. B. divaricatum**
4. Leaves elliptic, 1.3–2.3 × 1–1.5 cm; lip ovate, 3–4 mm long, transversely ridged . . . . **31.29. B. peyrotii**
5. Pseudobulbs not angular; flowers brown with darker spots . . . . **31.28. B. perreflexum**
5. Pseudobulbs 4–5-angled; flowers orange with red stripes and a yellow lip . . . **31.30. B. turkii**

32.25. B. coriophorum Ridl.   *(opposite, bottom right)*
(syn. *Bulbophyllum compactum* Kraenzl.; *B. crenulatum* Rolfe; *B. robustum* Rolfe; *B. mandrakanum* Schltr.)

Plant robust; pseudobulbs oblong-conical, 5–8 × 1.7–2.3 cm; leaves 2, narrowly oblanceolate, 13–18 × 2.7–4 cm; inflorescence up to 50 cm tall; rachis thickened and spongy; flowers dark red, finely warty, very numerous, inserted in cavities within rachis; lateral sepals 7–8 mm long; lip orbicular, 3 mm long, thick especially at the base, dotted with warts. Antananarivo, Antsiranana, Fianarantsoa, Toamasina; also in the Comoros. Humid, evergreen and dry forest; 95–1600 m. Fl. October–June.

32.26. B. divaricatum H.Perrier
Pseudobulbs ovoid-conical, 2.5–6 × 1–1.2 cm; leaves 2, oblanceolate, 9–11 × 1–1.5 cm; inflorescence slightly longer than the leaves; flowers papillose on the exterior, c. 7 mm long; sepals 4–6.5 mm long; lip glabrous, thick, elliptical, tongue-shaped, 2 × 1.5 mm, with 3 furrows in the lower half. Antsiranana. Lichen-rich forest; 1500–2000 m. Fl. April, October.

32.27. B. insolitum Bosser
Pseudobulbs flattened, ovoid-oblong, 5–7 × 3–4 mm; leaves 2, obovate, 4–6.5 × 2–3.5 mm; inflorescence short, 1-flowered; sepals 3–4.5 mm long, the laterals united, with lacerate wings; lip 3-lobed, 2.3–2.5 × 0.7–1 mm. Antsiranana. Humid, lowland forest. Fl. unknown.

32.28. B. perreflexum Bosser & P.J.Cribb
Pseudobulbs ovoid, 3–3.5 × 1–2 cm; leaves loriform, 9–12 × 0.7–1.2 cm; inflorescence distantly many-flowered, 30–35 cm long; rachis at angle to peduncle; bracts elliptic, reflexed, covering flowers; flowers strongly reflexed, brown, with darker spots; sepals 7.5–11 mm long, lateral sepals united, with erose broad apical wings; lip ligulate, acute; close to *B. coriophorum* and *B. divaricatum*, but the flowers of these species are very different, particularly in the shape of the lip. Antsiranana. Lower montane forest; 900–1300 m. Fl. December.

32.20

32.23

32.24

32.25

32.29. **B. peyrotii** Bosser (*opposite*)

(syn. *Bulbophyllum fimbriatum* H.Perrier non Rchb.f.; *B. flickingerianum* A.D.Hawkes; *B. mayae* A.D.Hawkes)

Pseudobulbs clustered, suborbicular, 13–25 mm across, bright yellow; leaves spreading, elliptic, 13–23 × 10–15 mm; inflorescence fleshy, 5–5.5 cm long, 10–12-flowered; flowers small, greenish-yellow scattered with warts and dark red dots; sepals 4 mm long; lateral sepals joined with deeply fimbriate-dentate margins; petals linear-oblong, 3 mm long; lip ovate, rounded in front, 3–4 mm long, transversely ridged in apical half. Antananarivo, Toamasina. Humid evergreen mossy forest, on tree trunks; 900–1000 m. Fl January–August.

32.30. **B. turkii** Bosser & P.J.Cribb

Pseudobulbs ovoid, 4–5-angled, 2–3 × 1–1.5 cm; leaves loriform, 10–12 × 0.7–0.8 cm; inflorescence 17–22 cm long; rachis at an obtuse angle to peduncle; bracts elliptic, reflexed, covering flowers; flowers reflexed, sepals orange striped with red, lip yellow; sepals 7.5–10 mm long; lateral sepals united, with broad erose winged margins; lip fleshy, ligulate, 4–4.5 mm long; close to *B. perreflexum* (p. 196) by its shape and size of the flower and the strong reflection of the ovary, but distinct by the shape of the lip and petals. Fianarantsoa. Moist, montane forest; 950–1150 m. Fl. March.

Ploiarium Group 4

Peduncle as broad at the apex as at the base; rachis dilated or not but not winged below each flower; keels on lateral sepals entire; spike short, less than 4 mm long and densely flowered.

Key to groups of species in Ploiarium Group 4

1. Inflorescence hidden under the leaf; petals terminated by a long acicular tip . . . .
. . . . **32.31. B. protectum**
1 Inflorescence not hidden by the leaves; petals lacking a terminal aciculate tip 2
2. Peduncle shorter than the pseudobulb or equalling it **32.32–32.36**
2. Peduncle more than twice as long as the pseudobulb 3
3. Rachis cylindrical, in the same line as the peduncle **32.37–32.39**
3. Rachis at an obtuse angle to the slender peduncle 4
4. Pseudobulbs 2 cm or less long, less than 3 times as long as broad;
leaves 5 times as long as broad or less **32.40–32.41**
4. Pseudobulbs more than 2 cm long; leaves more than 5 times as long as broad . . . .
. . . . **32.42–32.44**

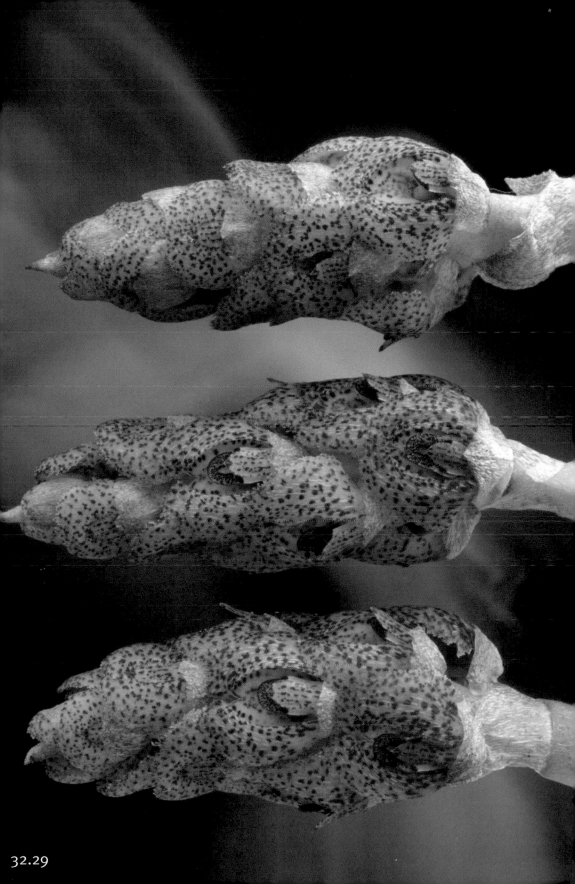

32.29

32.31. B. protectum H.Perrier

Rhizome fractiflex; pseudobulbs 2.5–3 cm long; leaves 2, elliptic, 6–7.5 × 2.2–2.6 cm, flowers c. 1 cm long; raceme thickened, white or pink; inflorescence shorter than the leaves and hidden under one; rachis 2–2.5 cm long; petals extended by a filiform acumen; sepals 3–3.5 mm long; lip like a tripod, truncate. Antananarivo, Toamasina. Humid, lowland, evergreen forest; c. 700 m. Fl. February, October.

32.32. B. cyclanthum Schltr.   *(opposite, top left)*

Plant 10–15 cm tall; pseudobulbs 4–5 cm apart, fusiform, 5–6 cm long; leaves 2, oblanceolate, 8–13.5 × 1.2–1.7 cm; rachis densely flowered; floral bracts rigid, almost as long as the flower; flowers small, finely papillose on the exterior; sepals 4–4.5 mm; lip orbicular, curved, very obtuse. Antsiranana, Toamasina. Humid, evergreen forest on plateau; c. 1300 m. Fl. December.

32.33. B. jackyi G.A.Fischer, Sieder & P.J.Cribb   *(opposite, top right)*

Plant creeping; pseudobulbs 3–5 cm apart, orbicular, 0.8–1 cm long; leaf 1, obovate-elliptic, emarginated, 4–6 × 2.5–3.8 cm; inflorescence clavate, 2.5–3.8 cm long, densely flowered; floral bracts rigid, longer than the flower; flowers dark purple; sepals 0.6–1.8 mm; petals ovate; lip ovate, recurved, papillose. Fianarantsoa. In low heath forest on white sand; sea level to 100 m. Fl. November.

32.34. B. leptochlamys Schltr.   *(opposite, bottom left)*

Plants 8–12 cm tall; pseudobulbs 1–3 cm apart, ovoid or cylindrical, 2–4.5 cm long; leaves linear-ligulate, 6–10 × 0.4–0.6 cm; inflorescence 4–6 cm long, 10–20-flowered; flowers glabrous, covered by the bracts; sepals 3.5–4 mm long; lip suborbicular, 2 mm long. Antsiranana. Humid, lichen-rich, evergreen forest on plateau; c. 1500 m. Fl. January.

32.35. B. paleiferum Schltr.

Plant 20–25 cm tall; pseudobulbs 4–5 cm apart, cylindrical, 5–12 cm long; leaves ligulate, 8–9 × 1.6–1.8 cm; inflorescence 8.5–9 cm long; flowers greenish yellow, covered by the bracts; sepals 3–5 mm long, the laterals finely ciliate; lip orbicular, 2 mm long. Toamasina. Humid, evergreen forest; 500–900 m. Fl. January–March.

32.32

32.33

32.34

32.36

32.36. **B. quadrialatum** H.Perrier *(p. 201, bottom right)*

Plant 9–12 cm tall; pseudobulbs 3–4 cm apart, elliptic-conical, 2.5–3 × 0.9–1.2 cm, 4-angled, yellow; leaves 2, narrowly elliptic, 5.5–7.5 × 1–1.7 cm; inflorescence 5–8 cm long; flowers white to rose-coloured; sepals 4–4.5 mm long; lip suborbicular, 1.8 mm long. Antsiranana. Mossy, montane forest; c. 1800 m. Fl. November.

32.37. **B. aggregatum** Bosser *(opposite)*

Pseudobulbs 1 cm apart, ovoid, 4-angled, 8–15 × 4–10 mm; leaves 2, elliptic or oblong, bilobulate, 1.8–2.5 × 1–1.1 cm; inflorescence 13–15 mm long, subcapitate; flowers 8–10, purple; sepals 4–4.5 mm long, the laterals papillose; lip fleshy, recurved, suborbicular, 2.4 × 2 mm. Antananarivo, Fianarantsoa. Humid, evergreen forest on plateau, mossy forest; 1300–1700 m. Fl. April–August.

32.38. **B. oreodorum** Schltr. *(p. 205, top left)*

Lithophyte 8–13 cm tall, compact; pseudobulbs 1.5–2 cm apart, ovoid, 2.8–4 cm long; leaves oblong-ligulate, 4.5–7 × 1.3–1.6 cm; inflorescence 11.5–12.5 cm long; peduncle fleshy, 2 or more mm in diameter; flowers up to 50, very dense, red, finely papillate; sepals 4 mm long; lip suborbicular, 2.5 mm long. Antsiranana, Fianarantsoa. Rocky outcrops amongst mosses; 2000–2400 m. Fl. April.

32.39. **B. perseverans** Hermans
(syn. *Bulbophyllum graciliscapum* H.Perrier)

Plant 14–20 cm tall; pseudobulbs cylindrical, 2.5–4 cm long; leaves lanceolate, 7–9.5 × 0.8–1.2 cm; inflorescence 15.5–21 cm long; peduncle slender; flowers glabrous; sepals 6–7 mm long; lip ovate, obtuse, 3 mm long, the base almost auriculate, with 2 distinct spots below the middle. Antsiranana (Tsaratanana), Toamasina. In lichenous montane forest; 2000 m. Fl. February.

32.40. **B. lemuraeoides** H.Perrier

Plant small, 4–9 cm tall; pseudobulbs ovoid, 1–1.5 cm long; leaves ovate, 1.3–3 × 0.8–1 cm; inflorescence slender, 4–7 cm long; rachis at angle to peduncle; flowers 4–10; sepals 4.5–7 mm long; lip triangular, obtuse, 2.3 × 1.2–2 mm, heart-shaped at the base. Antananarivo, Fianarantsoa. Humid, evergreen forest on plateau; 800–2000 m. Fl. November–February.

32.41. B. tampoketsense H.Perrier

Plant small, 5–7 cm tall; pseudobulbs ovoid, 1.5–2 cm long; leaves obovate or oblanceolate, 1.5–6 × 0.9–1.2 cm; inflorescence 7–7.5 cm long; flowers up to 15, red, 5 mm long, papillose on the exterior; sepals 4–6 mm long; lip ovate-ligulate, 3 × 2 mm, with 2 coloured spots below the middle. Antananarivo. Humid, evergreen forest on plateau; c. 1600 m. Fl. December.

32.42. B. ankaratranum Schltr.

Plant 9–13 cm tall; pseudobulbs 6–10 cm apart, ellipsoidal or ovoid, 2.5–5 cm long; leaves ligulate, 4–8 × 0.8–1.3 cm; inflorescence 10.5–14.5 cm long; flowers glabrous, greenish-yellow, with the extremities of the lip and the petals blackish-purple, red spots mainly on the sepals; sepals 5–6 mm long; lip very curved at the base, 3 mm long. Antananarivo. Mossy, montane forest; c. 2000 m. Fl. February–June.

32.37

32.43. B. antongilense Schltr.

Plant 8–13 cm tall; pseudobulbs 3.5–4.5 cm apart, cylindrical, 3–3.5 cm long; leaves linear, 5–8.5 × 0.5–0.7 cm; flowers 8–15, reddish, glabrous; sepals 3–3.25 mm long; lip ovate, obtuse. Close to *B. ankaratranum* (p. 203) but stronger in growth and with larger, broader pseudobulbs, longer leaves, differently coloured flowers and larger floral bracts. Toamasina. Humid, lowland, evergreen forest; c. 500 m. Fl. August.

32.44. B. namoronae Bosser

Pseudobulbs cylindrical, 7–11 × 0.7–1 cm; leaves 2, linear-oblong, 8–17 × 1.2–2.2 cm; inflorescence 25–30 cm tall, densely many-flowered; flowers fleshy; sepals 3.5–5 mm long; lateral sepals papillose; lip rounded, 2 × 1.8 cm violet, curved, concave at the base and the margins upright, within the lower part, on both sides, a central depression; stelids bidentate. Fianarantsoa. Humid, evergreen forest; 800–1500 m. Fl. January.

Ploiarium Group 5

Peduncle as broad at the apex as at the base; rachis dilated or not but not winged below each flower; keels on lateral sepals entire; spike lax and longer, usually less than 30-flowered, if more then peduncle more than twice as long as the spike; bracts 2 mm or more apart.

Key to groups of species in Ploiarium Group 5

1. Pseudobulbs and leaves usually less than 3 times as long as broad ... **32.45. B. brachyphyton**
1. Pseudobulbs and leaves more than 3 times as long as broad 2
2. Pseudobulbs not angled, conical, ovoid or cylindrical **32.46–32.53**
2. Pseudobulbs angled, sometimes winged 3
3. Inflorescence more than twice as long as the pseudobulb and leaves **32.54–32.56**
3. Inflorescence less than twice as long as the pseudobulb and leaves **32.57–32.67**

32.45. B. brachyphyton Schltr.

Plant 4 cm tall; pseudobulbs narrow, 4-angled, 1–1.5 × 0.3–0.5 cm; leaves oblong, 1.7–2 × 0.6–0.8 cm; inflorescence shorter than the leaves; flowers 5–8, glabrous; sepals 4–4.5 mm long; lip curved, 1.25 mm long, subrectangular above a contracted base. Toamasina. Humid, lowland forest; sea level–100 m. Fl. unknown.

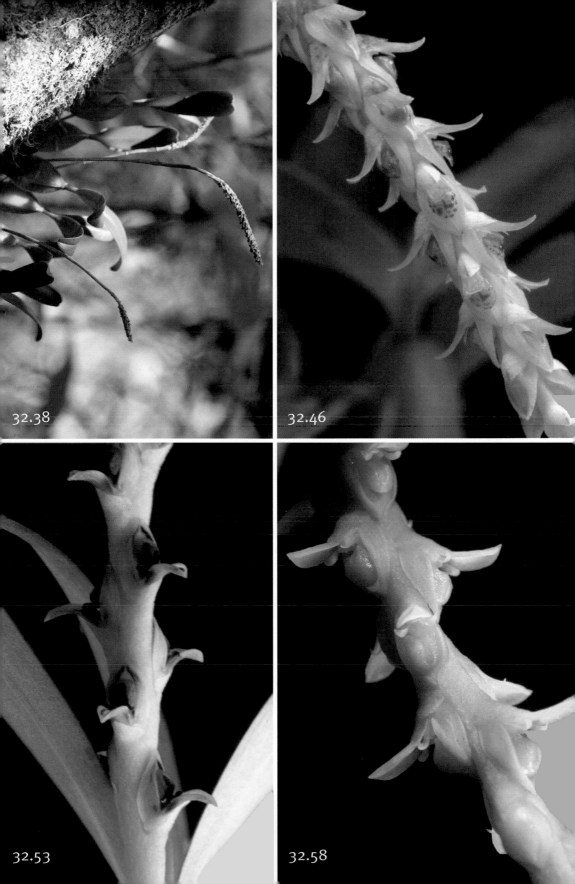

32.38

32.46

32.53

32.58

32.46. B. lucidum Schltr.   *(p. 205, top right)*

Plant 12–13 cm tall; pseudobulbs 4–5 cm apart, ovoid, 2.5–3 cm long; leaves leathery, oblong, obtuse, 4–5.5 × 1.5–2.3 cm; inflorescence 15–18 cm long; flowers 3–5 mm apart, greenish, glabrous; sepals 5–5.5 mm long; lip broadly ovate, 3 mm long. Antsiranana. Coastal forest; up to 200 m. Fl. October.

32.47. B. rictorium Schltr.

Plant 10–13 cm tall; pseudobulbs oblong to subglobose, 4-angled, 1.2–2.3 cm long; leaves oblong or narrowly oblong, 3.2–4 × 1.1–1.6 cm; inflorescence 13–16 cm long; bracts broadly ovate; flowers 8–15, glabrous; sepals 3–3.5 mm long; lip oblong, 2 mm long. Antsiranana. Mossy, evergreen forest; c. 1500 m. Fl. October.

32.48. B. rubiginosum Schltr.

Plant 4–5 cm tall; pseudobulbs 1–1.5 cm apart, narrowly ovoid, 1.5–2 cm long; leaves oblong or ligulate, 1.7–2.8 × 0.7–1 cm; inflorescence as long as the leaves or a little shorter; flowers 5–8, greenish; sepals 2.5–3 mm long; lip orbicular, obtuse, verrucose; related to *B. brachyphyton* (p. 204) but with a less leathery leaf, narrower petals and a differently structured lip which is verrucose. Toamasina. Humid, evergreen forest; 800–900 m. Fl. January.

32.49. B. auriflorum H.Perrier   *(opposite)*

Plant 9–15 cm tall; pseudobulbs 1.5–4.5 cm long; leaves oblanceolate, 2.5–8.2 × 0.7–1.2 cm; inflorescence 6–14 cm long; flowers orange-yellow, glabrous; sepals 5–7 mm long; lip obovate, 4 mm long. Toamasina. Humid, lowland, evergreen forest; 400–800 m. Fl. September–October.

32.50. B. hovarum Schltr.

Plant c. 15 cm tall; pseudobulbs 3–4 cm apart, cylindrical-conical, 2.5–3.6 cm long; leaves 2, petiolate, narrowly elliptic, 6–13 × 0.7–2 cm; inflorescence densely many-flowered; flowers violet-black, finely warty; sepals 3–4 mm long; lip suborbicular, 2 mm long. Antsiranana. Lichen-rich forest; 1000–2000 m. Fl. May.

32.49

32.51. B. metonymon Summerh.

(syn. *Bulbophyllum zaratananae* Schltr.; *B. schlechteri* Kraenzl.)

Plant up to 13 cm tall; pseudobulbs 4 cm apart, oblong-ovoid, 3 cm long; leaves ligulate, 7–8 × 0.7–1 cm; inflorescence equalling or passing the leaves; flowers dense, red; sepals 5–5.5 mm long; lip broadly ovate, 3 mm long; similar to *B. subclavatum* (p. 216) and *B. rubrolabium* (p. 210) but with larger red flowers and a thickened pale green rachis, the distinct lip. Antsiranana. Lichen-rich forest; c. 1200 m. Fl. January.

32.52. B. minax Schltr.

Plant up to 13 cm tall; pseudobulbs 2 cm apart, ovoid, 3 cm tall; leaves ligulate, 9–10 × 1 cm; inflorescence 18–19 cm tall; flowers 10, yellow; sepals 6 mm long; lip recurved, 2.3 mm long, with 2 pronounced ridges on top. Fianarantsoa. Mossy, montane forest; c. 1700 m. Fl. May.

32.53. B. teretibulbum H.Perrier

(p. 205, bottom left)

Plant 10–15 cm tall; pseudobulbs subcylindrical, 3–7.5 cm long; leaves narrowly elliptic, 3–7.5 × 0.5–1.2 cm; inflorescence longer than the leaves; peduncle 8–9 cm, laxly flowered; flowers 6–9, yellow; sepals 7–8 mm long; lip obovate, 4 × 2.5 mm, very obtuse. Toamasina. Humid, evergreen forest on plateau; 900–1080 m. Fl. October.

32.54. B. labatii Bosser

Epiphyte; pseudobulbs ovoid, 4-angled, 1.5–2 cm tall, borne on a long rhizome, yellow; leaves 2, oblong, 2.5–5 × 0.8–1.5 cm; inflorescence 42–63 cm long, many-flowered; flowers yellow; sepals 4.5–5 mm long, the lateral united; petals linear, 1.5–2 mm long; lip ovate, rounded at tip, bilobulate at base, 2.5 mm long. Toamasina. Humid lowland forest; 200 m. Fl. October.

32.55. B. myrmecochilum Schltr.

Plant 14–20 cm tall; pseudobulbs oblong-ovoid, 1.8–2.3 cm long; leaves oblong-ligulate, 5–7.5 × 1.2–1.5 cm; inflorescence very slender, up to 20 cm tall, longer than the leaves; flowers red with a green lip; sepals 6–7 mm long; lip recurved. Antsiranana. Lichen-rich forest; c. 2000 m. Fl. January.

32.56. B. sanguineum H.Perrier

Pseudobulbs slender, 4-angled, 2.5–3 cm long; leaves narrowly elliptic, 6–7 × 0.5–0.8 cm; inflorescence 17–23 cm long, red; flowers red; lodged in cavities in rachis; sepals 4–5 mm long; lip obovate, 2 mm long; close to *B. rhodostachys* (p. 210), but differs in having an inflorescence that is much longer than the leaves, thinner rachis, less than 2 times the thickness of the apex of the peduncle, smaller flowers, acute dorsal sepals, and entire stelids. Toamasina. Humid, evergreen forest on plateau; c. 1000 m. Fl. February.

32.57. B. acutispicatum H.Perrier

Characterised by its approximate flowers, set shallowly within the rachis, covered by the distichous, acute floral bracts, distant on the raceme. Antsiranana. Humid, evergreen forest on plateau; 1500–1700 m. Fl. December.

32.58. B. coccinatum H.Perrier   (*p. 205, bottom right*)

Epiphyte or terrestrial; pseudobulbs 2–4 cm apart, ovoid-conical, 2.5–3 cm long, 3–4-angled, yellow; leaves narrowly oblong or oblanceolate, 4.5–11 × 1–1.7 cm; inflorescence 10–14 cm long; rachis red, longer than the peduncle; flowers 15–30, red; sepals 4 mm long; lip ovate, 2.2 mm long, a little narrowed at the tip, biauriculate at the base with 2 swellings. Toamasina. Humid, lowland forest by lake; sea level–100 m. Fl. unknown.

32.59. B. ferkoanum Schltr.

Plant up to 25 cm tall; pseudobulbs 4.5 cm long; 4-angled; leaves ligulate, 12–14 × 1.4–1.8 cm; inflorescence longer than the leaves; flowers glabrous; sepals 5 mm long; lip oblong, 3 mm long; related to *B. sarcorhachis* (p. 211) but differs in its longer, 4-anguled pseudobulbs, longer spike, and somewhat larger flowers, with wider petals and a distinct lip and column. Toamasina. Humid, lowland forest. Fl. unknown.

32.60. B. lancisepalum H.Perrier

Plants 20–30 cm tall; pseudobulbs 5–8 cm apart, elongate, 4.5–6 cm long; leaves oblong, 10–13 × 2.3–2.5 cm; inflorescence 17–23 cm long; flowers yellow, papillose; sepals 5.5 mm long; lip broadly ovate, obtuse, 3 × 2 mm; differs from *B. minax* (p. 208) by the distinctly angular pseudobulbs, the yellow flowers, the anterior piece of the lateral sepals with winged ridges, the petals velvety at the top, and the flat lip. Toliara. Humid, lowland forest. Fl. unknown.

32.61. **B. obtusilabium** W.Kittr.

(syn. *Bulbophyllum rhizomatosum* Schltr.)

Plant 10–16 cm tall; pseudobulbs 1.5–2.5 cm apart, 4-angled, 3–5.7 cm long; leaves ligulate, 7–11 × 1–1.5 cm; inflorescence 9–11.5 cm long; flowers 6–10, dull red; bracts longer than the flowers; sepals 6–6.5 mm long; lip orbicular, 3 mm long; related to *B. ankaratranum* (p. 203), but the plant is more compact, the spike somewhat longer, and the lip longer with short parallel lamellae. Antsiranana, Toamasina. Humid, mossy, evergreen forest on plateau; 800–1600 m. Fl. January.

32.62. **B. pallens** (Jum. & H.Perrier) Schltr.

(syn. *Bulbophyllum ophiuchus* var. *pallens* Jum. & H.Perrier)

Plant 22–30 cm tall; pseudobulbs 2.5–3 cm apart, ellipsoidal, yellow, 3–3.5 cm long; leaves oblanceolate, 10–13 × 1–1.2 cm; inflorescence 19–26 cm long; flowers dull-coloured; sepals 1.8–2.2 mm long; lip 1.7 mm long. Mahajanga. Humid, mossy, evergreen forest on plateau by streams; c. 1000 m. Fl. August.

32.63. **B. ranomafanae** Bosser & P.J.Cribb

Pseudobulbs 7–12 × 6–9 mm, conical; leaves oblong, 2.3–3.5 × 0.4–0.7 cm; inflorescence 12–15 cm long; sepals 6–7 mm, the laterals united; lip fleshy, 3 × 1.3 mm, its margins finely denticulate, upper surface with 2 depressions at the base of the basal lobes, and 2 short fine central crests. Fianarantsoa. Moist, montane forest; 950–1150 m. Fl. March.

32.64. **B. rhodostachys** Schltr.

Pseudobulbs 3 cm apart, oblong-conical, 3.5–4 cm long; leaves linear-ligulate, 7–9 × 0.8–1.3 cm; inflorescence 12–17 cm long, red; flowers many, glabrous; sepals 5 mm long; lip ovate, obtuse, 3.5 × 2.5 mm; close to *B. rubrolabium*, but differs in the thickening of the rachis and the larger flowers. Antsiranana, Toamasina. Humid, lowland, evergreen forest, along the edge of streams; c. 300 m. Fl. November.

32.65. **B. rubrolabium** Schltr.

Plant up to 15 cm high; pseudobulbs 2–4 cm apart, ovoid, 2.3–3.5 cm long; leaves narrowly oblong, 6–9 × 0.6–0.9 cm; inflorescence shorter than the leaves; rachis thickened; flowers dense, 12–15, red; sepals 4–5 mm long; lip ovate, verrucose. Antsiranana, Toliara. Humid, evergreen forest; 1100–1500 m. Fl. November–December.

32.66. B. sarcorhachis Schltr.

Plant 10–12 cm tall; pseudobulbs 1–2 cm apart, oblong, 2–2.2 cm long, 4-angled; leaves elliptic-oblong, 5–6.2 × 1–1.6 cm; inflorescence 10–12 cm long; flowers greenish grey; sepals 4.5–6 mm long; lip 2 mm long; close to *B. rubrolabium* (p. 210) and *B. rhodostachys* (p. 210) but differs in its narrower petals and wider lip. Fianarantsoa, Toamasina. Humid, lowland, evergreen forest; sea level–200 m. Fl. September–December.

32.66a. B. sarcorhachis var. befaonense (Schltr.) H.Perrier
(syn. *Bulbophyllum befaonense* Schltr.)

Differs from the typical form by the narrower and longer pseudobulb, longer leaves, 9–14 cm × 8–12 mm, the inflorescence almost twice the length, 16–19 cm long, greenish yellow raceme, the slightly wider sepals and the more acute petals. Toamasina. Humid, lowland, evergreen forest; c. 700 m. Fl. September.

32.67

32.66b. B. sarcorhachis var. **flavomarginatum** H.Perrier ex Hermans

The habit and inflorescence of this variety are similar to those of var. *befoanense* (p. 211), but its pseudobulbs are wider and oval-oblong and its leaves are wider; raceme greyish-green; dorsal sepal slightly obovate, narrowed below the middle; lateral sepals narrower, with light yellow margins, contrasting with the brown colouration of its sepals. Antsiranana. On moss- and lichen-covered trees; 1100–2600 m. Fl. April–July.

32.67. B. toilliezae Bosser   *(p. 211)*

Plant up to 25 cm tall; pseudobulbs 2–3 cm apart on rhizome, ovoid, angled, 4–6 × 2 cm, rugose on angles; leaves oblong-ligulate, 13–17 × 2–2.3 cm; inflorescence 20–25 cm tall, 18–22-flowered; flowers yellow-green to slightly orange-yellow, fleshy; sepals 11–13.5 mm long; lip fleshy, much recurved, ovate c. 4 mm × 2 mm, 2 oval, deep depressions near the base, upper surface very papillose. Toamasina. Humid, evergreen forest; c. 900 m. Fl. March.

Ploiarium Group 6

Peduncle as broad at apex as at base; rachis dilated or not but not winged below each flower; keels on lateral sepals entire; spike dense or subdense, many-flowered, peduncle 2–3 times as long as the spike; bracts less than 2 mm apart.

Key to groups of species in Ploiarium Group 6

1. Rachis red 2	

1. Rachis red 2
1. Rachis yellow or green 3
2. Dorsal sepal much longer than the lateral sepals ....
   ....**32.68–32.69**
2. Dorsal sepal as long as the lateral sepals **32.70–32.74**
3. Petals glabrous **32.75–32.77**
3. Petals pubescent **32.78–32.82**

32.68. B. henrici Schltr.   *(opposite, top left)*

Plant up to 45 cm tall; pseudobulbs 7 cm apart, narrowly oblong, 9–10 × 2 cm; leaves oblong-ligulate, 15–18 × 3.5–3.8 cm; inflorescence up to 45 cm tall; rachis 11–12 cm long; flowers rugose, yellowish; sepals 3.5–5.5 mm long; lip suborbicular, 2.5 × 2 mm. Related to *B. ophiuchus* (p. 215) and *B. moramanganum* (p. 214), but distinguished by the broad leaves, thick spike and rugose flowers with long sepals. Toamasina. Humid, evergreen forest; c. 700 m. Fl. September.

32.68

32.68a

32.68a

32.72

32.68a. B. henrici var. **rectangulare** H.Perrier ex Hermans

(p. 213, middle and bottom left)

Differs from the typical variety by the synsepal that is bi-apiculate at the apex, the smaller subrectangular lip, lacking a circular depression, and the papillae which are less pronounced. Fianarantsoa. Humid, evergreen forest; c. 700 m. Fl. October.

32.69. B. moramanganum Schltr.

Plant up to 40 cm tall; pseudobulbs 4 cm apart, cylindrical-conical, up to 5.5 cm long; leaves oblong-ligulate, 13–17 × 1.7–2 cm; inflorescence up to 40 cm long; rachis 7 cm long; flowers red; sepals 4–6 mm long, verrucose; lip geniculate at base. Toamasina. Humid, evergreen forest on plateau; 900 m. Fl. October–November.

32.70. B. ankaizinense (Jum. & H.Perrier) Schltr.

(syn. *Bulbophyllum ophiuchus* var. *ankaizinensis* Jum. & H.Perrier)

Pseudobulbs 2.5–4.5 × 1.2–2 cm, 2-leaved; leaves narrowly elliptic, 7–10 × 1.5–3 cm; inflorescence 23–37 cm long; raceme red, densely flowered; flowers red; sepals 4 mm long; petals villous; lip suborbicular, 2 × 1.7 mm with a central depression. A number of local forms have been found. Antsiranana, Fianarantsoa, Mahajanga, Toamasina. Mossy, evergreen forest; 600–2000 m. Fl. October–January.

32.71. B. crassipetalum H.Perrier

Plant up to 24 cm tall; pseudobulbs oblong, 2.5–3.5 cm long; leaves oblanceolate, 7–8 × 1–1.2 cm; inflorescence 19–23 cm long; rachis 7–8 cm long, dark red, many-flowered; sepals 4 mm long; petals papillose-pubescent; lip curved, fleshy, suborbicular, 2 × 1.5 mm, papillose and with a wide speculum. Antsiranana, Fianarantosa. Humid, evergreen forest; 400–1750 m. Fl. November–December.

32.72. B. hirsutiusculum H.Perrier

(p. 213, right)

Plant 50–60 cm tall, pseudobulbs ovoid, 2.5–3 cm long, 6-angled; yellow; leaves elliptic or oblong, 4.5–10 × 1.6–3 cm; inflorescence 43–54 cm long; rachis cylindrical, 11–17 cm long, many-flowered; flowers dull red, covered by short stiff hairs; sepals 4–6 mm long; petals glabrous; lip broadly elliptic, 3 × 1.8 mm. Fianarantsoa. Humid, evergreen forest on plateau; 1300–1700 m. Fl. December.

32.73. **B. ophiuchus** Ridl.

(syn. *Bulbophyllum mangoroanum* Schltr.)

Pseudobulbs subcylindrical, 2.5–5 cm long, 4-angled, yellow; leaves lanceolate, 7–13 × 1.1–1.8 cm; inflorescence 50 cm or more long; rachis dull red, 12–15 cm long; flowers green, turning red on fading; sepals 4 mm long; lip 2 × 1 mm, subrectangular, 2 × 1 mm, with a broad central speculum surrounded by a wide, finely verrucose margin. Fianarantsoa, Toamasina. Humid, evergreen forest on plateau; 900–1400 m. Fl. November–February.

32.73a. **B. ophiuchus** var. **baronianum** H.Perrier ex Hermans

Differs from the typical variety by its smaller flowers, its dorsal sepal which is obtuse and as wide as the anterior part of the almost orbicular synsepal, its suborbicular lip and the foot of the column which is not dilated in front of the stigmatic surface. Toamasina. Fl. unknown.

32.74. **B. rubrigemmum** Hermans

(syn. *Bulbophyllum simulacrum* Schltr.)

Pseudobulbs 3–5 cm apart, narrowly ovoid, 3.5–4.5 cm long; leaves ligulate, 9–11 × 0.8–1.1 cm; Inflorescence more than 30 cm long; rachis red, up to 10 cm long; flowers red; sepals 3–3.4 mm long; petals glabrous; lip 1.6 × 1 mm; similar to *B. lucidum* (p. 206) but with smaller and narrower pseudobulbs, much smaller leaves, smaller flowers, more pointed petals, a different lip structure and shorter stelids. Toamasina. Humid, evergreen forest; c. 700 m. Fl. January.

32.75. **B. callosum** Bosser

(p. 217, top)

Pseudobulbs 4-angled, 5–6.5 × 2 cm, 4–5 cm apart on rhizome; leaves 2, oblong, 8–13 × 2.5–3.2 cm; inflorescence cylindrical, 35–44 cm tall, with fleshy rachis at angle to stalk, 5–9 cm long; flowers dark purple; sepals 4–4.5 mm long, papillate-puberulent; petals oblong; lip suborbicular, 2 mm across, verruculose. Antananarivo, Toamasina. Humid, mossy, evergreen forest on plateau; 1300–1400 m. Fl. September–October.

32.76. **B. nitens** Jum. & H.Perrier

(p. 219, top left)

Plant up to 8 cm tall; pseudobulbs 1.5–2 cm apart, oblong, 1.2–3 cm long; leaves lanceolate, 1.6–6 × 0.7–1.5 cm; inflorescence 25–36 cm long; rachis reflexed or horizontal, 5–6 cm long, densely many-flowered; flowers greenish yellow; sepals 3 mm long; lip 1.2 × 0.8 mm, with 2 slightly oblique raised ridges. Antananarivo, Antsiranana, Fianarantsoa, Toamasina. Lichen-rich, humid evergreen forest on plateau; 1000–2000 m. Fl. March, August–June.
LOCAL NAME *Tsiakondroakandro*.

32.76a. B. nitens var. intermedium H.Perrier

Flowers 5.5 mm long; dorsal sepal 1.5 mm broad at base; lip subrectangular, 2 × 1 mm, auriculate at base, papillose. Antsiranana. Mossy forest; 1800–2000 m. Fl. April.

32.76b. B. nitens var. majus H.Perrier

Pseudobulbs 4–5 × 1–1.5 cm; leaves 8–9 × 1.1–2 cm; flowers 7 mm long; lip suborbicular, non-papillose. Antananarivo. Mossy forest. Fl. September.

32.76c. B. nitens var. pulverulentum H.Perrier

Pseudobulbs less than 3 cm tall; flowers 6 mm long; lower bracts 4 mm long; sepals, petals and lip as in var. *majus*. Antananarivo. Evergreen forest; 900–1400 m. Fl. February.

32.77. B. subclavatum Schltr.   (*opposite, bottom*)

Plant 25–30 cm tall; pseudobulbs 2–3 cm apart, oblong, 2.5–4 cm long, 4-angled; leaves ligulate, 9–11 × 1–1.3 cm; inflorescence 18–25 cm long; rachis greenish, 6–9 cm long, densely many-flowered; flowers reddish; sepals 5.5–6.5 mm long; lip curved at the base, suborbicular, 2 × 2.3 mm and subapiculate. Antananarivo, Fianarantsoa. Humid, evergreen forest on plateau; 1200–1700 m. Fl. September–December.

32.78. B. subcrenulatum Schltr.

Plant 30–40 cm tall; pseudobulbs 6 cm apart, narrowly oblong, 5–6 cm long, 4-angled; leaves ligulate, 13–15 × 1.2–1.5 cm; inflorescence 30–38 cm long; rachis 10–13 cm long; flowers greenish, becoming red with age, c. 8 mm long; sepals 6 mm long; petals papillose; lip with the margins expanded, broadly ovate or slightly obovate, 2.5 × 1.8 mm. Antsiranana, Toamasina. Humid, evergreen forest; 800–900 m. Fl. August.

32.75

32.77

32.79. B. humbertii Schltr.   (*opposite, top middle*)

Plant up to 25 cm tall; pseudobulbs 3 cm apart, ovoid-conical, 2.5–4 cm long; leaves not known; inflorescence 25 cm long; rachis 7 cm long; floral bracts long and narrow; flowers greenish; sepals 3–4 mm long; lip broadly ovate, 2-keeled. Antananarivo, Fianarantsoa. Humid, evergreen forest on plateau; c. 1700 m. Fl. December.

32.80. B. masoalanum Schltr.   (*opposite, top right*)

Close to *B. henrici* (p. 212) but differs by the less robust plant; pseudobulbs 2.5–3 cm apart, 3 cm long; leaves 8–11 × 1.3–1.6 cm; rachis thinner, 7 cm long; flowers smaller and differently coloured, red-veined; sepals 4–4.5 mm long; column toothed at the edge. Antsiranana, Toamasina. Humid, lowland, evergreen forest; c. 300 m. Fl. October–November.

32.81. B. verruculiferum H.Perrier

Epiphyte 10–25 cm tall; pseudobulbs oblong, 2.5–3 cm long; leaves 2, narrowly ligulate-oblanceolate, 11–13 × 1–1.4 cm; inflorescence 13–25 cm tall, densely many-flowered; rachis 5–8 cm long; flowers dark red; sepals 3–5 cm long, the dorsal verrucose, the keels on the laterals verrucose; lip subquadrate, 2 × 1.8 mm. Antsiranana, Fianarantsoa. Lichen-rich forest; 1400–1700 m. Fl. November–December.

32.82. B. vulcanorum H.Perrier

Plant 32–47 cm tall; pseudobulbs 2–4.5 cm apart, ovoid, 2.5–3 cm long; leaves elliptic, 5–9.5 × 1.2–2 cm; inflorescence 32–47 cm; rachis 12–17 cm long, yellow; flowers yellow; sepals 5–7.5 mm long; lip 3 × 1.6 mm, lacking a speculum. Antsiranana. Humid evergreen forest on plateau; c. 1000 m. Fl. November.

VI. Section Alcistachys

Robust plants; pseudobulbs 2-leaved, often depressed, yellow to bright red; peduncle robust; bracts large; flowers large, more than 15 mm long; sepals free; lip glabrous, pubescent or ciliate; stelids elongate.

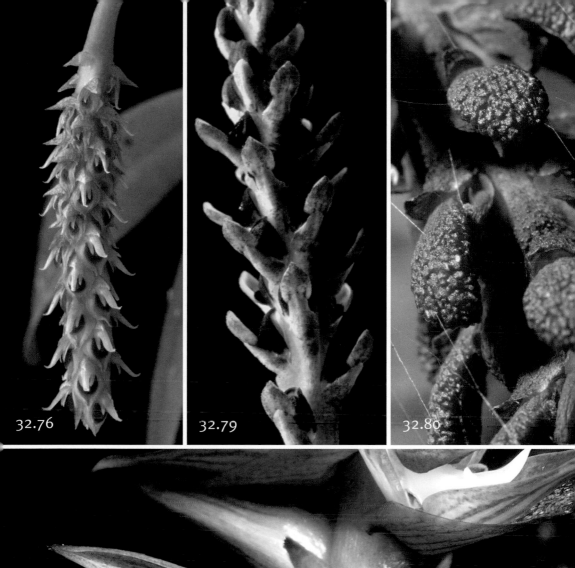

32.76   32.79   32.80

32.85

Key to species in Section Alcistachys
1. Pseudobulbs flattened and discoid .... **31.83. B. hamelinii**
1. Pseudobulbs compressed or not, but not discoid 2
2. Lip surface papillose of glabrous, never pubescent or ciliate 3
2. Lip surface pubescent of margins ciliate in lower part 4
3. Flowers whitish green with a purple lip; lip callus ridges papillose ... **31.84. B. bathieanum**
3. Flowers yellow finely marked with purple and with a purple lip; lip glabrous ....
   .... **31.86. B. occlusum**
4. Pseudobulbs not angular; leaves oblanceolate; bracts not yellow; flowers
   yellow striped with red-brown; lip ciliate in basal part .... **31.85. B. brevipetalum**
4. Pseudobulbs 4-angled; leaves lanceolate; bracts yellow; flowers green
   spotted with red and with a red pubescent lip .... **31.87. B. sulfureum**

32.83. B. hamelinii W.Watson   *(opposite)*

The largest *Bulbophyllum* in Madagascar; pseudobulbs flattened, discoid, up to 11 cm in diameter; leaves elliptic, up to 35 × 9 cm; inflorescence densely many-flowered; bracts up to 2.3 cm long, covering flowers; flowers in an 8–15 cm long tight raceme, pinkish-purple and malodorous. Toamasina. Evergreen, humid forest; 800–1000 m. Fl. February.

32.84. B. bathieanum Schltr.

Pseudobulbs compressed, ovoid, 4 cm long; leaves oblong-ligulate, acuminate, 16–21 × 3–4 cm; inflorescence 35 cm or more long; flowers whitich green with a dark red lip; lip c. 12 mm long, very curved-folded, the margins of the upper two-thirds roundly recurved, folded into a furrow in the middle, flanked by 2 prominent lines of papillae. Antananarivo, Antsiranana, Toamasina. Humid, evergreen forest at intermediate elevation; 500–1400 m. Fl. November–December.

32.85. B. brevipetalum H.Perrier   *(p. 219, bottom)*
(syn. *B. brevipetalum* subsp. *majus* H.Perrier; *B. brevipetalum* subsp. *speculiferum* H.Perrier)

Plant 30–45 cm tall; pseudobulbs depressed, 2.5–3 cm long, bright yellow; leaves oblanceolate, 13–18 × 2.5–3.2 cm; inflorescence up to 45 cm long; flowers yellow, striped with brownish-red, 15–18 mm long; lip oblong, 10 × 4 mm, ciliate in the lower part only, the margins folded back and the wings erect and ciliate. Antananarivo, Fianarantsoa. Humid, evergreen forest on plateau; c. 1500 m. Fl. February, September.

32.83

32.86. B. occlusum Ridl. *(opposite, top left)*

Plant large; pseudobulbs compressed, orbicular or broadly ovoid, 5–6 cm long; leaves oblanceolate, 17–35 × 3.5–5 cm; peduncle c. 20 cm long, almost completely covered by papery sheaths, covering the purplish flowers; sepals 12 mm long; lip broad; column-teeth long. Antananarivo, Antsiranana, Fianarantsoa, Toamasina; also in the Mascarenes. Humid, mossy, evergreen forest; 600–1600 m. Fl. August–March.

32.87. B. sulfureum Schltr. *(opposite, top right)*

Related to *Bulbophyllum occlusum*; plant 30–35 cm tall; pseudobulbs ovoid, 4-angled, 2–2.5 cm long; leaves lanceolate, acute, 12–19 × 2.6–3 cm; inflorescence up to 32 cm long; rachis 12 cm long; bracts yellow, c. 3 cm long; flowers green, spotted with red and with a dull red lip; sepals 15 mm long; lip ovate-oblong, 10 mm long; petals and the whole surface of the lip hairy; stelids large and blunt. Antananarivo, Antsiranana, Fianarantsoa, Toamasina. Humid, evergreen forest; 800–1700 m. Fl. December–February.

VII. Section Kainochilus

Large plants; pseudobulbs 2-leaved, rarely 1-leafed; flowers large, 8 mm or more long; sepals free; lip with large tooth on upper margin, very bearded, often with long caducous ligules in front; column longer than wide; stelids large.

Key to groups of species in Section Kainochilus

1. Pseudobulbs with a single leaf **32.88–32.89**
1. Pseudobulbs with 2 leaves 2
2. Lip hairy but lacking elongate apical appendages **32.90–32.93**
2. Lip with many c. 4 mm long, ligulate appendages in the apical part 3
3. Petals linear, as long as the column **32.94–32.95**
3. Petals short, triangular, much shorter than the column **32.96–32.99**

32.88. B. lemurense Bosser & P.J.Cribb

(syn. *Bulbophyllum clavigerum* H.Perrier non *B. clavigerum* F.Muell.)

Pseudobulbs cylindrical-conical, 2–3 cm tall; leaf narrowly oblong, 6–8 × 0.6–1 cm; inflorescence 20–30 cm tall, 10–20-flowered; flowers dull purple; sepals 10–15 mm long; petals linear, 1.5 mm long; lip claviform, 5–6 mm long, with short hairs laterally and linear appendages at the tip. Antsiranana. On rocks, humid, medium-elevation forest; c. 1400 m. Fl. March.

32.86

32.87

32.90

32.91

32.89. B. viguieri Schltr.

Pseudobulbs narrowly ovoid, 2–4.5 × 1–1.5 cm, 2.5–3 cm apart on 3 mm diam.; leaf ligulate, up to 12 × 0.8–1 cm; inflorescence longer than the leaf; cylindrical, densely many-flowered; flowers pale green, 10 mm long; sepals 6 mm long; petals less than 1 mm long; lip 6 mm long, bicostate at base, margins longly and densely ciliate-barbate. Antananarivo. Humid, evergreen forest on plateau; c. 1500 m. Fl. November.

32.90. B. alexandrae Schltr. *(p. 223, bottom left)*

Large epiphyte or lithophyte; pseudobulbs ovoid, 4-angled, 4–6 cm long, yellow or reddish brown; leaves oblong, 11–25 × 2.5–3.5 cm; inflorescence 20–35 cm tall; flowers deflexed, green with a yellow lip; sepals 15–25 mm long; lip fleshy, oblong, 5 mm long, margins longly ciliate. Antananarivo, Toamasina, Toliara. Seasonally dry, deciduous forest and woodland; humid, evergreen forest on plateau; on granite; 700–1500 m. Fl. September–February.

32.91. B. anjozorobeense Bosser *(p. 223, bottom right)*

Epiphyte; pseudobulbs cylindrical-conical, 3.5–6 × 2.5–2 cm; leaves oblanceolate, 12–22 × 0.8–1.5 cm; inflorescence 24–43 cm tall, many-flowered; bracts short; flowers green flushed with purple, lip yellow; sepals reflexed, 11–12 mm long; lip fleshy, entire, with 2 longitudinal ridges, margins densely longly ciliate; close to *B. imerinense*, but a more robust plant that is also distinct by the presence of caducous, black glands on the ovary and the back of the floral bracts and sepals. Antananarivo, Toamasina. Humid, evergreen forest remnants; 1000–1200 m. Fl. October.

32.92. B. cylindrocarpum Frappier ex Cordem. var. andringitrense Bosser

Epiphyte; pseudobulbs 4-angled, olive to reddish; leaves 2, narrowly oblong, 10–15 × 0.8–1.2 cm; inflorescence 25 cm tall, 15–20-flowered; flowers green with a yellow-green lip, glandular; sepals 10 mm long; petals 1 mm long; lip 6–7 mm long, with brownish red hairs on the margins. Fianarantsoa. Evergreen forest remnants; 1500–1800 m. Fl. April. *B. cylindrocarpum* var. *cylindrocarpum* is from La Réunion.

32.93. B. imerinense Schltr. *(opposite, left)*

Epiphyte; pseudobulbs ovoid, angular; leaves narrowly oblong, 10–25 × 0.7–1 cm; inflorescence 20–30-flowered; flowers small; sepals 6–7 mm long; petals 0.5 mm long; lip ligulate, 3.5–4 mm long, with marginal white hairs. Antananarivo, Fianarantsoa, Toamasina. Humid, evergreen forest on plateau; 1000–1300 m. Fl. September.

32.94. B. edentatum H.Perrier

(below, right)

Epiphyte; pseudobulbs ovoid, 4-angled, 2.5–5 cm tall; leaves oblong, 4.5–15 × 1.2–3.5 cm; inflorescence 13–35 cm tall, 20–40-flowered; flowers sparsely glandular; sepals 15–16 mm long; petals linear, as long as the column; lip narrow, claviform, with branching hairs on lateral margins and long appendages in the apical part. Fianarantsoa. Humid, evergreen forest on plateau; 1200–1800 m. Fl. November–December.

32.95. B. erythroglossum Bosser

Pseudobulbs 4-angled, 3–5 cm tall; leaves oblong, 5–15 × 1.5–2.5 cm; inflorescence 20–25 cm tall, up to 15-flowered; flowers spreading, green with a dull red lip; sepals spreading, 10–11 mm long; petals linear, 4 mm long; lip obovate, 5.5 mm long, cilate at base and with long appendages in apical part; stelids broadened at apex. Toamasina. Humid forest; 900–1000 m. Fl. May.

32.93

32.94

32.96. B. horizontale Bosser

Pseudobulbs leathery, tetragonal, 1.8–2.5 × 1–1.8 cm; 2–4 cm apart on rhizome, shiny red-brown; leaves horizontal, oblong to oblong-ovate, 3–6 × 1.5–2.5 cm; inflorescence 6–9 cm tall, densely many-flowered; flowers red, furfuraceous; sepals 8 mm long; petals triangular; lip dark red, very fleshy, with a central rounded ridge on top, the edges are raised into 2 wings at the base, extended by 2 papillose cushions on either side, lateral margins hairy, the tip has longer ligulate hairs. Antananarivo, Toamasina. Humid, evergreen and mossy forest; c. 900 m. Fl. March–July.

32.97. B. mirificum Schltr.

Epiphyte; pseudobulbs ovoid, roundly angular, 3–3.5 cm long; leaves linear-oblong, 10–15 × 1.5–2 cm; inflorescence 15–20-flowered; flowers not reflexed, non-resupinate; sepals reflexed, 12–13 mm long; petals minute; lip 4–5 mm long, spathulate, base narrow, with 2 short, rounded raised keels, a central keel extending to the apex, the sides shortly densely hairy, the apical part bearing many ligulate, mobile, caducous appendages. Toamasina. Eastern forest; sea level–100 m. Fl. February.

32.98. B. multiligulatum H.Perrier

Epiphyte; pseudobulbs 4-angled, 2–5 cm tall; leaves narrowly oblong, 12–20 × 1–2 cm; inflorescence laxly 20–40-flowered, raceme longer than the peduncle; flowers green with white petals and a white lip spotted with red; sepals 11–14 mm long; petals 0.6–0.7 mm long; lip fleshy, clavate, 6–7 mm long, with a lateral row of branched hairs and below the apex a transversal row of ligulate appendices, mobile, caducous. Antsiranana, Mahajanga. Humid evergreen forest on plateau; 1000–1500 m. Fl. April.

32.99. B. reflexiflorum H.Perrier

(syn. *Bulbophyllum inauditum* Schltr. (1925) non Schltr. (1913); *B. bosseri* K.Lemcke)

Epiphyte up to 20 cm high; pseudobulbs ovoid, 4–5-angled, 2–4 cm tall; leaves narrowly oblong, 10–18 × 1–2 cm; inflorescence 20–37 cm long, with 15–20 flowers; bracts persistent; flowers pendent, violet-red; sepals 15 mm long; lip fleshy, oblong, 6.5–7 × 2.5 mm, with many long ligulate appendages at apex, the upper surface has 2 short upright rounded keels at the back, with a slightly raised central keel, sides rounded, papillose, covered in caducous hairs, and with somewhat long hairs on the margins. Antsiranana, Antananarivo. Montane, ericaceous scrub; lichen-rich forest on moss- and lichen-covered trees; 2000–2400 m. Fl. March–April, September.

32.99a. B. reflexiflorum subsp. **pogonochilum** (Summerh.) Bosser (syn. *Bulbophyllum pogonochilum* Summerh.; *B. comosum* H.Perrier in Notul. Syst. (Paris) 6 (2):94 (1937), non Collett & Hemsley (1890))

Epiphyte similar in general shape to the typical subspecies but lip more elongate, and the petals longer and of a different shape; plant up to 20 cm tall; raceme laxly 5–8-flowered; rachis slightly verrucose; flowers 22 mm long, dark red, verrucose on the exterior; lip carrying long, mobile, linear appendages. Fianarantsoa. Humid evergreen forest on plateau; 1200–1700 m. Fl. November–December.

VIII. Section Trichopus

Small plants, less than 10 cm tall; pseudobulbs 1-leafed; inflorescence 6–many-flowered; peduncle and rachis sinuous, with distant, regularly alternating flowers; peduncle longer than the rachis; sepals free; lip ciliate, rarely not; stelids short.

Key to groups of species in Section Trichopus

1. Lip not ciliate .... **32.100. B. muscicolum**
1. Lip ciliate or papillose 2
2. Leaf oblong or oblong-spatulate, less than 6 times as long as broad **32.101–32.103**
2. Leaf linear, oblanceolate or linear-lanceolate, more than 10 times as long as broad **32.104–32.106**

32.100. B. muscicolum Schltr.

Pseudobulbs lenticular, 3 mm across; leaf ovate-lanceolate, 3.5 6 × 2–4 mm; inflorescence 1–2-flowered; flowers violet, the lip darker; sepals 2–3 mm long; lip oblong-linguiform, 1.5–2 mm long. Antsiranana. In dry forests. Fl. March.

32.101. B. andohahelense H.Perrier

Pseudobulbs globose, 5 mm across; leaf oblanceolate, 15–35 × 2.5–6 mm; inflorescence filiform, 3–7-flowered, longer than the leaf; flowers dark violet; petals longly ciliate; lip 2.5 mm long, margins ciliate. Fianarantsoa, Toliara. Montane, ericaceous scrub; 1800–2000 m. Fl. November–January.

32.102. **B. hyalinum** Schltr.

Pseudobulbs ovoid or subglobose, 6–8 × 4–6 mm; leaf oblong, 16–23 × 4.5–5 mm; inflorescence 2–3 times as long as the leaf, slender, 8–15-flowered; flowers white; sepals 3 mm long; lip oblong, 3 mm long, ciliate. Antsiranana; also in the Comoros. Humid, evergreen forest; 400–1200 m. Fl. January–May.

32.103. **B. obscuriflorum** H.Perrier

Pseudobulbs globose, 8–12 mm across; leaf oblanceolate, 15–35 × 4–6 mm; inflorescence 4–5 cm long, laxly 8–10-flowered; flowers dark red, 5 mm long; sepals 3.5–4 mm long; lip narrowly spatulate, ciliate. Antsiranana. Seasonally dry, deciduous forest; c. 400 m. Fl. January.

32.104. **B. boiteaui** H.Perrier

Close to *Bulbophyllum nigriflorum*, but differs mainly by the flowers which are twice as large and have a differently shaped lip. Pseudobulbs orbicular, 10–12 mm across; leaf oblanceolate, up to 62 × 6 mm; inflorescence up to 20 cm long, slender, 14–22-flowered; flowers violet-black, 8–9 mm long; sepals 6–7 mm long; petals ciliate; lip auriculate at base, very narrow, bent in middle, 2.5 mm long, ciliate. Antsiranana, Antananarivo, Fianarantsoa. Mossy forest; 1200–2000 m. Fl. January–February.

32.105. **B. ciliatilabrum** H.Perrier

Pseudobulbs depressed globose, 3–4 mm across; leaf oblanceolate-linear, 15–20 × 1.5–2 mm; inflorescence 2–3 times as long as the leaf, slender, 8–15-flowered; flowers deep red, small; sepals 3 mm long; lip oblanceolate, 1.2 mm long, with an obtuse protruding callus on top, the upper part somewhat enlarged-spathulate, ciliate with long white hairs at the margins. Toamasina. Humid, evergreen forest; 800–900 m. Fl. January.

32.106. **B. nigriflorum** H.Perrier

Pseudobulbs orbicular, 4–6 mm across; leaf linear-oblanceolate, 30–50 × 3–4 mm; inflorescence 2–3 times as long as the leaf, slender, 6–21-flowered; flowers nearly black, finely pubescent; sepals 3–4 mm long, ciliolate; petals circular, ciliolate; lip oblanceolate, 2 mm long, very obtuse, ciliate with long bristles. Antananarivo, Toamasina. Humid, evergreen forest; c. 1000 m. Fl. February.

IX. Section Pantoblepharon

Plants small; pseudobulbs unifoliate; inflorescence stubby, shorter than or longer than the leaf, 1–8-flowered; flower parts ciliate; sepals free; lip ciliate; stelidia absent.

Key to species of Section Pantoblepharon

1. Pseudobulbs discoid; lip clavate at tip . . . . **31.107. B. discilabium**
1. Pseudobulbs globose to ovoid; lip not clavate at tip 2
2. Inflorescence 5–8-flowered 3
2. Inflorescence 1–3-flowered 4
3. Inflorescence 8–10 cm long, twice as long as the leaf; sepals 6 mm long . . . .
. . . . **31.109. B. maudeae**
3. Inflorescence less than 4.5 cm long, shorter than the leaf . . . . . . **31.111. B. pantoblepharon**
4. Inflorescence 1-flowered; sepals 5 mm long; lip recurved, red . . . . . . **31.108. B. latipetalum**
4. Inflorescence 2–3-flowered; sepals 3.5 mm long or less 5
5. Lip obovate, pale coloured . . . . **31.110. B. onivense**
5. Lip oblong-ligulate, reddish . . . . **31.112. B. pleurothallopsis**

32.107. **B. discilabium** H.Perrier

Plant small; pseudobulbs approximate, discoid, 3–4 × 1.5–1.8 cm; leaf linear-oblanceolate, 18–30 × 2–3 mm; inflorescence 20–35 mm tall, 2–3-flowered; sepals 6 mm long; lip with the base thickened then clavate, 4 mm long, apex longly ciliate. Antsiranana. Humid, evergreen forest on plateau; 1500–2000 m. Fl. December–March.

32.108. **B. latipetalum** H.Perrier

Plant small, 4–5 cm tall; pseudobulbs broadly ovoid, 8–10 mm long; leaf oblong, 1.5–3 × 0.4–0.6 cm; inflorescence short, 1-flowered; sepals 5 mm long; lip incurved, 3 × 2 mm, red, with short hairs at the apex. Antsiranana. Humid, lichen-rich, evergreen forest on plateau; c. 1500 m. Fl. December.

32.109. **B. maudeae** A.D.Hawkes

(syn. *B. nigrilabium* H.Perrier)

Plant 8–10 cm tall; pseudobulbs small, green; leaf ligulate-oblanceolate, 3–4 cm long; inflorescence 8–10 cm long, 3 times the length of the plant, 5–8-flowered; sepals 6 mm long; lip black, narrow, 3 mm long, base much thickened, from there extended and spathulate, apex with black hairs. Antsiranana. Lichen-rich forest; c. 1700 m. Fl. December.

32.110. B. onivense H.Perrier *(opposite, top left)*

Epiphyte less than 4 cm tall; pseudobulbs orbicular, 5–6 mm across; leaf lanceolate, petiolate, 7.5–17 × 3–5 mm; inflorescence 2–3-flowered, as long as the leaf; flowers reddish, veined with red with a white lip; sepals 3.5 mm long; lip obovate; similar to *B. pleurothallopsis* but has more acute floral bracts, shorter and more acute petals, and a pale lip. Antananarivo, Toamasina. Humid, evergreen forest on plateau; 1300–1500 m. Fl. February.

32.111. B. pantoblepharon Schltr. *(opposite, top right)*

Epiphyte less than 5 cm tall; pseudobulbs globose, 5–11 mm across; leaf oblanceolate, 3–4.5 × 0.5–0.7 mm; inflorescence much shorter than the leaf, densely 5–8-flowered; flowers purplish black; sepals 4.5 mm long; lip lanceolate, ciliate, 1.5–2 mm long. Antananarivo, Antsiranana. Lichen-rich forest; c. 1600 m. Fl. January.

32.111a. B. pantoblepharon Schltr. var. vestitum H.Perrier ex Hermans

Differs from the typical variety by its peduncle which is completely covered by 5–6 thick, lax sheaths, which open at the apex and are keeled at the back, its elliptical sepals, its smaller, 0.6 mm long, completely orbicular petals, the 2 small, triangular, thick bracteoles on the outside of the perianth, the acute stelids that are the same height as the erect part of the column, and the anther that lacks an apicule. Antananarivo, Fianarantsoa. Humid, evergreen forest on plateau; 900–1400 m. Fl. unknown.

32.112. B. pleurothallopsis Schltr. *(opposite, middle left)*

Small epiphyte, less than 3 cm tall; pseudobulbs subglobose, 3–4 mm across; leaf narrowly oblanceolate, 8–20 × 2.5–4 mm; inflorescence 2–3-flowered; flowers small, greenish or reddish; sepals 3–3.5 mm long, hyaline, more or less ciliate; petals ovate, ciliate; lip oblong-ligulate, 2 mm long, ciliate. Antsiranana, Toamasina. In lichen-rich forest; c. 1400 m. Fl. January–February.

X. Section Lupulina

(formerly Section Calamaria sens. Perrier and including Section Humblotiorchis). Following the treatment outlined by Fischer, G. (2007). Evolution of the orchid genus *Bulbophyllum* in Madagascar. PhD thesis, University of Vienna, Austria.

Robust plants; pseudobulbs 1–2-leaved; inflorescence many-flowered; peduncle and rachis usually not sinuous; sepals free; lip ciliate, rarely papillate; stelids acicular.

Key to groups of species in Section Lupulina

1. Lip glabrous . . . . .32.113. **B. humblotii**
1. Lip ciliate 2

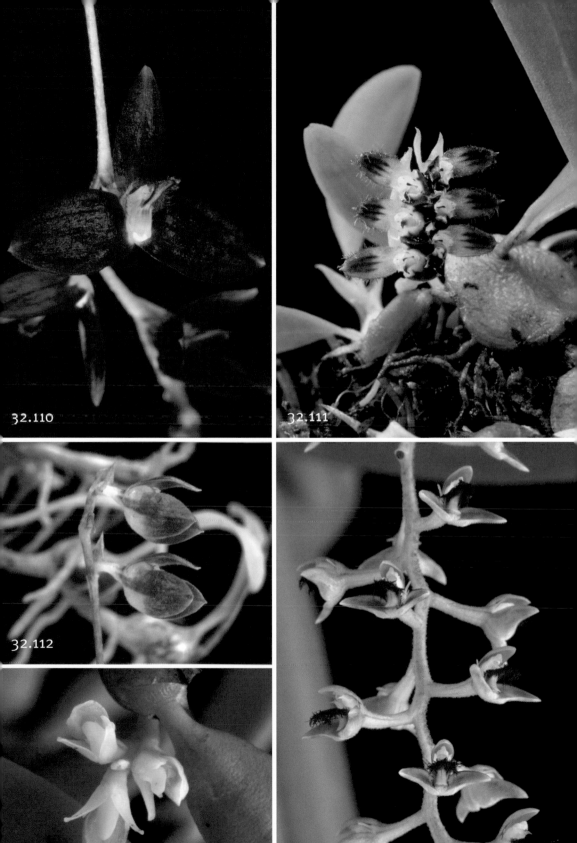

32.110

32.111

32.112

32.113

32.114

2. Rachis slender and sinuous; ovary 4.5–5 mm long, narrow .... **32.114. B. hildebrandtii**
2. Rachis fleshier than the apex of the peduncle: ovary less than 4 mm
   long, fleshy 3
3. Bracts very large, more than 10 mm long, covering entirely the flowers
   and rachis; pseudobulbs always bifoliate **32.115–32.120**
3. Bracts small, less than 10 mm long, not covering entirely the flowers and
   rachis; pseudobulbs uni- or bifoliate **32.121–32.130**

32.113. B. humblotii Rolfe   *(p. 231, bottom left)*

(syn. *Bulbophyllum album* Jum. & H.Perrier; *B. laggiarae* Schltr.; *B. luteolabium* H.Perrier; *B. linguiforme* P.J.Cribb)

Small epiphyte; pseudobulbs 2-leaved; inflorescence 1.5–3.5 cm long; flowers creamy-white with a yellow lip; lip fleshy, tongue-like, 2.5 × 1–1.5 mm, geniculate at base, the tip rounded, the margins entire; stelids well-developed. Antsiranana, Toamasina, Toliara; also in the Seychelles and E Africa. Seasonally dry, deciduous forest and woodland; humid, lowland, evergreen forest; sea level–1300 m. Fl. March–October.

32.114. B. hildebrandtii Rchb.f.   *(p. 231, bottom right)*

(syn. *Bulbophyllum maculatum* Jum. & H.Perrier; *B. madagascariense*; *B. melanopogon* Schltr.)

Pseudobulbs 3–4 cm apart on rhizome, 4-angled, 2–3 × 1–1.8 cm, yellow; leaves 2, lanceolate to oblong-ligulate, 6–7 × 1.5–1.8 cm; inflorescence 30–40 cm tall, never pendulous; flowers 7–10 mm, greenish-yellow with a red lip; sepals 5 mm long; petals linear, 1.5 mm long; lip 3 mm long, red-ciliate on margins and at tip. Antsiranana, Fianarantsoa, Mahajanga, Toamasina. On *Eugenia* and other large trees by rivers; seasonally dry, deciduous forest and woodland; sea level–600 m. Fl. July–December.

32.115. B. bicoloratum Schltr.   *(opposite, top left)*

(syn. *Bulbophyllum theiochlamys* Schltr.; *B. coeruleum* H.Perrier)

Pseudobulbs 2–3 cm apart, ovoid, 4-angled, 12–17 × 7–10 mm; leaves oblong, 3–7 × 0.7–1.5 cm; inflorescence up to twice as long as the leaves; rachis 3–7 cm long, 1 cm in diameter; flowers small, pale green to reddish with a red lip with a yellow tip, hidden underneath the imbricate red floral bracts; sepals 4–4.5 mm long; petals linear, 2 mm long; lip thickened, 2.5–4 mm long, densely ciliate with black hairs. Antananarivo, Antsiranana, Fianarantsoa, Toliara. Humid, lowland, evergreen forest; 300–1000 m. Fl. February–May.

32.116. B. cirrhoglossum H.Perrier   *(opposite, top right and bottom)*

Pseudobulbs compressed-discoid, 2–2.3 cm across; leaves lorate, 4.5–5 × 1.7–2 cm; inflorescence up to 20 cm long; spike conical; flowers in 3 rows, sunk into rachis; sepals 8–8.5 mm long; lip saddle-shaped, 3 mm long, dark red, with long violet hairs in the middle. Toamasina. Humid, evergreen forest on plateau; 900–1200 m. Fl. unknown.

32.115   32.116

32.116

32.117. B. luteobracteatum Jum. & H.Perrier (*opposite*)

Pseudobulbs 8–12 mm apart, ovoid, 2–3 × 1–1.5 cm, 5-angled; leaves oblong, 5–8 × 1.3–1.5 cm; inflorescence c. 30 cm tall; bracts bright yellow, 3–4 cm long; flowers c. 16 mm long, yellowish and spotted by dull brown, the lip red; sepals 6 mm long; petals linear, 2 mm long; lip papillose, curved and truncate at the base, then extended into a thick lump. Antsiranana. Lichen-rich forest; c. 1500 m. Fl. May.

32.118. B. obtusatum Schltr.

Pseudobulbs 10–15 mm apart, oblong, 3.5–4 × 2.3–2.5 cm, 6-angled, yellow; leaves narrowly elliptic, 10–13 × 2–2.5 cm; inflorescence slightly longer than the leaves; bracts in 4 rows, 15–18 mm long; flowers red; sepals 5 mm long; lip papillate at base, narrow, very thick and round at tip, the edges recurved, ciliate with many violet-red hairs in the lower half. Antsiranana, Mahajanga. Humid, evergreen forest; dry forest; up to 1400 m. Fl. April–May.

32.119. B. occultum Thouars (*p. 237, top left*)

Pseudobulbs 2.5–3 cm apart, ovoid, 4-angled, 3 × 2 cm; leaves oblong, 5–12 × 1.8–3 cm; inflorescences 15–25 cm tall; bracts in 3 rows, 15 mm long; flowers reddish; sepals 4 mm long; lip oblong, 3 × 1 mm, ciliate. Antananarivo, Antsiranana, Fianarantsoa, Mahajanga, Toliara; also in the Mascarenes. Humid, evergreen, coastal and lowland forest; sea level–1500 m. Fl. September–May.

32.120. B. quadrifarium Rolfe (*p. 237, bottom*)

Pseudobulbs orbicular to ovoid, 2.5–4 × 2.8–2.9 cm; leaves elliptic-oblong, 7–8 × 2–2.5 cm; inflorescence 30–40 cm tall; rachis 7–15 cm long, 1.5–2 cm in diameter; bracts in 5 rows, the lowermost 1 cm long, reddish yellow; flowers cream edged with purple; sepals 4 mm long; lip semi-circular, 3 mm long, long-ciliate on the margins. Fianarantsoa, Toamasina, Toliara. Coastal forest; sea level–500 m. Fl. October.

32.121. B. cardiobulbum Bosser

Pseudobulbs clustered, biconvex, heart-shaped, 3–5 cm across, shiny yellow-brown; leaves 2, oblong, 3.5–7 × 1.6–2 cm; inflorescence 20–30 cm tall, 12–24-flowered; stalk strongly compressed; flowers green spotted with red, the petals white, the lip pale yellow spotted with purple; sepals 13–15 cm long; petals triangular, 2.5 mm long; lip lanceolate, acute, 12 × 3.5 mm, the base recurved, with 2 ridges, the basal lobe erect, papillate-ciliolate, ciliate. Antananarivo, Fianarantsoa. Humid, mossy, evergreen forest on plateau; 1300–1400 m. Fl. November.

32.117

32.122. B. cryptostachyum Schltr.

Pseudobulbs 1–2 cm apart, ovoid, 3–3.5 × 1.7–1.9 cm, 4-angled; leaves 2, oblong or ligulate, 5–6 × 1.5–1.8 cm; inflorescence arcuate, reflexed, 20–25 cm long; bracts elliptic-lanceolate, 10 mm long; flowers furfuraceous, 6 mm long; sepals 4 mm long; lip curved, oblong, very obtuse, c. 2 mm long, base with a small swelling, densely ciliate on basal margins. Antsiranana, Mahajanga. Humid, evergreen forest; c. 1000 m. Fl. August–October.

32.123. B. elliotii Rolfe   *(opposite, top right)*

(syn. *Bulbophyllum malawiense* B.Morris)

Pseudobulbs well-spaced, suborbicular, 1–1.5 cm across; leaves broadly oblong, 11.8–2.5 × 0.8–1.5 cm; inflorescence 5–10 cm tall, 10–20-flowered; bracts ovate-triangular, 3.5–4.5 mm long; flower dull maroon; sepals 3–3.5 mm long; lip strongly recurved, linear, obtuse, pectinate-fimbriate. Antananarivo, Toamasina, Toliara; also in E and S-C Africa. Humid, evergreen forests; sea level–1200 m. Fl. April–June.

32.124. B. erectum Thouars   *(opposite, middle right)*

(syn. *Bulbophyllum lobulatum* Schltr.; *B. calamarioides* Schltr.)

Pseudobulbs 1–1.5 cm apart, oblong-ovoid, 1.5–3 × 0.6–1 cm, green, sometimes finely spotted red; leaf ligulate, 4.5–13 × 0.9–2 cm; inflorescence 30–37 cm tall; inflorescence 15–50-flowered; bracts lanceolate, almost as long as the flowers; flowers greenish-brown, purple-spotted, lip reddish or orange; sepals 7–8 mm long; lip linguiform, 4–5 mm long, ciliate and a little narrowed towards the obtuse apex. Antsiranana, Toamasina; also in the Mascarenes. Coastal dry and evergreen forests, not far from the sea, on *Sarcolaenaceae* and *Chlaenaceae*; sea level–400 m. Fl. February–April, August, October.

32.125. B. lecouflei Bosser   *(p. 239, left)*

Pseudobulbs disc-like, bilaterally flattened, 2 × 2 cm; leaves 2, oblong, 5.5–6 × 1.8–2.2 cm; inflorescence up to 20 cm long, pendent, distantly many-flowered; bracts pale yellow; flowers rose-violet with a darker lip; sepals 6–7 mm long; lip 3-lobed, side lobes erect, ciliate at tips; midlobe tongue-like; stelidia elongate, recurved. Antsiranana. Coastal and lowland forest; sea level–100 m. Fl. September–November.

32.126. B. pusillum Thouars

Pseudobulbs not flattened, unifoliate; leaf elliptic-ovate; inflorescence covered with black rugosities; flowers yellow or greenish brown with black verrucosities, the lip yellow or orange with red ciliae; lip compressed in middle, rounded at apex, ciliate on lateral margins only; stelds strongly falcate. Antananarivo, Antsiranana, Fianarantsoa. Montane forest; 600–1600 m. Fl. unknown.

32.119

32.123

32.124

32.120

32.127. B. sambiranense Jum. & H.Perrier   *(opposite, top right)*

Pseudobulbs slightly flattened, 2–2.5 × 1–1.5 cm, 1–1.5 cm apart on rhizome; leaves 2, ovate-lanceolate, 5–7 × 1.4–1.7 cm; inflorescence 15–20 cm long; floral bracts and flowers rugose, red; flowers small, 7.5–8 mm long; sepals 5 mm long; lip curved, ligulate, obtuse, 2 mm long, ciliate with thick hairs. Antananarivo, Antsiranana, Mahajanga. Seasonally dry, deciduous forest and woodland; humid, evergreen forest on plateau; 100–1900 m. Fl. September–April.

32.128. B. rubrum Jum. & H.Perrier   *(opposite, bottom right)*
(syn. *Bulbophyllum ambongense* Schltr.)

Pseudobulbs 1–2 cm apart, broadly ovoid, 1.5–2.5 cm across, 4-faced; leaves oblong to oblong-ligulate, 3–7 × 1.2–1.8 cm; inflorescence 15–27 cm tall; bracts distichous, lanceolate, acute, 8 mm long; flowers dark reddish brown with a dull red lip; sepals 4 mm long; petals lanceolate, 2.5 mm long; lip 2.5 mm long, strongly ciliate in the middle. Antsiranana, Mahajanga. On coastal trees and in lowland forest; up to 300 m. Fl. July–Nov.

32.129. B. ruginosum H.Perrier

Pseudobulbs 0.5–1 cm apart, suborbicular, 2–2.5 × 1.5–2 cm, bifoliate. Leaves broadly elliptic, 2.5–4 × 1.2–2 cm; inflorescence 15–25 cm long, up to 50-flowered; flowers dark brownish red or dull purple, furfuraceous on the exterior; sepals 7–9 mm long; lip ligulate, 3 mm long, papillose all over. Antsiranana (Mt Ambre). Montane forest; 1000 m. Fl. November.

32.130. B. trifarium Rolfe

Pseudobulbs ovoid-oblong, 1.5 × 5 cm, bifoliate; leaves elliptic-oblong, 4–9 × 1–2 cm; inflorescence 13–15 cm long, 30–50-flowered; rachis recurved; flowers dull lurid purple with minute darker spots; sepals 6–7 mm long; lip elliptic, narrowed in middle, densely cilate. Toliara. Dry or subhumid forest, on tsingys; sea level–300 m. Fl. October–November.

XI. Section Loxosepalum

Small to medium-sized plants; pseudobulbs unifoliate, yellow, surrounded by cottony white fibres; inflorescence usually longer than the leaf, laxly to densely many-flowered; flowers glabrous; sepals free, except in *B. subsecundum* (p. 246) and *B. ventriosum* (p. 244); lip glabrous, curved-geniculate; stelids short.

32.125

32.127

32.128

Key to groups of species in Section Loxosepalum

1. Pseudobulbs bifoliate .... **32.159. B. rutenbergianum**
1. Pseudobulbs unifoliate 2
2. Leaf semi-terete; pseudobulbs finely warty .... **32.131. B. leandrianum**
2. Leaf dorsiventrally flattened; pseudobulbs smooth or rough but not warty 3
3. Flowers 6 mm or more long 4
3. Sepals 5.5 mm or less long 5
4. Pseudobulbs ovoid or globose, usually less than 1.5 cm tall **32.132–32.137**
4. Pseudobulbs conical-cylindrical, usually more than 1.5 cm tall **32.138–32.145**
5. Pseudobulbs ovoid or discoid, usually less than 1.5 cm tall **32.146–32.153**
5. Pseudobulbs conical or cylindrical, often more than 1.5 cm tall **32.154–32.158**

32.131. **B. leandrianum** H.Perrier (*opposite*)

Pseudobulbs subglobose, 7–10 × 10–14 mm, finely warty; leaf linear, sub-terete, 10–22 × 1–1.5 mm; flowers greenish yellow; sepals 5 mm long; lip curved wider towards the base, the edges very finely denticulate. Antananarivo, Toamasina. Humid, evergreen forest; c. 900 m. Fl. December–January.

32.132. **B. ambrense** H.Perrier

Pseudobulbs ovoid, 6–8 mm tall; leaf oblanceolate, 7.5–10 × 5–7 mm; sheaths woolly; inflorescence 5.5–9 cm tall; flowers 7 mm long; sepals 5–5.5 mm long; lip small, 2.6 × 2 mm, slightly emarginate at the tip, with a thick rounded heel, the lobes suborbicular and the margins denticulate. Antsiranana. Mossy, evergreen forest; c. 1200 m. Fl. September.

32.133. **B. marojejiense** H.Perrier

Plant small; pseudobulbs conical-terete; inflorescence 6–10-flowered; flowers small; sepals 6–7 mm; lip curved, 4 × 2 mm. Antsiranana. Humid, lichen-rich, evergreen forest on plateau; 1000–1600 m. Fl. November–December.

32.134. **B. marovoense** H.Perrier

Plant 18–30 cm tall; pseudobulbs narrowly conical, 1.8–2.7 cm long; leaf narrowly lanceolate, 8–12 × 1–1.5 cm; inflorescence slender, pendent or recurved, 25–35 cm long; sepals lanceolate-linear, 6–6.5 mm long; petals very short, semi-rounded; lip 1.5× 1 mm; stelids triangular, acute. Toamasina. Humid, evergreen forest on plateau; 900–1200 m. Fl. unknown.

32.131

32.135. B. melleum H.Perrier *(opposite)*

Pseudobulbs ovoid, 10–16 × 8–10 mm; leaf elliptic, 2–4 × 1–1.4 mm; inflorescence slender, 12–15 cm long, laxly flowered; flowers yellow, honey-scented; sepals 10–12 mm long; with denticulate-crenulate margins. Fianarantsoa, Toamasina. Humid, evergreen forest on plateau; 1000–1400 m. Fl. October–November.

32.136. B. rienanense H.Perrier

Pseudobulbs ovoid, 10 mm tall, clustered; leaf elliptic to ovate, 10–22 × 7–12 mm; inflorescence 7–10 cm long, 2–7-flowered; flowers 13 mm long; sepals 10 mm long. Fianarantsoa. Humid, evergreen forest on plateau; c. 1200 m. Fl. November.

32.137. B. sphaerobulbum H.Perrier

Pseudobulbs globose, 10 mm across; leaf oblong, 4–5 cm long; inflorescence 14–17 cm long; flowers 6–8 mm long; sepals 5 mm long; lip very small, 1 mm long, with the edges a little expanded in the lower half. Toamasina. Humid, lowland, evergreen forest; c. 400 m. Fl. October.

32.138. B. approximatum Ridl.

Pseudobulbs conical, up to 25 mm long; leaf elliptic-spatulate, up to 5 × 1.8 cm; inflorescence 12–15 cm long; flowers 10 mm long; sepals 7–8 mm long; oblong, emarginated. Antsiranana, Fianarantsoa. Humid, highland forest. Fl. September.

32.139. B. kainochiloides H.Perrier

Pseudobulbs conical, 3–3.5 cm tall; leaf ligulate, 10 × 1.2 cm; inflorescence 20–30 cm tall; peduncle c. 12 cm long; flowers 7 mm long; sepals 4.5 mm long; lip curved, the margins truncate and a little expanded in the upper third and with a thick heel; stelids narrow, triangular, acute. Antananarivo. Humid, evergreen forest on plateau; 1200–1400 m. Fl. October, January.

32.135

32.140. B. minutilabrum H.Perrier

Pseudobulbs conical, 2.5–3 cm tall; leaf 12–18 × 2–2.2 cm; inflorescence 30–35 cm tall; flowers 6–7 mm long; sepals 6 mm long; lip expanded at the base. Antsiranana. Lichen-rich, evergreen forest; c. 2400 m. Fl. April.

32.141. B. papangense H.Perrier

Pseudobulbs conical, 8–25 × 3–4 mm; leaf oblanceolate, 4–8.2 × 0.4–0.8 cm; inflorescence 11–16 cm long, many-flowered; flowers yellowish-white, 6–7 mm long; sepals 4–5 mm long; lip very recurved, 2 mm long, a little indented at the apex and the margins reflexed. Fianarantsoa, Toliara. Humid, evergreen forest on plateau; ericaceous scrub; c. 1500 m. Fl. November.

32.142. B. subapproximatum H.Perrier

Pseudobulbs conical, 1.3–2.5 cm tall, slightly rugulose, blackish; leaf ligulate of oblong, 1.8–6.5 × 0.8–1.6 cm; inflorescence 12–15 cm long; flowers 6–8 mm long; sepals 4.5–6 mm long; lip subrectangular. Antsiranana. Humid, evergreen forest on plateau; c. 1000 m. Fl. May.

32.143. B. trichochlamys H.Perrier

Pseudobulbs conical, c. 12 mm long; leaf ovate, 4 × 1.4 cm; inflorescence 6–9 cm tall; flowers 8 mm long; sepals 6 mm long; lip relatively large, 3 mm long. Antananarivo. Humid; evergreen forest on plateau; c. 1200 m. Fl. February.

32.144. B. ventriosum H.Perrier

Pseudobulbs slender, yellow; leaf oblong or oblanceolate, 6–13 × 1–2 cm; inflorescence twice as long as the leaf; flowers pale yellow; sepals 5–6 mm long; distinct by the wide, rounded mentum which is formed by the fusion of the 2 lateral sepals and by its lip with spreading margins; lip auriculate. Antananarivo, Antsiranana. Lichen-rich, evergreen forests; c. 2000 m. Fl. April, August–September.

32.145. B. zaratananae Schltr. *(p. 247, top left)*

Pseudobulbs cylindrical-conical, 17–30 × 7–10 mm; leaf ligulate, 5–15 × 1–1.3 cm; inflorescence 12–25 cm tall; flowers white to yellowish, 8 mm long; sepals 5–6.5 mm long; lip recurved; related to *B. baronii* (p. 248) but with a thicker rhizome and longer leaves, also a distinct lip. Antsiranana. Humid, evergreen and lichen-rich forest, also in ericaceous scrub; 400–2000 m. Fl. November–January.

32.145a. B. zaratananae subsp. **disjunctum** H.Perrier ex Hermans

The variety has slightly larger flowers up to 9 mm long, with the margins of the column straight below the rostellum, a column-foot that is flat at the front and acute stelids. Fianarantsoa. Humid, mid-elevation forest. Fl. October.

32.146. B. ambatovense Bosser *(p. 247, middle left and bottom)*

A small epiphyte; pseudobulbs dorsiventrally flattened, discoid, 4–7 mm in diam.; leaf narrowly oblong, 15–25 × 1.5–2.5 mm; inflorescence 2.5–4.5 cm tall, up to 12-flowered; flowers white or pale pink with a green lip; sepals 3–3.5 mm long; petals 1.4–1.5 mm long; lip fleshy, 0.8–1 mm long; side lobed oblong, erect; midlobe emarginated. Toamasina. Humid, evergreen forest; 700 m. Fl. September–December.

32.147. B. curvifolium Schltr.

Pseudobulbs ovoid, 10–15 cm across; leaf slightly recurved, linear-ligulate, 40–70 × 3.5–5 mm; inflorescence longer than the leaf, 10–18-flowered; flowers 5.5–6 mm long; sepals 4 mm long; lip 2.5 mm long; related to *B. subsecundum*, but with shorter inflorescence and differently shaped flowers; lip 2.5 mm long, very obtuse, rectangular-suborbicular, the base narrower. Antsiranana. Humid, evergreen forest; c. 500 m. Fl. October.

32.148. B. decaryanum H.Perrier

Pseudobulb ovoid, 1.5–2.2 × 1–1.3 cm; leaf oblanceolate, 7.7–10 × 1.2–1.7 cm; inflorescence 10–12 cm long, densely many-flowered; flowers small, c. 5 mm long; sepals 4–4.5 mm long; lip 0.8 mm long. Toamasina. Humid, evergreen forest; 900–1000 m. Fl. unknown.

32.149. B. lineariligulatum Schltr.

Pseudobulbs ovoid, 8–10 mm tall; leaf linear or linear-ligulate, 8–10 × 5–7 mm; inflorescence more or less as long as the leaf, 10–15-flowered; flowers 5 mm long; sepals 4 mm long; lip ovate, obtuse, 3 mm long; closely related to *Bulbophyllum curvifolium* (p. 245), but differs by its smaller pseudobulbs, shorter leaves, the blunter and differently shaped petals and the lip which is the same width at the apex and base but narrowed in the middle. Antsiranana, Fianarantsoa, Toamasina. In lichen-rich, evergreen forest and in ericaceous scrub; 1000–2000 m. Fl. September–April.

32.150. B. subsecundum Schltr.

Pseudobulbs ovoid, 12–15 mm tall; leaf oblong-ligulate, 4–8 × 0.8–1.5 cm, longly narrowed towards the base; inflorescence 13–18 cm tall; flowers yellowish-white, 4–5 mm long; sepals 3.5 mm long; lip obtuse, 2 mm long. Antsiranana, Toamasina. Humid, evergreen forest; 800–1400 m. Fl. November–January.

32.151. B. subsessile Schltr.

Pseudobulbs ovoid, 8–10 × 5–6 mm; leaf oblong-ligulate, 2.4–4 × 0.4–0.6 cm; inflorescence up to 10 cm long, 6–15-flowered; stalk setiform; flowers 5 mm long; sepals 4 mm long; lip wide and rounded. Toamasina. Humid, lowland, evergreen forest; 500–1000 m. Fl. September.

32.152. B. vakonae Hermans   (*opposite, top right*)
(syn. *Bulbophyllum ochrochlamys* Schltr. (1924) non Schltr. (1913))

Pseudobulbs ovoid, 10–14 mm tall; rhizome thick; leaf ligulate 4–6.5 × 1–1.3 cm; inflorescence 10–13 cm tall, densely flowered; flowers yellow, 5 mm long; sepals 4 mm long; lip very short, broadly oblong. Antananarivo, Toamasina. Humid, evergreen forest; 800–900 m. Fl. January.

32.153. B. xanthobulbum Schltr.

Pseudobulbs broadly ovoid, 10 × 9 mm; leaf elliptic, 17–23 × 10–13 mm; inflorescence 5–6 mm long; flowers 4.5–5.5 mm long; sepals 4–4.5 mm long; lip 1.3 mm long, very recurved, with a wide heal, the margins of the lower part upright and those of the upper part reflexed. Toamasina. Coastal and humid, evergreen forest; sea level–800 m. Fl. January.

32.145

32.146

32.152

32.146

32.154. **B. alleizettei** Schltr.

(opposite, top left)

Pseudobulbs subcylindrical, 10–20 × 3–5 mm, the sheaths woolly; leaf ligulate, 25–50 × 2.5–5 mm; inflorescence shorter than the pseudobulb, arcuate; flowers small, c. 3 mm long, yellow; sepals 1.75 mm long; lip ovate, obtuse. Antananarivo, Toamasina. Humid, evergreen forest on plateau; 1000–1500 m. Fl. February, October.

32.155. **B. baronii** Ridl.

(opposite, top right and bottom)

Pseudobulbs elongate, ovoid-conical, 10–16 × 4–8 mm, yellow or sometimes red; leaf narrowly oblong or oblanceolate, 2–7.5 × 0.5–0.9 cm; inflorescence 8–12 cm long; flowers pale yellow, 4–5 mm long; sepals 4 mm long; lip small, rounded at tip. Antananarivo, Antsiranana, Fianarantsoa, Toamasina, Toliara. Humid, lichen-rich, evergreen forest; on rocks; 800–2200 m. Fl. November–June.

32.156. **B. florulentum** Schltr.

Pseudobulbs cylindrical-conical, 15–30 mm long, rugulose; leaf linear or ligulate, 3–10 × 0.4–0.8 cm; inflorescence 11–20 cm high, densely many-flowered; flowers small, 5 mm long, yellowish-white; sepals 4 mm long; lip very curved, suborbicular, with the lower half wide and with the edges folded back. Antsiranana. Lichen-rich, evergreen forest; 1500–2000 m. Fl. January, October.

32.157. **B. leptostachyum** Schltr.

Pseudobulbs cylindrical-conical, 10–20 mm long; leaf linear or linear-lanceolate, 2–6.5 × 0.2–0.9 cm; inflorescence 6–14 cm high, 15–25-flowered; flowers small, 5–7 mm long, yellowish-white; sepals 3–5 mm long; lip very curved, oblong-linguiform at apex from a broad base. Antananarivo, Fianarantsoa, Toamasina. Humid, evergreen and mossy forest; 400–1600 m. Fl. October–March.

32.158. **B. multiflorum** Ridl.

(p. 251, top left)

(syn. *Bulbophyllum ridleyi* Kraenzl.)

Pseudobulbs conical, 1.5–3 cm tall; leaf lanceolate, obtuse, 4–8 × 1.2–1.5 cm; inflorescence 18–25 cm tall, many-flowered; flowers white, 5–6 mm long; sepals 5 mm long; lip very curved, hardly 2 mm long. Antananarivo, Antsiranana, Fianarantsoa, Toamasina. Humid, evergreen forest on plateau, covering tree trunks; 900–1500 m. Fl. March, October.

32.154

32.155

32.155

32.159. B. rutenbergianum Schltr.

(syn. *Bulbophyllum peniculus* Schltr.; *B. spathulifolium* H.Perrier; *B. coursianum* H.Perrier)

Pseudobulbs ovoid, 11–20 × 6–10 mm, 1–1.5 cm apart on rhizome; leaves oblong-ligulate, 20–50 × 5–10 mm; inflorescence 6–9 cm tall, many-flowered; flowers whitish yellow; sepals 3.5–5 mm long; petals narrowly oblong, 1–12 mm long; lip recurved, 2 mm long, emarginated. Antananarivo, Antsiranana, Fianarantsoa. Lichen-rich, evergreen forest; 1400–2500 m. Fl. October–February.

XII. Section Lyperocephalum

Peduncle often as long as the straight or slightly recurved, often fleshy, rachis; flowers fleshy. Stelids with a fleshy extension at the base, the latter differing from the stelids in consistency, form and colour; inflorescence dense or subcapitate, 6–12-flowered.

32.160. B. lyperocephalum Schltr.   *(opposite, top right)*

Small epiphyte on a creeping rhizome; flowers greenish, 5 mm long, glabrous; lip 2 × 1.25 mm, folded-curved, with a thick keel, the lower half sunk and with 2 thick, pronounced horns, the upper part rounded. Antsiranana, Toamasina. Humid, evergreen forest; 800–900 m. Fl. February.

XIII. Section Lyperostachys

Inflorescence 10 cm or more in length; rachis bent or straight, more or less fleshy; peduncle as long as the rhachis. Flowers fleshy; stelids simple or toothed on the lower margin; dorsal sepal not the largest segment of the flower.

Key to species of Section Lyperostachys

1. Leaf 1.2–1.7 cm broad . . . .**32.161. B. lyperostachyum**
1. Leaf 2.5 cm broad . . . .**32.162. B. thompsonii**

32.161. B. lyperostachyum Schltr.

Pseudobulbs conical-cylindrical, 2–3 × 0.6–0.8 cm, rugose; leaf ligulate, 8.5–14 × 1.2–1.7 cm; inflorescence shorter than the leaf, 9–11 cm long, 6–20-flowered; flowers yellow with darker orange-brown lines, 6.5–7 mm long; sepals 4.5 mm long; lip 3.5 × 2 mm, curved, obtuse, glabrous, a small obsolete swelling in the middle; stelids well-developed, bidentate at the apex. Antsiranana. Lichen-rich, evergreen forest; c. 2000 m. Fl. January.

32.158

32.160

32.163

32.165

32.162. B. thompsonii Ridl.

Pseudobulbs ovoid-globose, rugose, shiny; leaf oblanceolate, 7.5 × 2.5 cm; inflorescence up to 20-flowered; sepals triangular, acute; lip oblong-lanceolate, acute. Madagascar. Fl. unknown.

XIV. Section Pachychlamys

Plants 2-leaved; rachis more fleshy than the top of the peduncle; flowers fleshy; sheaths of the young pseudobulbs often very fleshy, hard and long; rachis recurved.

Key to groups of species in Section Pachychlamys

1. Sepals 8 mm or more long **31.163–31.166**
1. Sepals 7 mm or less long **31.167–31.172**

32.163. B. molossus Rchb.f.   *(p. 251, bottom left)*

Pseudobulbs conical, 2–5 cm long, with papery sheaths when young; leaves oblanceolate, 4–7 × 1.2–2 cm; inflorescence shorter than or slightly longer than the leaves, 2–5-flowered; flowers red, glabrous, 12–13 mm long; sepals 10–11 mm long; mentum large; lip recurved, obtuse at the apex and auriculate at the base, 2 mm long and wide. Antsiranana, Antananarivo, Fianarantsoa, Toamasina. Humid, evergreen forest on plateau; 1200–1500 m. Fl. November–February.

32.164. B. moratii Bosser

Epiphyte, up to 29 cm tall; rhizomes stout, elongate; pseudobulbs, cylindrical, 7–9 cm long; leaves 2, narrowly obovate, 17–20 × 5–6 cm; inflorescence densely many-flowered; bracts as long as flowers, yellow; flowers with yellow-rose sepals spotted with violet, white petals edged with violet and a yellow lip; sepals 9–11 mm long; lip long, narrow fleshy, erose on lower margins; one of the largest species on the island. Antsiranana, Toamasina. Humid, evergreen montane and plateau forest; c. 1000 m. Fl. May.

32.165. B. pachypus Schltr.   *(p. 251, bottom right)*

Pseudobulbs ovoid-cylindrical, 3–5 × 1–1.5 cm; leaves oblong-ligulate or ligulate, 6.5–14 × 1–1.7 cm; inflorescence 11–20 cm tall, 4–8-flowered; flowers 11–13 mm long, tinted violet with a few darker stripes; sepals 9–10 mm long; lip oblanceolate, 8 mm long, red and yellow. Antananarivo, Fianarantsoa, Toamasina. Humid, evergreen forest; 800–1400 m. Fl. January–April.

32.166

32.166

32.170

32.171

32.172

32.166. B. septatum Schltr. *(p. 253, top left and top right)*
(syn. *Bulbophyllum serratum* H.Perrier; *B. ambreae* H.Perrier)

Pseudobulbs oblong-conical, 1.7–2 × 0.9–1.2 cm; leaves oblong-ligulate, 3–4.5 × 0.7–1.2 cm, shortly contracted at the base; inflorescence 6–10 cm tall, 5–10-flowered; flowers yellowish, c. 1 cm long; sepals 8 mm long, pointed, with a distinct mentum; lip tongue-shaped, 6 mm long. Antsiranana. Lichen-rich and mossy forests; 1500–2000 m. Fl. November–January.

32.167. B. ikongoense H.Perrier

Pseudobulbs narrowly elongate, 4–6 × 0.8–1 cm, 4-winged; leaves 2, narrowly oblanceolate, 12–15 × 0.8–1.1 cm; inflorescence up to 20 cm tall; raceme strongly reclining, 4.5–6 cm long, not very densely many-flowered; flowers striped with brown; sepals 3 mm long; lip very small, 1 × 1 mm, with 5 obtuse lobules, and a narrow transverse callus. Fianarantsoa. Humid, lowland forest. Fl. October–December.

32.168. B. longivaginans H.Perrier

Pseudobulbs ovoid, 2.5–3 cm long, with papery sheaths at the base of the pseudobulbs; leaves oblanceolate-linear, 20–22 × 0.8–1 cm; inflorescence 12.5–15 cm long, 5–10-flowered, shorter than the leaves; flowers red, glabrous, c. 7 mm long; sepals 6 mm long; lip recurved, obtuse at the apex and auriculate at the base. Antananarivo, Toamasina. Humid, evergreen forest on plateau; c. 1000 m. Fl. January–February.

32.169. B. multivaginatum Jum. & H.Perrier

Pseudobulbs oblong-conical, 3–4 × 1.5–2 cm, rugulose; leaves elliptic, 2–3 × 1.5–2 cm; inflorescence 18–23 cm tall; flowers thick, c. 8 mm long; sepals 5–6 mm long, reddish brown; lateral sepals greenish on the outside, red-brown on the inside; disc of the lip purple-black; lip 2.5 mm long, curved, obtuse, velvety. Antsiranana, Toamasina. Humid, lowland, evergreen forest; 1100–1600 m. Fl. May, October.

32.170. B. perrieri Schltr. *(p. 253, middle right)*

Pseudobulbs conical, 2–23.5 × 0.6–0.8 cm; leaves oblanceolate, 15–20 × 1–1.8 cm; inflorescence equalling or longer than the leaves, 15–20-flowered; flowers 7 mm long; sepals 4 mm long; mentum very pronounced, protruding by 4 mm; petals orbicular; lip 3.5 mm long. Antsiranana. Humid, moss- and lichen-rich, evergreen forest on plateau; 1500–1600 m. Fl. January.

32.171. B. sandrangatense Bosser *(p. 253, bottom left)*

Pseudobulbs conical, 2–2.5 × 1.3–1.5 cm, fibres enveloping the pseudobulbs; leaves 2, suberect, ligulate-oblanceolate, 10–18 × 0.7–1 cm; inflorescence equalling or longer than the leaves, 5–13-flowered; flowers red; sepals 4–5 mm long; petals subquadrate, 2 mm long; lip triangular, subcordate with a papillose-ciliate surface. Toamasina. Humid, mossy, evergreen forest; 900–1100 m. Fl. February.

32.172. B. vestitum Bosser *(p. 253, bottom right)*

Pseudobulbs cylindrical-conical, 2.5–4.5 cm long; leaves linear, 15–20 × 0.7–1 cm; inflorescence 20–25 cm tall, 15–25-flowered; flowers white flushed red; sepals 5–6 mm long; lip fleshy, arcuate, 3 × 2.2 mm; close to *B. sandrangatense* with similar fibres at the base of the bulb, but different in floral morphology, especially in the lip; also similar to *B. longivaginans* (p. 254) but differs in the plant and shape of the petals and lip. Antsiranana, Toamasina. Humid, mossy, evergreen forest; 500–1100 m. Fl. June–September.

32.172a. B. vestitum var. meridionale Bosser

Petals ovate-lanceolate, acute at tip, much wider and larger, c. 3 × 2 mm; inflorescence laxly few-flowered. Toliara. Evergreen forest. Fl. unknown.

XV. Section Lepiophylax

(including Section Micromonanthe sensu Perrier)

Plants very small; pseudobulbs uni- or bifoliate; peduncle setiform, as long as or much longer than the leaves; stelidia very short; free part of column-foot elongate.

Key to groups of species in Section Lepiophylax

1. Tiny plants; pseudobulbs 1-leafed; inflorescence 1–3-flowered **31. 173–31.178**
1. Small plants; pseudobulbs 2-leaved; inflorescence up to 14-flowered **31.179–31.180**
1. Larger plants; inflorescence 15–30-flowered . . . . **31.181. B. sciaphile**

32.173. B. amoenum Bosser *(p. 257, top left)*

Pseudobulbs ovoid-subglobose, 6–8 mm across, yellow, flushed with red; leaf elliptic, 10–21 × 6–9 mm; inflorescence erect, slender, 4.7–5.4 cm tall, 6–9-flowered; flowers pale rose with red veins and a pale yellow lip; sepals 10–11 mm long; petals oblong, 2.5 mm long; lip obovate, 2 mm long, glabrous. Antananarivo, Toamasina. Humid, mossy, evergreen forest on plateau; 1200–1400 m. Fl. September–December.

32.174. B. bryophilum Hermans　　　　　　　*(opposite, top right)*
(syn. *Bulbophyllum muscicola* Schltr. (1913), non Rchb.f (1872))
Small epiphyte, carpeting its substrate, 1-flowered on a 12 mm peduncle; lip 2 × 1 mm, fleshy, oblong-ligulate, a little narrowed in the lower half, obtuse, blackish-violet. Antsiranana. Seasonally dry, deciduous woodland. Fl. March.

32.175. B. calyptropus Schltr.　　　　　　*(opposite, bottom left)*
Epiphyte; pseudobulbs ovoid, 2.5–4 mm across, 5–8 mm apart on rhizome; leaf oblong, 8–12 × 3.5–7 mm; inflorescence 1-flowered; flower glabrous, almost transparent, finely streaked with red and with a purple lip; sepals 7 mm long, obtuse; lip suborbicular from a narrow base, 5 mm long. Antsiranana. Lichen-rich forest; c. 2000 m. Fl. January.

32.176. B. conchidioides Ridl.　　　　　*(opposite, bottom right)*
(syn. *Bulbophyllum pleurothalloides* Schltr.)
Epiphyte or lithophyte, up to 2.5 cm tall; pseudobulbs ovoid or subglobose, 3 mm across; leaf oblong or elliptic, 5–10 × 2.5–5 mm; inflorescence 1–5-flowered; flower pale with a pinkish-red lip, small on a setiform peduncle; sepals 5–6 mm long; lip 3 mm long, ovate. Antananarivo, Toamasina. Humid evergreen forest on plateau; in moss- and lichen-rich forest; 900–1400 m. Fl. October–February.

32.177. B. johannis H.Wendl. & Kraenzl.
Pseudobulbs irregularly globose, 2–3 mm across; leaf oblong, acute, 3 × 2 mm; inflorescence 1-flowered; flower white; sepals 3 mm long; lip ovate with a linear claw. Madagascar. Fl. unknown.

32.178. B. moldekeanum A.D.Hawkes
(syn. *Bulbophyllum microglossum* H.Perrier)
Plant small, 5–7-flowered; pseudobulbs sub-confluent, flattened; lip very small, ovate-lanceolate, ciliate fimbriate on the lower side, 4–6 × 1.5 mm. Antsiranana. Lichen-rich forest; 1500–1700 m. Fl. November–December.

32.173

32.174

32.175

32.176

32.179. B. jumelleanum Schltr.

Pseudobulbs 8–12 mm apart on rhizome, globose, 4–6 mm across, 2-leaved; leaves oblong, 5–11 × 3–6.5 mm; inflorescence 7 cm tall, 5–8-flowered; flowers small, white, the lip greenish or yellow; sepals 2 mm long; lip 2 mm, pandurate, furrowed, obtuse and shortly indented at the apex. Antsiranana. Humid, evergreen forest on plateau, on tree trunks; c. 1200 m. Fl. October.

32.180. B. pervillei Rolfe   *(opposite, top left)*

Pseudobulbs flattened, suborbicular; leaves elliptic, 18–30 × 10–15 mm; peduncle subfiliform, 10–15 cm long; flowers small; sepals 5 mm long; lip long and narrow, 4 × 0.6 mm, papillose and with 2 ridges on the upper surface. Antsiranana, Fianarantsoa, Toamasina, Toliara. Coastal forest; sea level–1000 m. Fl. October–July.

32.181. B. sciaphile Bosser   *(opposite, top right)*

Pseudobulbs clustered, subglobose, 6–8 mm across; leaves 2, spreading, elliptic to suborbicular, 6–7 × 4.5–5.5 mm; inflorescence 4.5–6 cm long, with 15–30 flowers; flowers 8–9 mm, yellow, furfuraceous on the exterior; sepals 6–6.5 mm long; petals obovate, 4 mm long; lip oblong, 4 mm long. Antananarivo. Humid, mossy, evergreen forest on plateau; 1300–1400 m. Fl. August.

XVI. Section Lemuraea

Small plants; pseudobulbs 2-leaved, usually well-spaced on rhizome; peduncle often rigid, fleshy; rachis usually short and at angle, 5–15-flowered; sepals free; lip not ciliate; stelids short.

Key to groups of species in Section Lemuraea

1. Peduncle shorter than the rachis .... 32.182. B. andrangense
1. Peduncle longer than the rachis 32.183–32.187

32.182. B. andrangense H.Perrier

Pseudobulbs conical-cylindrical, 18–25 × 5–6 mm, 2–3 cm apart on rhizome; leaves narrowly oblanceolate, 7–9.5 × 0.7–1.1 cm; inflorescence shorter than the leaves, 4–5-flowered; sepals 8–10 mm long; petals oblong, 2 mm long; lip very recurved, 6 × 1.5 mm. Antananarivo, Toamasina. Humid evergreen forest; 1200–1400 m. Fl. October.

32.180

32.181

32.183

32.184

32.185

32.183. B. brachystachyum Schltr.   *(opposite and p. 259, middle left)*
(syn. *Bulbophyllum pseudonutans* H.Perrier)

Plant small, 5–6 cm high, similar in shape to *B. nutans*; pseudobulbs oblong, 12–16 × 5–6 mm; pseudobulbs oblong, 12–18 × 5–7 mm, 2–3 cm apart on rhizome; leaves 2, oblong, 20–35 × 5–10 mm; inflorescence subcapitate, with 3–10 flowers; flowers 7 mm long, greenish yellow; sepals 5–6 mm long; lip flat, spathulate at the base, with a rather protruding triangular callus. Antsiranana, Manongarivo. Dry forest; humid, evergreen forest on plateau on moss- and lichen-covered trees; 1200–1600 m. Fl. January, April.

32.184. B. liparidioides Schltr.   *(p. 259, bottom left)*

Resembles some species of *Liparis* in habit; pseudobulbs conical, 10–15 × 5–7 mm; leaves 2, narrowly oblanceolate, 2.6–6 × 0.5–0.9 cm; inflorescence slightly longer than the leaves, 6–10-flowered; flowers 5–6 mm long; sepals 2–4 mm long; lip recurved, 1.5–2 mm long, obtuse and glabrous. Toamasina. Humid, evergreen forest; 800–1200 m. Fl. February.

32.185. B. nutans (Thouars) Thouars   *(p. 259, bottom right)*
(syn. *Bulbophyllum andringitranum* Schltr.; *B. tsinjoarivense* H.Perrier; *B. chrysobulbum* H.Perrier)

Pseudobulbs ovoid or oblong, 10–20 × 6–10 mm, yellow, 1–3 cm apart on rhizome; leaves ovate or oblong, 14–30 × 3–12 mm; inflorescence longer than the leaves, 5–14-flowered; flowers hyaline, whitish yellow; sepals 3–7 mm long; petals 2.5–4 mm long; lip recurved, 2.5–3.6 × 2–3 mm, motile, sides undulate-plicate. Antananarivo, Fianarantsoa; also in the Comoros and Mascarenes. On shaded rocks in humid forest; 800–2200 m. Fl. December–June.

32.185a. B. nutans var. variifolium (Schltr.) Bosser
(syn. *Bulbophyllum variifolium* Schltr.; *B. ambohitrense* H.Perrier)

Distinct from the typical variety by the more slender inflorescence, longer peduncle and leaves that are contracted at the base into a longer petiole; the flowers are similar but a little larger, whereas the shape of the stelids is slightly different. Antsiranana. Mossy and lichen-rich, evergreen forest; 1500–2000 m. Fl. January.

32.186. B. ormerodianum Hermans
(syn. *Bulbophyllum abbreviatum* Schltr. non Rchb.f.)

Plant small, similar to *B. nutans*; inflorescence as *B. brachystachyum*; flowers c. 5 mm long, yellowish; lip 3-lobed, curved, the margins folded downwards. Toamasina. Humid, evergreen forest; 700–1000 m. Fl. February.

32.183

32.187. B. trilineatum H.Perrier

Dwarf epiphyte, less than 5 cm tall; pseudobulbs narrowly conical, 4–6 mm tall; leaves 2, elliptic, 8–16 × 3–4 mm; inflorescence 2–4 cm long, filiform, 1–3-flowered; flowers pinkish-white; sepals 4.5 mm long; lip oblong-elliptic; stelids very small and obtuse. Toliara. Humid, evergreen forest on plateau; c. 1500 m. Fl. November.

XVII. Section Bifalcula

Plants small; pseudobulbs 2-leaved; inflorescence slender, many-flowered; rachis often sinuous; sepals free; lip glabrous, obovate-spatulate, with a basal tooth; stelids long, bidentate at tip.

Key to species of Section Bifalcula

1. Inflorescence axis zigzag, 14–15-flowered; flowers dark red .... **32.190. B. minutum**
1. Inflorescence axis not zigzag; flowers yellow or yellow-green 2
2. Pseudobulbs ovoid, 2.5–3 mm in diameter; flowers 2–3, yellowish green ....
.... **32.188. B. capuronii**
2. Pseudobulbs discoid, 8–10 mm in diameter; flowers yellow with red veins ....
.... **32.189. B. complanatum**

32.188. B. capuronii Bosser

Pseudobulbs 3–4 × 2.5–3 mm; leaves 2, oblong, 2–2.7 × 0.4–0.6 cm; inflorescence 1.5 cm long, 2–3-flowered; flowers greenish-yellow; sepals 5–6 mm long; lip linguiform, 4 × 1 mm; several characteristics distinguish this species from *B. minutum*, especially the short, few-flowered inflorescence and the very different shape of the lip. Fianarantsoa. Coastal forest; up to 100 m. Fl. unknown.

32.189. B. complanatum H.Perrier
(syn. *Bulbophyllum sigilliforme* H.Perrier)

Pseudobulbs 1–2 cm apart, discoid, 8–10 mm across; leaves 2, broadly ovate, 8–12 × 5–6 mm; inflorescence 12–15 mm long, with 5–7 flowers; flowers 8 mm, hyaline, yellowish, red-veined; sepals 4 mm long; petals oblong, 2.5 mm long; lip suborbicular, 2.5–3 × 1.5–2 mm long. Antsiranana, Toamasina. Seasonally dry, deciduous and humid, evergreen forest and woodland; sea level–1600 m. Fl. February–April, August–September.

32.190. B. minutum Thouars   (opposite)
(syn. *Bulbophyllum implexum* Jum. & H.Perrier)

Pseudobulbs ovoid or subglobose, 5–8 × 5–6 mm; leaves 2, broadly elliptic, 10 × 8 mm; inflorescence 9–12 cm long; rachis zigzag shaped, with 14–15 small dark red flowers; stelids subulate. Antananarivo, Antsiranana. Semi-deciduous forest, on *Chlaenaceae*, in seasonally dry deciduous woodland; sea level–800 m. Fl. September–April.

32.190

33. ACAMPE

A genus of some 10 species, predominantly Asian, with a single species in tropical and South Africa and Madagascar. Epiphytes with stout leafy stems and boot-lace-like branching aerial roots. Leaves distichous, very leathery, V-shaped in cross-section. Inflorescences axillary, with clustered flowers in capitate heads. Flowers very fleshy, somewhat cup-shaped. Sepals and petals free, subsimilar, obovate. Lip very fleshy. Column short, fleshy; pollinia 2, attached by a stalk to the viscidium.

33.1. A. pachyglossa Rchb.f.   (*opposite*)
(syn. *Acampe renschiana* Rchb.f.; *A. madagascariensis* Kraenzl.; *A. pachyglossa* subsp. *renschiana* (Rchb.f.) Senghas)

Sepals and petals greenish cream, with transverse crimson or reddish brown stripes and spots; lip white, spotted with red. Antsiranana, Fianarantsoa, Mahajanga; also in the Comoros, Seychelles, E and S Africa. Deciduous, seasonally dry forest, woodland and scrub, on tamarind trees in tropical, western forests, on mangroves, coastal cliffs on rock, secondary evergreen forest, limestone formations; on tree trunks, large branches or rocks; up to 1000 m. Fl. throughout the year.

LOCAL NAME *Kisatrasatra*.

34. MICROCOELIA

A genus of about 30 species in tropical and South Africa, Madagascar, the Comores and Mascarene Islands. 11 species in Madagascar. Small, twig or branch epiphytes. Stem short to elongate. Roots clustered or scattered along stem, photosynthetic. Leaves scale-like, not photosynthetic. Inflorescences racemose, short, few–many-flowered. Flowers small, semi-translucent, white, often with green markings on the lip, or orange. Sepals and petals similar, not widely spreading. Lip entire or obscurely lobed, spurred at base.

Key to species of *Microcoelia*

1. Stem elongate; roots scattered along stems 2
1. Stems very short; roots clustered 3
2. Lip 3-lobed; spur incurved . . . . **34.1. M. aphylla**
2. Lip orbicular, emarginated; spur tapering to tip . . . . **34.2. M. cornuta**
3. Flowers white, tinged with green or carmine-rose 4
3. Flowers orange 10
4. Spur of lip globose; sepals 1–1.5 mm long . . . . **34.6. M. exilis**
4. Spur of lip cylindrical to clavate; sepals 2 mm or more long 5
5. Spur appressed to ovary 6
5. Spur perpendicular to ovary 7
6. Flowers white tinged with carmine-rose; lip midlobe triangular, concave; spur straight, calvate, rounded at tip, 5 mm long . . . . **34.5. M. dolichorhiza**
6. Flowers whitish tinged with green; lip midlobe broadly obovate, cup-shaped; spur cylindrical, gradually inflated to an obtuse tip, 4 mm long . . . . **34.4. M. decaryana**
7. Sepals 6.3–7.7 mm long; lip broadly obovate, emarginate, 6.3–9.3 × 5–11.6 mm; spur incurved, triangular in cross-section . . . . **34.7. M. macrantha**

33.1

7. Sepals 5.5 mm or less long; lip not broadly obovate, emarginated, less
    than 5.5 mm long; spur perpendicular, not incurved 8
8. Lip midlobe tongue-shaped; spur cylindrical with an inflated tip, 7.5–9.5
    mm long . . . . **34.9. M. physophora**
8. Lip midlobe ovate to obovate; spur 6.3 mm or less long 9
9. Spur 1.3–2 mm long; apically inflated and acute at tip; sepals 3.9–5.2 mm
    long . . . . **34.3. M. bispiculata**
9. Spur 5.4–6.3 mm long, cylindrical, inflated at tip; sepals 1.7–2.8 mm long . . **34.8. M. perrieri**
10. Spur geniculate at base, appressed to ovary; sepals 2.6–3.6 mm long . . **34.11. M. gilpinae**
10. Spur not geniculate, perpendicular to ovary, 10 mm long; sepals 2–2.5 mm
    long . . . . **34.10. M. elliotii**

34.1. M. aphylla (Thouars) Summerh.   *(opposite, top)*
(syn. *Angraecum aphyllum* Thouars; *Solenangis aphylla* (Thouars) Summerh.)
Flowers white marked with pink; lip obscurely 3-lobed; spur c. 5 mm long,
incurved. Antsiranana, Toamasina, Toliara; also in Comoros, Mascarenes,
Kenya, Mozambique. In humid forest; sea level–30 m. Fl. July–December.

34.2. M. cornuta (Ridl.) Carlsward   *(opposite, bottom left)*
(syn. *Gussonea cornuta* Ridl.; *Solenangis cornuta* (Ridl.) Summerh.)
Differs from *M. aphylla* in its fewer-flowered inflorescences, its more or less
orbicular lip, retuse at the apex, its spur tapering towards the apex, and its
more acute rostellum lobes. Antsiranana, Mahajanga; also in Comoros. In
dry, semi-deciduous forest on calcareous rocks and sand; sea level–300 m.
Fl. November–December.

34.3. M. bispiculata L.Jonsson
Inflorescences up to 10 cm long, sparsely up to 15-flowered; lip obscurely 3-
lobed at base, with mouth of the spur thickened, the midlobe obovate,
3.3–5.6 mm long; spur 4.6–6 mm long, perpendicular or apically inflexed and
inflated. Antsiranana, Toamasina. In coastal, humid, evergreen forest; sea
level–100 m. Fl. February–April.

34.4. M. decaryana L.Jonsson
Inflorescences up to 5 cm long, up to 23-flowered; lip large, obovate, concave,
with a distinct ridge-like thickening at each side of the spur opening, c. 3 mm
long; spur cylindrical-clavate, c. 4 mm long, parallel to ovary. Mahajanga. In
seasonally dry, deciduous forest, woodland, and scrub, on *Dalbergia*,
Commiphora and *Hildegardia*; c. 300 m. Fl. October–December.

34.1

34.2

34.6

34.5. **M. dolichorhiza** (Schltr.) Summerh.

Inflorescences up to 1.5 cm long, 10–12-flowered; lip triangular with a plate-like, transverse lamella on each side at the mouth of the spur; spur straight, 5 mm long, parallel to ovary. Antsiranana, Mahajanga. In dense rain-forest; c. 1000 m. Fl. October–December.

34.6. **M. exilis** Lindl.   (*p. 267, bottom right*)
(syn. *Gussonea exilis* (Lindl.) Ridl.)

Inflorescences up to 25 cm long, up to 50-flowered or more; flowers very small; sepals and petals 1–1.5 mm long; spur almost globular, up to 1 mm across. Toliara; also in tropical and South Africa. In coastal forest; seasonally dry, deciduous woodland and scrub, on small branches of understorey trees and bushes; sea level–2000 m. Fl. January–December.

34.7. **M. macrantha** (H.Perrier) Summerh.   (*opposite, top*)
(syn. *Gussonea macrantha* H.Perrier)

Inflorescences up to 7 cm long, up to 13-flowered but usually fewer; flowers large, c. 13 mm long; lip flabellate-obovate, 6.5–9.5 cm long, notched at apex; spur slightly incurved, 5.6–8.7 mm long. Antsiranana, Fianarantsoa, Toamasina, Toliara. In evergreen forest, mostly on smaller branches and twigs, often near water courses; sea level–1000 m. Fl. January–April.

34.8. **M. perrieri** (Finet) Summerh.   (*opposite, bottom left*)
(syn. *Rhaphidorhynchus perrieri* Finet)

Inflorescences many, up to 9.5 cm long, up to 26-flowered; characterised by its stout stem, long slender rachis and hair-like pedicels, its almost cylindrical, 5–6 mm long spur with a swollen apex, and 3-lobed lip with an orbicular midlobe. Antsiranana, Mahajanga, Toliara. In seasonally dry, deciduous woods, and scrub and open woodland, on *Didiereaceae*; up to 500 m. Fl. April–October.

34.9. **M. physophora** (Rchb.f.) Summerh.
(syn. *Angraecum physophorum* Rchb.f.)

Inflorescences 1–1.5 cm long, usually up to 6-flowered; distinguished by its flattened roots, its 1.5–3 mm long sepals and petals, its 7.5–9.5 mm long spur which is thickened at the apex and its narrow, ligulate, erect and concave lip, up to 2.7 mm long. Antsiranana, Mahajanga; also in E Africa. In seasonally dry, deciduous woods and open woodland; up to 500 m. Fl. September–December.

34.7

34.8

34.11

34.10. **M. elliotii** (Finet) Summerh.

Inflorescences up to 6 cm long, up to 15-flowered; flowers orange, like *M. gilpinae* but distinguished by the more or less folded, 3–3.5 mm long, broadly elliptical lip with a row of tubercles at the spur opening, and by the long, slightly recurved, 8–10 mm long spur. Antsiranana, Mahajanga, Toliara. In dense rainforest; up to 1300 m. Fl. January–April.

34.11. **M. gilpinae** (Rchb.f. & S.Moore) Summerh.   *(p. 269, bottom right)*
(syn. *Angraecum gilpinae* Rchb.f. & S.Moore)

Flowers bright orange; distinguished from *M. elliotii* by the sigmoid, 6.3–8.8 mm long spur, parallel to the ovary, and the slipper-shaped, 2.5–3.5 mm long lip. Antananarivo, Antsiranana, Fianarantsoa, Mahajanga, Toamasina, Toliara. In dense humid evergreen forest and in secondary and disturbed vegetation on various understorey shrubs and trees, seems to prefer shaded habitats; up to 1800 m. Fl. throughout the year.

35. BECLARDIA

A genus of two species endemic to Madagascar and the Mascarenes. Epiphytes, normally with a well-developed stem, leaves distichous, ligulate. Lip 3–4-lobed, the terminal lobe more or less indented at the tip, spur wide and funnel-shaped at the base.

Key to species of *Beclardia*

1. Leaves 27–30 × 3–4 cm; sepals 18–24 mm long; lip 25–32 mm long; spur
   15 mm long . . . . **35.1. B. grandiflora**
1. Leaves 8–14 × 1.2–2 cm; sepals 7–17 mm long; lip 12–20 mm long; spur
   7–12 mm long . . . . **35.2. B. macrostachya**

35.1. **B. grandiflora** Bosser

Stem 27–30 cm tall; leaves loriform, 27–30 × 3–4 cm; inflorescence up to 40 cm long, 10–15-flowered. Flowers white with a green mark in throat of lip; sepals 18–24 mm long; petals spatulate, 25–27 mm long; lip 4-lobed, 25–32 mm long; side lobes rounded, 12–13 mm across; midlobes oblong-rounded, spreading, 8–10 × 12 mm; spur 15 mm long, funnel-shaped in mouth, tapering abruptly to a subcalvate tip. More robust than *B. macrostachya* (p. 272), with longer and wider leaves and larger flowers; lip 4-lobed, with the terminal lobes larger than the laterals and with the margins undulate, not crispate. Toamasina, Toliara. In shady forest; 900–1000 m. Fl. December–March.

35.2

36.1

36.2

35.2. **B. macrostachya** (Thouars) A.Rich. *(p. 271, top)*
(syn. *Beclardia brachystachya* (Thouars) A.Rich.)

Plants medium-sized with flabellate leaves; flowers white, often with a green-tipped spur and a hairy yellow disk on the lip; lip 1.2–2 × 1–1.8 cm; lateral lobes rounded; terminal lobe sub-entire, bilobed at the tip, the edges undulate-crispate; spur 7–12 mm long, the base funnel-shaped, tip suddenly contracted in a cylinder. Antananarivo, Antsiranana, Fianarantsoa, Toamasina, Toliara; also in the Mascarenes. In coastal, lowland and montane humid, evergreen forest; seasonally dry, deciduous forest and woodland; 400–2000 m. Fl. December–June.

36. AERANGIS

A genus of about 50 species in tropical and South Africa, Madagascar and the Comoros, one species in Sri Lanka. 21 species in Madagascar. Epiphytes or rarely lithophytes with a short, or rarely elongate, stem and a fan of obovate or elliptic, deeply bilobed leaves. Inflorescences pendent, arcuate or spreading, short to long, simple, few–many-flowered. Flowers stellate, usually white, rarely pink or flushed with pale brown on sepals and spur. Sepals and petals spreading, usually reflexed. Lip entire, usually with a long cylindrical spur at base; sometimes with short ridges either side of mouth of spur. Column short, fleshy; rostellum elongate; pollinia 2, waxy, attached by a slender linear stalk to a small viscidium.

Key to species of *Aerangis*

1. Plants small; stems very short; leaves less than 10 cm long; inflorescences with many small flowers or a few large ones 2
1. Plants of medium to large size; stems short or elongated; leaves usually 10 cm or more in length; inflorescences usually with many flowers of medium to large size 11
2. Spur less than 3 cm long 3
2. Spur more than 5 cm long 6
3. Flowers white or creamy white 4
3. Flowers green, bronze or buff 5
4. Flowers opening widely to lie flat; lip and petals obovate-flabellate .... **36.1. A. citrata**
4. Flowers not opening widely; lip and petals ovate-lanceolate .... **36.2. A. hyaloides**
5. Flowers green, spur cylindrical, 21–23 mm long .... **36.3. A. pallidiflora**
5. Flowers buff to bronze; spur clavate, 18 mm long .... **36.4. A. seegeri**
6. Flowers pale green to pale pinkish; spur clavate 5–6 cm long .... **36.7. A. macrocentra**
6. Flowers white, sometimes with pale yellowish, buff, pink or brown sepals and petals; spur cylindrical, 6 cm or more long 7
7. Spur in a spiral .... **36.8. A. monantha**
7. Spur not spiralled 8
8. Roots flattened, verrucose; leaves finely black-spotted .... **36.10. A. punctata**
8. Roots not flattened and verrucose; leaves not spotted 9
9. Flowers 1–5; spur 7 cm or more long 10
9. Flowers 12–20; spur 6–7.2 cm long .... **36.9. A. pulchella**

36.15

36.1. **A. citrata** (Thouars) Schltr.   (*p. 271, bottom left*)
(syn. *Angraecum citratum* Thouars)

Stem up to 6 cm long; leaves 2–8, dark glossy green, 4–16 × 1–4 cm; inflorescence up to 30 cm long; flowers 15–60, white to cream-coloured; sepals oblanceolate, 5–10 mm long; petals and lip obovate; 8–12 × 6–10 mm; spur subclavate, 25–30 mm long. Throughout E Madagascar. Humid, evergreen forest from coast to plateau, usually on twigs or small trees; up to 1500 m. Fl. August–May.
LOCAL NAME *Manta*.

36.2. **A. hyaloides** (Rchb.f.) Schltr.   (*p. 271, bottom right*)
(syn. *Aerangis pumilio* Schltr.)

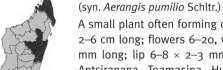

A small plant often forming clumps; leaves 2–6 × 0.7–2.2 cm; inflorescence 2–6 cm long; flowers 6–20, white, hyaline, do not open widely; sepals 5–7 mm long; lip 6–8 × 2–3 mm; spur clavate, 5–12 mm long. Antananarivo, Antsiranana, Toamasina. Humid, evergreen forest; forest, mainly in the shade on moss- and lichen-rich-covered small trees. Often on twigs and small branches; up to 1100 m. Fl. September–January.

36.3

36.4

36.5

36.6

36.3. A. pallidiflora H.Perrier *(p. 275, top left)*

(syn. *Angraecum ramulicolum* H.Perrier)

Stem up to 3 cm long; leaves 2–6, elliptic to broadly elliptic, 3–7 × 1–2 cm; inflorescences, pendent, up to 25 cm long, old ones persisting; flowers 7–9, yellowish green; sepals 7–8 mm long; lip lanceolate, 8–9 × 1–2 mm; spur 2.1–1.3 cm long. Similar to *A. seegeri* but the shape of the tepals and the spur are different. Antananarivo, Antsiranana, Fianarantsoa, Toamasina. Coastal and plateau, humid, mossy, evergreen forest; on branches and twigs; up to 1500 m. Fl. December–March, August.

36.4. A. seegeri Senghas *(p. 275, top right)*

Stem up to 2 cm long; 2–4 leaves, 3–4 × 2–2.5 cm; inflorescence up to 25 cm long; flowers 6–12, pale green to yellow-green or buff; sepals 8 mm long; lip lanceolate, 8–9 mm long; spur, clavate, acute, 18 mm long. Similar to *A. pallidiflora* but flowers more star-shaped and the lower part of the spur thickened and flattened. Antananarivo, Toamasina. Coastal forest and evergreen forest on plateau; up to 1500 m. Fl. June.

36.5. A. fastuosa (Rchb.f.) Schltr. *(p. 275, bottom left)*

(syn. *Angraecum fastuosum* Rchb.f.)

Plant small; leaves rounded, succulent, up to 7.5 cm long; flowers 1–3, large, white, with, a 7–8 cm long spur and a 5–13 mm long rostellum. Antananarivo, Fianarantsoa, Toamasina, Toliara. Humid, evergreen forest; on twigs and small branches in sclerophyllous forest; 900–1500 m. Fl. September–October.

36.6. A. fuscata (Rchb.f.) Schltr. *(p. 275, bottom right)*

(syn. *Aerangis umbonata* (Finet) Schltr.)

Small plants; stem 0.5–6 cm long; leaves 1.5–9 × 1–3 cm; inflorescence 1–5-flowered; flowers yellow-green to pink with a white lip; sepals 18–32 mm long; lip lanceolate, 20–40 × 7–12 mm; spur 9–13 cm long, broad at mouth; rostellum short. Antananarivo, Antsiranana, Toamasina, Toliara. On twigs and branches in a range of habitats, on shrubs and small trees; up to 1500 m. Fl. November.

36.7. A. macrocentra (Schltr.) Schltr. *(opposite, top left)*

(syn. *Aerangis clavigera* H.Perrier)

Stem up to 6 cm long; 4–6 leaves, 7–10 × 2–5 cm, greyish or dark green; inflorescence 8–40 cm long; flowers 12–30, white, tinged with green or pinkish brown; perianth parts 5–7 mm long, blunt, reflexed; lip 8–10 × 3–4 mm; spur 5–6 cm long, inflated towards the tip. Antananarivo, Antsiranana, Fianarantsoa, Toamasina. Humid, evergreen forest on plateau; lichen-rich and mossy forest; up to 2200 m. Fl. March–June.

36.7

36.8

36.10

36.11

36.8. A. monantha Schltr.   *(p. 277, top right)*

Small plants with 1–2-flowered racemes; 3–7 leaves, 2.5–6 × 1.5 cm, dull green with a red margin; flowers relatively large, with pale brown sepals and petals and a white lip; sepals 2–2.4 cm long; lip obovate or ovate, acuminate, 2–2.4 × 1 cm; spur 13 cm long, in a spiral, funnel-shaped at the base; rostellum very long, Antsiranana, Fianarantsoa, Toamasina. On twigs and branches in a range of habitats, on shrubs and small trees; 800–1200 m. Fl. November.

36.9. A. pulchella (Schltr.) Schltr.

Stem up to 10 cm long; leaves 5, 12–15 × 2.5–5 cm, greyish green; inflorescence 25–30 cm long; flowers 12–20, white; perianth parts 8–10 mm long, obtuse to apiculate; lip elliptic, 8–9 × 4–4.5 mm; spur 6–7.2 cm long. Mahajanga. Seasonally dry woodland and forest; sea level–200 m. Fl. March–June.

36.10. A. punctata J.Stewart   *(p. 277, bottom left)*
(syn. *Aerangis curnowiana* auct. non (Rchb.f.) H.Perrier)

Roots flattened, verrucose; stem 1–2 cm long; 2–4 leaves, 2–3.5 × 0.6–1.2 cm, with tiny pale dots on the surface; inflorescence 1–4-flowered; flowers with pale yellow-buff sepals and petals and white lip; sepals 14–20 mm long; lip ovate or obovate, acute, 16–22 × 7–9 mm; spur 10–13 cm long, broad at the mouth; flowers similar to those of *Angraecum curnowianum* (p. 334). Antananarivo. Humid, evergreen forest on plateau; 1000–1500 m. Fl. February.

36.11. A. articulata (Rchb.f.) Schltr.   *(p. 277, bottom right)*
(syn. *Angraecum articulatum* Rchb.f.; *Aerangis venusta* Schltr.; *A. calligera* (Rchb.f.) Garay)

Stem up to 30 cm long; 2–11 leaves, 15–23 × 3–5 cm; inflorescence pendent, up to 30 cm or more long; flowers up to 25, white, stellate; sepals, petals and lip are similar; sepals 15–30 mm long; lip oblong-lanceolate, acute, 15–30 × 6–9 mm; spur 10–15 cm long; column with a prominently beaked anther-cap. Antananarivo, Antsiranana, Fianarantsoa, Toamasina, Toliara; also in the Comoros. Humid, evergreen forest from coast to plateau, on trunks and major branches of large forest trees; up to 2000 m. Fl. September–February.

36.12

36.14

36.13

36.12. **A. modesta** (Hook.f.) Schltr. *(p. 279, top left)*

(syn. *Angraecum sanderianum* Rchb.f.; *Aerangis crassipes* Schltr.; *A. fastuosa* var. *angustifolia* H.Perrier)

Similar to *A. articulata* (p. 278) but smaller in all respects and flowers all a similar size along the inflorescence; stem up to 20 cm long; leaves up to 14 × 2–4 cm; inflorescence up to 30 cm long; flowers 7–24, white; sepals 15–22 mm long; lip 20–24 × 8–10 mm; spur 4–8 cm long. Antananarivo, Antsiranana, Fianarantsoa, Mahajanga, Toamasina; also in the Comoros. Humid, evergreen forest; often in forest remnants on trunks and small branches of trees; 100–1500 m. Fl. October–January.

36.13. **A. concavipetala** H.Perrier *(p. 279, bottom)*

Stem up to 3 cm long; leaves 8–11 × 2.8–3.5 cm; inflorescence 20–30 cm long; flowers 12–15, white with clear pink venation; sepals similar, 6–7 × 5 mm, less than twice as long as broad; petals 5 mm long, broadly ovate to suborbicular and concave; lip concave, ovate, apiculate, 6–8 × 5 mm; spur 7–9 cm long. Mahajanga. On trunks in humid forest; up to 300 m. Fl. June.

36.14. **A. mooreana** (Rolfe) P.J.Cribb & J.Stewart *(p. 279, top right)*

(syn. *Aerangis ikopana* Schltr.; *A. anjoanensis* H.Perrier; *A. pulchella* (Schltr.) Schltr.; *A. karthalensis* R.Neirinck & M.Herremans)

Stem up to 6 cm long; 2–6 leaves, 7–12 × 2–3.5 cm; inflorescence 15–30 cm long, flowers 10–22, white with a pinkish spur; sepals up to 15 mm long; lip obovate, obtuse, 8–12 × 3–6 mm; spur 7–12 cm long. Recognised by the obovate lip and short, reflexed dorsal sepal which is smaller than the other perianth parts. Flowers often pale pink. Antananarivo, Mahajanga, also in the Comoros. Lowland, humid, evergreen forest; seasonally dry, deciduous forest and woodland on shrubs; sea level–600 m. Fl. July–October.

36.15. **A. ellisii** (B.S.Williams) Schltr. *(p.273 and opposite, top left)*

(syn. *Aerangis buyssonii* God.-Leb.; *A. caulescens* Schltr.; *A. platyphylla* Schltr.; *A. alata* H.Perrier; *A. ellisii* var. *grandiflora* J.Stewart)

Epiphyte or lithophyte; stem up to 60 cm tall; leaves 4–20, thick, succulent, 7–25 × 2–6 cm; inflorescence ascending or spreading, up to 30 cm long; sepals and petals 15–30 mm long, strongly reflexed; lip deflexed, oblanceolate or obovate, 12–22 × 8–11 mm, with two basal callus ridges; spur 13–27 cm long. Plants from forest are different from those found in more exposed rocky habitats but flower shape and size are variable in both. Antananarivo, Antsiranana, Fianarantsoa, Toliara. Humid, evergreen forest; plateau, rocky outcrops of basalt, quartz, granite and gneiss; amongst xerophytic vegetation; 300–1800 m. Fl. October–May.

36.15

36.16

36.18

36.20

36.16. A. spiculata (Finet) Senghas   *(p. 281, top right)*

Plants large; stem up to 10 cm long; 2–7 leaves, 10–17 × 1.5–5.5 cm, greyish or dull green; inflorescence pendent, 30–75 cm long; flowers white tinged with pink on the tips and spur; sepals 20–28 mm long; lip reflexed, oblanceolate to ligulate-oblong, abruptly acuminate, 18–35 × 10–15 mm; spur 16–27 cm long, with minute teeth in the mouth. Antsiranana, also in the Comoros. Humid, lowland, evergreen forest; 80–1000 m. Fl. March.

36.17. A. cryptodon (Rchb.f.) Schltr.

(syn. *Angraecum cryptodon* Rchb.f.; *Rhaphidorhynchus stylosus* Finet; *Aerangis malmquistiana* Schltr.)

Stem up to 8 cm long; leaves up to 15 × 4.5 cm; inflorescence arching to pendent; flowers 8–20, brownish sepals and white petals and lip; sepals 15–19 mm long; lip deflexed, obovate, 12–16 × 5–9 mm; spur 10–14 cm long; often confused with forms of *A. ellisii* (p. 280) but differs in having 2 denticulate wings that stand upright above the anther at the apex of the column and the backward directed teeth at the edge of the curved part of the lip where it joins the spur. Antananarivo, Fianarantsoa, Toamasina. Humid, evergreen forest and rocky outcrops on basalt slopes; 200–1800 m. Fl. April–August.

36.18. A. decaryana H.Perrier   *(p. 281, bottom left)*

Stem up to 15 cm long; leaves succulent, 4–10 × 0.8–2 cm, with undulate margins, greyish-green; flowers white, tinged with pink; sepals 9–20 mm long; lip deflexed, ligulate to narrowly pandurate, 10–22 × 4–7 mm; spur 6–12 cm long, flattened at the tip and pale pink. Antsiranana, Fianarantsoa, Toliara. On *Alluaudia*, *Euphorbia*, tamarind trees and other spiny forest trees; in deciduous, dry, southern forest and scrubland, also in transition forest with evergreen, humid forest. 60–900 m. Fl. December–March.

36.19. A. rostellaris (Rchb.f.) H.Perrier

(syn. *Aerangis avicularia* (Rchb.f.) Schltr.; *A. buchlohii* Senghas)

Stem 2–10 cm long; leaves 2–10, thick, up to 15 × 5 cm; inflorescence 20–30 cm long; flowers 6–15, white; sepals 15–20 mm long, distinctly incurved; lip narrowly pandurate, cuspidate, reflexed, 16–18 × 5–6 mm; spur 9–11 cm long; column 4 mm long, with an elongated rostellum. Antsiranana, Fianarantsoa; also in the Comoros. Humid, evergreen forest on plateau; 1000–1300 m. Fl. September–May.

36.20. A. stylosa (Rolfe) Schltr.   *(p. 281, bottom right)*

(syn. *Angraecum fournierae* André)

Stem up to 5 cm long; leaves 2–6, 9–18 × 2.5–6 cm, dull bluish or greyish green; inflorescence 30–60 cm long; flowers 9–20, white; sepals 18–24 mm long; lip reflexed, obovate, longly acuminate, 18–22 × 6–8 mm; spur 8–15 cm long. Similar to *A. rostellaris* but column 6–8 mm long. Antananarivo, Antsiranana, Fianarantsoa, Toamasina; also in the Comoros. Humid, evergreen forest from coast to plateau and in mossy forest; up to 1400 m. Fl. September–May.

37. ANGRAECOPSIS

A small genus of about 15 species; mostly in tropical Africa, one in Madagascar. Small epiphytes with short leafy stems. Leaves distichous, ligulate to oblanceolate, often twisted at the base to lie in one plane. Inflorescences axillary, usually distantly several-flowered. Flowers semi-translucent, white, yellow or pale green. Dorsal sepal and petals subsimilar; lateral sepals very obliquely triangular. Lip 3-lobed, rarely entire. Column short with a 3-lobed rostellum; pollinia 2, each attached by a separate stipe to a viscidium.

37.1. A. parviflora (Thouars) Schltr. *(below left)*

Petals obliquely ovate, strongly expanded at the front; lobes of the lip broad. Antananarivo, Antsiranana, Fianarantsoa, Toamasina; also in the Mascarenes and tropical Africa. Humid, evergreen forest on plateau; 700–1100 m. Fl. March and July.

37.1

38.2

38. MICROTERANGIS

A small genus of about 4 species, endemic to Madagascar and the Mascarenes. Small epiphytes with short stems and a fan of ligulate to obovate-oblong leaves, unequally bilobed at apex, twisted to lie in one plane. Inflorescences axillary, unbranched, many-flowered. Flowers tiny. Sepals and petals spreading, free, 2–3 mm long. Lip entire, spurred at base. Column short, fleshy; rostellum entire; pollinia 2, attached to a single stipe and a small viscidium.

Key to species of *Microterangis*

1. Inflorescence 8–15-flowered; sepals and petals 2.5 mm or more long; lip ovate-lanceolate, acute, 2.5 mm long .... **38.3. M. oligantha**
1. Inflorescence 20- or more-flowered; sepals and petals 1.75 mm or less long; lip suborbicular or rhombic, less than 2 mm long 2
2. Inflorescence laxly several-flowered; sepals and petals 1 mm long; lip subrhombic .... **38.1. M. coursiana**
2. Inflorescence densely many-flowered; sepals and petals 1.75 mm long; lip suborbicular .... **38.2. M. divitiflora**

38.1. M. coursiana (H.Perrier) Senghas
(syn. *Chamaeangis coursiana* H.Perrier)

Differs from *M. hariotiana* in having smaller leaves, shorter inflorescences and a broadly rhombic lip. Toamasina. Humid, lower montane forest; c. 900 m. Fl. September.

38.2. M. divitiflora (Schltr.) Senghas   *(p. 283, bottom right)*
(syn. *Angraecum divitiflorum* Schltr.; *Chamaeangis divitiflora* (Schltr.) Schltr.)

Distinguished by its dense, many-flowered inflorescence, 1.75 mm long sepals and petals, suborbicular lip 1.75 × 2 mm and 1.25 mm long, cylindrical, obtuse spur. Antananarivo, Mahajanga, Toamasina. Humid, highland forest. Fl. July–September.

38.3. M. oligantha (Schltr.) Senghas
(syn. *Angraecum oliganthum* Schltr.; *Chamaeangis oligantha* (Schltr.) Schltr.)

Plant small; inflorescence 8–15-flowered; 2.5 mm long sepals and petals; lip ovate-lanceolate, acute, 2.5 × 1.5 mm; spur obtuse, 1.25 mm long. Antsiranana. Dry, semi-deciduous western forest; up to 200 m. Fl. June.

39. AERANTHES

A genus of 40 species in Madagascar, the Mascarenes, Comoros and tropical E Africa; 37 species in Madagascar. Epiphytes with short to long, leafy stems. Leaves leathery, linear. Inflorescences wiry, short to long, usually bearing flowers in succession, rarely simultaneously. Flowers somewhat translucent, white, pale yellow or green. Sepals often with slender tails. Petals smaller than sepals. Lip attached to the column-foot, entire, often with a tail. Column short with the foot inflated, usually spurred; rostellum trifid with a central short tooth; pollinia 2, each with its own viscidium.

Key to species of *Aeranthes*

1. Spur absent or very obscure . . . . **39.1. A. ecalcarata**
1. Spur or spurs present 2
2. Column-foot with 3 spurs . . . . **39.2. A. tricalcarata**
2. Column-foot with a single spur 3
3. Inflorescence many-flowered; flowers opening simultaneously . . . . **39.3. A. polyanthemus**
3. Inflorescences 1–4-flowered, or more flowered but flowers opening successively 4
4. Spur noticeably dilated or bifid near apex 5
4. Spur cylindrical or tapering to tip 18
5. Sepals 5 cm or more long . . . . **39.4. A. grandiflora**
5. Sepals 2–5 cm or less long 6
6. Flowers white . . . . **39.5. A. albidiflora**
6. Flowers pale to dark green 7
7. Leaves V-shaped in cross-section; inflorescence erect . . . . **39.8. A. strangulata**
7. Leaves flat in cross-section; inflorescence ascending, spreading or pendent 8
8. Inflorescence much shorter than the leaves . . . . **39.10. A. leandriana**
8. Inflorescence as long as or longer than the leaves 9
9. Lip margins denticulate 10
9. Lip margins smooth 11
10. Leaves 16–22 × 3.5–4 cm; inflorescences 16–20 cm long; sepals 22–26 mm
    long; lip ovate, acute, 22–28 × 18–20 mm; spur 8–10 mm long . . . . **39.7. A. denticulata**
10. Leaves 2.5–5.5 × 1–1.8 mm; inflorescences 5–10 cm long; sepals 17–19 mm
    long; lip subrhombic, acute, 10–12 × 6–7 mm; spur 6–6.5 mm long . . . . . . **39.17. A. tenella**
11. Sepals 10–12 mm long; leaves 10–12 × 0.5–1 cm . . . . **39.12. A. parvula**
11. Sepals 18 mm or more long; leaves 15 cm or more long 12
12. Leaves linear, acute, less than 1.4 cm broad 13
12. Leaves oblong, ligulate or rarely linear, more than 1.8 cm broad 14
13. Plant pendent; sepals 30–38 mm long; lip oblong-subpanduarte, 22–28 ×
    8–12 mm; spur 8–12 mm long . . . . **39.13. A. peyrotii**
13. Plant spreading; sepals 24 mm long; lip broadly ovate, 19 × 10 mm; spur
    23 mm long, globular at tip . . . . **39.16. A. sambiranoensis**
14. Leaves 11–12; lip orbicular, apiculate, 15–20 × 16–21 mm . . . . **39.11. A. orthopoda**
14. Leaves 7 or less; lip not orbicular, not apiculate 15
15. Leaves linear; lip ovate, 18 × 12 mm; spur 10 mm long . . . . **39.9. A. filipes**
15. Leaves oblong to ligulate; lip obovate to rhombic, 22 mm or more long 16
16. Leaves 2–3 cm broad; sepals 21–24 mm long; lip broadly obovate; spur
    12–13 mm long . . . . **39.6. A. carnosa**
16. Leaves 2.7–6 cm broad; sepals 22–40 mm long; lip obovate or rhombic;
    spur 14–15 mm long 17

17. Lip obovate, 40 × 20 mm; inflorescence up to 30 cm long, 1–3-flowered . . . .
    . . . . **39.14. A. ramosa**
17. Lip rhombic, 20–30 mm long; inflorescence up to 50 cm long, up to 10-flowered . . . .
    . . . . **39.15. A. robusta**
18. Flowers large with sepals 2.7 cm long or more, acumen more than 1.5 cm long 19
18. Flowers small; sepals 2.5 cm long or less; acumen less than 1.5 cm long 24
19. Spur 12 cm or more long . . . . **39.18. A. schlechteri**
19. Spur 1.5 cm or less long 20
20. Sepals 9 cm or more long . . . . **39.19. A. antennophora**
20. Sepals 7 cm or less long 21
21. Leaves 12 cm or less long 22
21. Leaves 20 cm or more long 23
22. Leaves grey-green, leathery; lip oblong, acuminate 18 × 8 mm; spur 15 mm
    long . . . . **39.22. A. crassifolia**
22. Leaves green, not very leathery; lip ovate, acuminate, 16–17 × 6–6.5 mm;
    spur conical 6–7 mm long . . . . **39.23. A. moratii**
23. Sepals 35–38 mm long; lip ovate, acuminate, 40–45 × 14–16 mm . . . **39.20. A. angustidens**
23. Sepals up to 60 mm long; lip obtrullate, acuminate, 40 × 20 mm . . . . **39.21. A. caudata**
24. Inflorescences erect or suberect-spreading 25
24. Inflorescences pendent 31
25. Inflorescences shorter than the leaves 26
25. Inflorescences as long as or longer than the leaves 28
26. Leaves fleshy, 2–3.4 cm broad; inflorescences 1–4-flowered, up to 10 cm long . . . .
    . . . . **39.28. A. laxiflora**
26. Leaves 0.5–1 cm broad; inflorescences 1-flowered, 2.5–5 cm long 27
27. Leaves 15–20 cm long; sepals 21–24 mm long; lip suborbicular,
    17 × 12 mm . . . .**39.29. A. nidus**
27. Leaves 4.5–7 cm long; sepals 5–6.5 mm long; lip braodly obovate,
    5–6 × 3.5–4.5 mm . . . . **39.24. A. adenopoda**
28. Spur conical 29
28. Spur cylindrical or tapering 30
29. Inflorescence 20–25 mm long; spur 1.5 mm long; sepals 7.5–8 mm long . . . .
    . . . . **39.26. A. ambrensis**
29. Inflorescence 6–15 mm long; spur 2.5–5 mm long; sepals 14–16 mm long . . . .
    . . . . **39.27. A. bathieana**
30. Leaves 3–5 cm long; spur 2–3 mm long . . . . **39.30. A. setiformis**
30. Leaves 6 cm or more long; spur 8 mm or more long . . . . **39.25. A. aemula**
31. Inflorescences 15 cm or more long 32
31. Inflorescences 14 cm or less long 33
32. Leaves 12–20 cm long; sepals 27–33 mm long; lip rhombic, mucronate,
    14–16 × 10–2 mm . . . . **39.31. A. multinodis**
32. Leaves 9–10 cm long; sepals 15–17 mm long; lip subrectangular, acute,
    15 × 14–15 mm . . . . **39.32. A. neoperrieri**
33. Leaves 10–17 × 1.5–1.9 cm; inflorescences 2–2.5 cm long, 1–4-flowered . . . .
    . . . . **39.36. A. tropophila**
33. Leaves less than 8 cm long, 9 cm broad; inflorescences 3 cm or more long 34
34. Sepals 7–10 mm long; lip broader than long . . . . **39.33. A. orophila**
34. Sepals 14 mm or more long; lip longer than broad 35
35. Spur conical; lip broadly rhombic, shortly acuminate, 10 × 6 mm . . . . **39.34. A. setipes**
35. Spur cylindrical; lip subrectangular, apiculate, 5.5 × 4.5–5 mm . . . . . . **39.35. A. subramosa**

39.4

39.1. A. ecalcarata H.Perrier   *(opposite, top left)*

Plant and flowers small; lip tiny, spur reduced to a slight swelling. Antananarivo, Antsiranana, Fianarantsoa, Toamasina, Toliara. Humid, evergreen, mossy forest, on tree trunks and branches; 1200–1400 m. Fl. December–July.

39.2. A. tricalcarata H.Perrier

Epiphyte or lithophyte; lip base and column-foot expanded into rounded lobes; the 2 lateral ones shorter than the middle one; this is almost certainly an abnormal form of a species that has more typical flowers. Toliara. Shaded, humid rocks. c. 1200 m. Fl. December.

39.3. A. polyanthemus Ridl.   *(opposite, bottom)*

Plant small; stem up to 5 cm long; leaves narrowly ligulate, up to 17.5 × 1.2 cm; inflorescence 12–13 cm long, pendent; flowers 7–9, yellow; sepals 12–16 mm long; lip lanceolate, 8 × 5 mm; spur 6–7 mm long, obtuse and incurved. Antananarivo, Fianarantsoa, Toamasina. Mossy, montane forest, on trees and rocks; 1100 m. Fl. October–January.

39.4. A. grandiflora Lindl.   *(p. 287)*
(syn. *Aeranthes brachycentron* Regel)

Stem short; leaves 5–7, narrowly oblong, 15–25 × 3–3.5 cm; inflorescence pendent, 10–60 cm long; flowers large, showy, white, yellowish white or greenish white; sepals 5 cm long, caudate; lip oblong-elliptic, acuminate, 4 × 2 cm; spur 1.5 cm long, abruptly dilated in apical half; distinguished from *A. ramosa* (p. 292) by its flowers with shortly caudate segments and cylindrical spur which is dilated above and bent forward beneath the lip. Antsiranana, Toamasina, Toliara; also in the Comoros. Humid, evergreen forest and littoral and plateau forest, on trunks; sea level–1200 m. Fl. July–January.

39.5. A. albidiflora Toill.-Gen., Ursch & Bosser

Stem short; leaves 6–7, ligulate, 17–20 × 1.4–1.6 cm; inflorescence pendent, up to 30 cm long; flowers white; sepals 26–30 mm long; lip broadly obovate, shortly apiculate, 21–24 × 14–15 mm; spur incurved, 12–13 mm long, strongly globular at the tip; close to *A. sambiranoensis* (p. 293) but differs by its pendent, 1-flowered inflorescence and slender peduncle, broader lip of a different shape, and shorter strongly incurved spur. Antananarivo, Fianarantsoa. Mossy, evergreen forest; 30–1400 m. Fl. November–April. **LOCAL NAME** *Velomihata.*

39.1   39.8

39.3

39.6. A. carnosa Toill.-Gen., Ursch & Bosser

Stem short; leaves 4, ligulate, 15–32 × 2–3 cm; inflorescence pendent, simple or branched, 20–70 cm long; flowers green, fleshy; sepals 21–30 mm long; lip broadly ovate, subapiculate, 23–30 × 13–18 mm, base short and narrow; spur to 7–8 mm long. Toamasina. Humid, mossy forest; up to 300 m. Fl. unknown.

39.7. A. denticulata Toill.-Gen., Ursch & Bosser

Leaves 4–5, oblong-ligulate, 16–22 × 3.5–4 cm; inflorescence pendent, 16–22 cm long, simple or few-branched; flower green with yellow tips; sepals 22–26 mm long; lip ovate, acute, 22–28 × 18–20 mm, margins denticulate, carrying 2 characteristic diverging keels at the base; spur 8–10 mm long. Toamasina. Humid, evergreen, eastern forests; 1000–1500 m. Fl. July–August.

39.8. A. strangulata Frapp. ex Cordem.   *(p. 289, top right)*

(syn. *Aeranthes longipes* Schltr.; *A. rigidula* Schltr.; *A. erectiflora* Senghas)

Stem 2 cm long; leaves 3–8, linear, V-shaped in cross-section, 17–23 × 1.1–2 cm; inflorescence erect or ascending, simple or up to 2-branched, rigid, up to 20 cm tall, peduncle covered by pale sheaths; flowers pale greenish yellow or green; sepals 19–20 mm long, longly acuminate; lip rhombic, shortly apiculate, 15 × 7–10 mm; spur incurved, 9–13 mm long, slightly bilobed at tip. Antananarivo, Fianarantsoa. Humid, evergreen montane forest; 1400–2000 m. Fl. February–March, July, September.

39.9. A. filipes Schltr.   *(opposite, top left)*

Stem short, slender; leaves 6–7, linear-ligulate, 20–25 × 1.8–4 cm; inflorescence 11–20 cm long, pendent, wiry; flowers green; sepals 18–34 mm long; lip broadly ovate, acuminate, 18–24 × 10–15 mm, with 2 calli at base; spur 12–15 mm long. Antsiranana, Toamasina, Toliara. Riverine forest; humid, evergreen forest; 1000–1400 m. Fl. March.

39.10. A. leandriana Bosser

Leaves 5–6, linear, 10–20 × 0.5–0.8 cm, glaucous green; inflorescence pendent, 2.5–10 cm long; flowers pale green, translucent; sepals 25–32 mm long; lip subrectangular, apiculate, 20 × 12–14 mm, papillose above; spur 10–12 mm long. Antananarivo. Humid, evergreen forest on plateau; 1400–1500 m. Fl. November–December.

39.9

39.12

39.13

39.13

39.11. **A. orthopoda** Toill.-Gen., Ursch & Bosser

Leaves 11–12, linear-oblong, leathery, 18–22 × 2.2–2.9 cm; inflorescence pendent, simple or branched, 30–70 cm long; flowers green; sepals 18–22 mm long; lip orbicular, shortly apiculate, 15–20 × 16–21 mm; spur 10–12 mm long. Antananarivo. Humid, evergreen, mossy forest on plateau, on moss- and lichen-covered trees. Fl. unknown.

39.12. **A. parvula** Schltr.   *(p. 291, top right)*

Plant small; leaves 6–7, linear, 10–12 × 0.5–1 cm; inflorescence short, erect, 14 cm long; flowers greenish white with brown thickened tips; sepals 10–12 mm long; lip subrectangular, acuminate, 8 × 5 mm; spur incurved, 5 mm long. Antsiranana. On tamarind trees; up to 200 m. Fl. February–May.

39.13. **A. peyrotii** Bosser   *(p. 291, bottom left and right)*

Plant pendent; sometimes carrying plantlets at the internodes of the spike and sheaths; leaves long and often glaucous; flowers large; lip oblong, sub-pandurate; spur club-shaped at the end. Antananarivo, Toamasina. Humid, evergreen forest; 850–1200 m. Fl. January–March.

39.14. **A. ramosa** Rolfe   *(opposite, left)*

(syn. *Aeranthes vespertilio* Cogn.; *A. brevivaginans* H.Perrier)

Plant large; leaves with a very long, slender, branching raceme; flowers large and variable in colour and shape; lip strongly auriculate at the base, then broadly obovate; spur club-shaped and narrowed at the base. *A. ramosa* is a very variable species both in shape and size. Antananarivo, Mahajanga, Toamasina. Humid, evergreen, mossy forest, on the trunks of *Philippia*; 1000–1500 m. Fl. May–August, October, December–February.

39.14a. **A.** cf. **ramosa**   *(opposite, right)*

39.15. **A. robusta** Senghas

Pendent epiphyte; stem short, 2 cm long; leaves congested, broadly linear, up to 27 × 4 cm, symmetrically bilobed at tip; inflorescences pendent, up to 50 cm long, up to 10-flowered; bracts dark brown; flowers green; dorsal sepal ovate, acuminate, 27 × 16 mm; petals obliquely ovate, acuminate, 30 × 16 mm; lip strongly recurved, rhombic, obtuse, 20–30 mm long; spur incurved, clavate, 15 mm long. Antananarivo, Toamasina. Montane forest; 1300–1600 m. Fl. July, September.

39.16. **A. sambiranoensis** Schltr.

Leaves 3–5, narrowly linear, 23–30 × 0.9–1.4 cm; inflorescence erect, simple or few-branched, up to 30 cm long; flowers greenish; sepals 24 mm long; lip broadly ovate, 3-lobed in front, 19 × 10 mm; spur 23 mm long, globose at tip. Antsiranana. Humid, evergreen forest at intermediate elevation; seasonally dry, deciduous forest and woodland; c. 800 m. Fl. January.

39.17. **A. tenella** Bosser

Leaves 4–6, linear, fairly thick, 3.5–6 × 0.7–0.9 cm; inflorescence spreading or pendent, simple or few-branched, up to 14 cm long; flowers whitich green; sepals up to 14 mm long; lip subrectangular, apiculate, 5.5 × 4.5–5 mm, with 2 ridges at base; spur 6 mm long. Antsiranana, Toamasina. Humid, mossy, evergreen forest. Fl. unknown.

39.14 39.14a

39.18. A. schlechteri Bosser *(opposite)*

(syn. *Neobathiea gracilis* Schltr.; *N. sambiranoensis* Schltr.)

Leaves 4–6, ligulate, 6–12 × 1–1.8 cm; deep glossy green; inflorescence pendent, slender, 2–3 times longer than the leaves; sepals 4.5–6 cm long; lip hastate, very longly acuminate, 3.2–4 × 0.8 cm, with 3 pubescent calli near base; spur filiform, up to 13 cm long. Antsiranana, Mahajanga. Deciduous and semi-deciduous forest, epiphyte on twigs, especially on *Erythroxylon*; 90–1250 m. Fl. February–March.

39.19. A. antennophora H.Perrier

Leaves 6–7, broadly ligulate, 30–40 × 2–3.5 cm; inflorescence slender, pendent, up to 30 cm long, sometimes branched; sheaths at most two-thirds the length of the inter-nodes of the peduncle; flowers greenish white; sepals 9–12 cm long, longly acuminate; lip broadly ovate, acuminate, 3.8–4.5 × 2 cm; spur 13–15 mm long. Antsiranana, Toamasina. Mossy forest; c. 1000 m. Fl. February.

39.20. A. angustidens H.Perrier

Leaves linear, 25–30 × 2–3 cm; inflorescences pendent, very slender, branches 2–3-flowered, can remain on the plant for several years, extend and become more branched as they get older; flowers green; sepals 35–38 mm long; lip ovate, acuminate, 40–45 × 14–16 mm; spur 12 mm long. Antananarivo(?), Fianarantsoa, Toamasina. Humid, evergreen forest; c. 700 m. Fl. January–February.

39.21. A. caudata Rolfe *(p. 297, top left)*

(syn. *Aeranthes imerinensis* H.Perrier)

Stem up to 15 cm long; leaves 6–8, ligulate, 20–30 × 2–4 cm; inflorescences pendent, branching, up to 45 cm long; flowers translucent green; sepals up to 6 cm long, tips filiform; lip broadly obtrullate, acuminate, 40 × 20 mm; spur 10 mm long. Antananarivo, Antsiranana, Toamasina, Fianarantsoa; also in the Comoros. Humid, mossy, evergreen forest; 700–1500 m. Fl. January–April.

39.22. A. crassifolia Schltr. *(p. 297, top right)*

Leaves 5–6, thick and leathery, 7–10 × 1.3–2.3 cm; inflorescences erect, simple or few-branched; flowers greenish; sepals 3 cm long; lip oblong, acuminate, 18 × 8 mm; spur 15 mm long. Antsiranana. On tamarind trees; up to 200 m. Fl. February.

39.18

39.23. **A. moratii** Bosser   *(opposite, bottom left)*

Leaves 4–5, thick, grey-green, linear-oblong, 10–12 × 1.5–2 cm; inflorescences pendent, simple or few-branched near base, 30–35 cm long; flowers 4–5, green with yellow tips; sepals 27–30 mm long, acuminate; lip ovate, acuminate, 16–17 × 6–6.5 mm; spur narrowly conical, dorsi-ventrally compressed, 6–7 mm long. Antsiranana. Humid, lowland forest; up to 200 m. Fl. unknown.

39.24. **A. adenopoda** H.Perrier   *(opposite, bottom right)*

Leaves 7–10, linear-lanceolate, 4.5–7 × 0.5–1 cm; inflorescences erect, half length of leaves; flowers white or greenish white; sepals 5–6.5 mm long; lip broadly obovate, subapiculate, 5–6 × 3.5–4.5 mm; spur 3.5–6.5 mm long. Antananarivo, Toamasina. Humid, montane forest; 1000–1300 m. Fl. unknown.

39.25. **A. aemula** Schltr.   *(p. 299, top left)*

(syn. *Aeranthes biauriculata* H.Perrier)

Leaves 4–6, linear or ligulate, 6–17 × 0.5–1.4 cm; inflorescence erect or spreading, shortly branched, up to 40 cm long; flowers white with yellowish tips; sepals 20–25 mm long; lip broadly ovate-cordate, obtuse, 9–10 × 6–8 mm; spur 8 mm long. Antananarivo, Antsiranana, Toamasina. Lichen-rich and mossy forest; 450–2000 m. Fl. January–February.

39.26. **A. ambrensis** Toill.-Gen., Ursch & Bosser

Leaves 5, thin-textured, ligulate, 10–12 × 1–1.1 cm, unequally bilobed at tip, with slightly protruding dorsal veins; inflorescence simple or few-branched, spreading or pendent, 20–25 cm long, covered on sheaths; flowers 7–9, hyaline; sepals 7.5–8 mm long; lip recurved, broadly ovate, acute, 7 × 4 mm, with a short basal callus; spur conical, 1.5 mm long. Antsiranana (Mte d'Ambre). Humid, evergreen forest; 800–1200 m. Fl. July.

39.27. **A. bathieana** Schltr.

Leaves up to 12, oblong-ligulate or ovate-elliptic, 5–6 × 1.3–1.7 cm; inflorescence erect, rigid, 1–4–flowered, 6–15 cm long; flowers white with yellow tips; sepals 14–16 mm long; lip ovate to subrectangular, obtusely apiculate, 5 × 4 mm, with 2 short callus ridges; spur conical-cylindrical, 2.5–5 mm long. Antananarivo (Ankaratra Massif). Mossy, montane forest; c. 2000 m. Fl. January–February.

39.21

39.22

39.23

39.24

39.28. A. laxiflora Schltr.

Leaves 4–5, ligulate, fleshy, 11–17 × 2–3.4 cm; inflorescences short, hidden among the leaves, 3–4-flowered, up to 10 cm long; flowers white to whitich green; sepals 23–25 mm long; lip broadly rhombic, shortly acuminate, 12–19 × 8 mm; spur 7.5–13 mm long. Antsiranana, Fianarantsoa. Montane forest; 1600–1800 m. Fl. January–February.

39.29. A. nidus Schltr.   *(opposite, bottom)*

(syn. *Aeranthes pseudonidus* H.Perrier)

Plant dense, with grass-like foliage; stems tufted; leaves narrowly linear, 15–20 × 0.8–1 cm; inflorescences 5–10, erect, 1-flowered, 2.5–3 cm long; flowers small, white with yellow tips; sepals 21–24 mm long; lip cordate at the base, suborbicular, 17 × 12 mm, with 2 basal calli; spur 10 mm long, straight. Antananarivo, Antsiranana, Fianarantsoa. Mossy, evergreen forest, encircling tree stems to catch debris, forming big clumps; 1000–2000 m. Fl. January–February, May.

39.30. A. setiformis Garay

(syn. *Aeranthes pusilla* Schltr.)

Plant small; leaves 4–7, linear-ligulate, 3–5 × 0.3–0.7 cm; inflorescences erect or spreading, as long as the leaves or longer; flowers white with a green tinge; sepals 7.5 mm long; lip suborbicular, obtuse, 3.5 × 3 mm; spur 2–3 mm long. Antananarivo, Fianarantsoa. Mossy, montane forest; c. 2000 m. Fl. October.

39.31. A. multinodis Bosser

Leaves 4, linear-oblong, 12–20 × 2.5–4 cm; inflorescences pendent, 15–35 cm long, 1–3-flowered; peduncle with many nodes, and sheaths that are longer than the internodes; flowers pale green, fleshy; sepals 27–33 mm long; lip rhombic, mucronate, 14–16 × 10–12 mm; spur 7–8 mm long. Close to *A. longipes*. Antananarivo. Humid, evergreen, forest on plateau; 1300–1400 m. Fl. unknown.

39.32. A. neoperrieri Toill.-Gen., Ursch & Bosser

Leaves 9–10, linear, 9–10 × 2.5 cm; inflorescence pendent, slender, 3–4-branched, 50–70 cm long; flowers green with a darker spur; sepals 15–17 mm long; lip subrectangular, acute, 15 × 14–15 mm; spur 10 mm long. Toamasina. Mossy, evergreen, montane forest; 800–1500 m. Fl. October.

39.25

39.33

39.29

39.33. A. orophila Toill.-Gen. *(p. 299, bottom right)*

Plant small with copious roots; leaves grass-like, 4–6 × 0.3–0.5 cm; inflorescences pendent, 3–6 cm long; flowers greenish white; sepals 7–10 mm long; lip rectangular, acuminate, 6–7 × 7–9 mm; spur 3–5 mm long. Antananarivo, Antsiranana. Mossy, evergreen, montane forest; 2000–2500 m. Fl. November–January.

39.34. A. setipes Schltr.

Plant small, slender; leaves 5–7, linear, leathery, 4.5–8 × 0.4–0.6 cm; inflorescence spreading or pendent, simple or few-branched, 6–14 cm long; flowers greenish white; sepals 15–16 mm long; lip broadly rhombic, somewhat acuminate, 10 × 6 mm, with 2 obscure basal calli; spur conical, 5 mm long; plantlets can develop along the inflorescence. Antsiranana. Mossy, montane forest; c. 1800 m. Fl. January.

39.35. A. subramosa (Schltr.) Garay

(syn. *Aeranthes gracilis* Schltr.)

Leaves 4–6, fairly thick, linear, 3.5–6 × 0.7–0.9 cm; inflorescences longer than the leaves, spreading or pendent, up to 14 cm long; flowers whitish green; sepals 14 mm long; lip subrectangular, apiculate, 5.5 × 4.5–5 mm, with 2 oblique ridges at the base; lip incurved, 6 mm long. Antsiranana. Mossy and lichen-rich forest. Fl. January.

39.36. A. tropophila Bosser

Leaves 5–7, linear-oblong, 10–17 × 1.5–1.9 cm; inflorescences 3–5, spreading, very short, 2–2.5 cm long, 1–4-flowered; flowers green or white; sepals 11–16 mm long; lip ovate or oblong, acute or apiculate, 7–8 × 4.5 mm; spur 5–6 mm long. Antsiranana, Mahajanga. Dry, deciduous scrubland, and in tropical xerophilous forests. Fl. September–October.

40.1

40. ERASANTHE

A monotypic genus endemic to Madagascar. Epiphyte with stout roots and a short stem. Leaves leathery, elliptic-obovate, obliquely bilobed at apex. Flowers large, white with green on lip. Sepals and petals lanceolate, acuminate. Lip entire, lacerate on margins, with a slender elongate spur much longer than the lip. Column with a prominent saccate foot.

40.1. **E. henrici** (Schltr.) P.J.Cribb, Hermans & D.L.Roberts   *(p. 301)*
(syn. *Aeranthes henrici* Schltr.)

Stem short; leaves leathery, ligulate, 9–24 × 1.5–5.5 cm, with undulate margins; inflorescence pendent-arching, 3–7-flowered; flowers white or greenish white with green on the lip; sepals 9–11 cm long; lip 10 × 3.8 cm, long acuminate, denticulate-fimbriate on margin; spur 11–16 cm long. Antananarivo, Antsiranana, Mahajanga. In humid forests on the plateau to the N and NE; 800–1000 m. Fl. March–April.

40.1a. **E. henrici** (Schltr.) P.J.Cribb, Hermans & D.L.Roberts subsp. **isaloensis** (H.Perrier ex Hermans) P.J.Cribb & D.L.Roberts
(syn. *Aeranthes henrici* Schltr. var. *isaloensis* H.Perrier ex Hermans)

Differs from the typical subspecies in its triangular, acute floral bracts and smaller flowers, which have a lip with an elliptic-transversal basal lamina extending into a square blade and a shorter acuminate apex, a shorter spur, shorter pedicel and ovary, and a column that is longer and distinct in structure. Fianarantsoa, Toliara. Low down on the trunks and branches of trees near water-courses in the deep sandstone canyons and on small trees and shrubs in the dry forest; 250–750 m. Fl. March.

41. LEMURORCHIS

A monotypic genus endemic to Madgascar. Robust, stemless or short-stemmed leafy epiphyte. Leaves distichous, arranged in a fan, linear-ligulate. Inflorescences axillary, unbranched, cylindrical, densely many-flowered. Flowers small, fleshy. Sepals free. Petals similar to dorsal sepal. Lip short, enclosing the column, obscurely 3-lobed in front; spur with a narrow mouth. Column short, lacking a foot; pollinia each with a rigid stipe and indistinct viscidium.

41.1. **L. madagascariensis** Kraenzl.   *(opposite, top left)*

Plant large, stemless; leaves 6–15, ligulate, 25–45 × 2.2–3 cm; inflorescence arcuate, cylindrical, 20–30 cm long, densely many-flowered; flowers up to 100, white to yellowish; sepals 3–4 mm long; lip 3-lobed at tip; spur cylindrical, 8 mm long. Antananarivo, Fianarantsoa, Toamasina. Mossy, montane forest; c. 2000 m. Fl. February–March.

41.1

42.1

42.2

42.5

42. NEOBATHIEA

A genus of 5 species endemic to Madagascar and the Comores. Small epiphytes or rarely lithophytes. Stem very short and with few leaves in a fan or rarely elongate and leafy in apical half. Leaves obovate to oblanceolate, often twisted at base to lie in one plane. Inflorescences 1–few-flowered. Flowers relatively large, white or pale green with a white lip. Sepals spatulate. Lip entire to 3-lobed; spur filiform with a wide mouth. Column short; pollinia 2.

Key to species of *Neobathiea*

1. Lip entire 2
1. Lip 3-lobed 3
2. Leaves 4–4.8 × 0.8–1.2 cm; flowers white; sepals and petals lanceolate ....
   .... **42.1. N. grandidieriana**
2. Leaves 5.5–6 × 2–2.5 cm; flowers greenish white; sepals and petals
   spatulate .... **42.2. N. keraudrenae**
3. Spur more than 7 cm long; lip side lobes ovate, acute, much smaller than
   midlobe .... **42.5. N. perrieri**
3. Spur less than 5 cm long; lip side lobes obovate, midlobe obovate to
   obtriangular 4
4. Stem short; leaves 5–8.5 × 1.5–1.8 cm; inflorescence 13–20 cm long; flower green
   with a white lip; sepals 15–18 mm long; lip 3-lobed, lobes obovate or sub-rectangular,
   the midlobe larger than the laterals; spur incurved, 3–3.2 cm long .... **42.3. N. hirtula**
4. Lithophyte or epiphyte; stem 3–21 cm long; leaves 2–3.5 × 1–1.5 cm;
   inflorescence 2.5–8 cm long, 1–7-flowered; flowers white; sepals spathulate,
   12–14 mm long; lip deeply 3-lobed, side lobes obovate, midlobe narrowly
   obtriangular; spur 4–5 cm long .... **42.4. N. spatulata**

42.1. **N. grandidieriana** (Rchb.f.) Garay   *(p. 303, top right)*
(syn. *Aeranthus grandidierianus* Rchb.f.; *Neobathiea filicornu* Schltr.)

Epiphyte; 4–5 leaves, 4–4.8 × 0.8–1.2 cm; flower white; sepals and petals lanceolate, acute, 11–13 mm long; lip entire, ovate, acute, 20 × 10 mm; spur up to 14 cm long. Antananarivo, Antsiranana, Fianarantsoa, Toliara; also in the Comoros. Humid, evergreen and lichen-rich forest on branches; 1000–1650 m. Fl. September–December.

42.2. **N. keraudrenae** Toill.-Gen. & Bosser   *(p. 303, bottom left)*

Epiphyte; stem up to 10 cm long; leaves 4–5, in a fan, obovate or oblong-spatulate, 5.5–6 × 2–2.5 cm; inflorescence 1–2-flowered; flowers greenish white; sepals and petals 11–12 mm long; spathulate at the tip, obtuse; lip truncate, broadened at the front, 20 × 12 mm; spur pubescent within, 13 cm long. This may be a local variant of *N. grandidieriana*. Antananarivo, Toamasina. Humid, mossy, evergreen forest on plateau. Fl. May.

42.3. N. hirtula H.Perrier

Epiphyte; 4–5 leaves, 5–8.5 × 1.5–1.8 cm; inflorescence 13–20 cm long; flower with green sepals and petals and a white lip; sepals 15–18 mm long; lip 3-lobed, lobes obovate or sub-rectangular, the midlobe larger than the laterals; spur incurved, 3–3.2 cm long. Mahajanga. Humid forest and semi-deciduous, western forest, on branches. Fl. December–February.

42.3a. N. hirtula H.Perrier var. floribunda H.Perrier ex Hermans

Differing from the typical variety in its larger size, longer stem, larger leaves, up to 11 × 2 cm, 6–10 inflorescences per stem each 30–50 cm long and carrying 6–10 flowers, and flowers that are slightly larger with an 18 mm long dorsal sepal and 21–22 mm long lateral sepals; the mid-lobule of the rostellum of the column does not exceed the wings. Mahajanga. Semi-deciduous western forest. Fl. April.

42.4

42.4. **N. spatulata** H.Perrier   *(p. 305)*

Lithophyte or epiphyte; stem 3–21 cm long, many-leaved towards apex; leaves elliptic, 2–3.5 × 1–1.5 cm; inflorescence 2.5–8 cm long, 1–7-flowered; flowers white; sepals spathulate, rounded or obtuse at the end, 12–14 mm long; lip deeply 3-lobed, 2.8 × 2 cm; side lobes obovate; midlobe narrowly obtriangular; spur 4–5 cm long. Antsiranana. Calcareous rocks and deciduous, western forest; sea level–300 m. Fl. December–January.

42.5. **N. perrieri** (Schltr.) Schltr.   *(p. 303, bottom right)*
(syn. *Aeranthes perrieri* Schltr.)

Epiphyte; 4–6 leaves, 3.5–7 × 1–1.9 cm; inflorescence 6–12 cm long, 2-flowered; sepals subspathulate, 22 mm long; lip 3-lobed, 2 cm long, side lobes ovate, acute, the midlobe much longer and acute at the tip; spur 7–12 cm long. Antsiranana, Mahajanga. Evergreen forest and semi-deciduous western forest, on shrubs; sea level–350 m. Fl. January–June.

43. JUMELLEA

A genus of about 41 species in Madagascar and the Comores and Mascarenes, and 2 species in tropical Africa. Epiphytic or lithophytic plants with short to elongate leafy stem. Leaves 4 to many, distichous, borne in a fan or in apical part of stem, articulate to a sheathing base, relatively thin-textured to leathery, rarely terete, bilobed at apex. Inflorescences 1-flowered, axillary. Flowers white, stellate. Sepals and petals subsimilar. Lip entire, broader than other segments, usually dilated above base, spurred at base, sometimes with a keel-like longitudinal callus at base; spur usually as long as or longer than the lip, rarely shorter, often geniculately bent near base. Column short; pollinia 2, each attached to its own viscidium.

Key to species of *Jumellea*

1. Spur shorter than the pedicel and ovary or, if slightly longer, then less than 2 cm long 2
1. Spur much longer than the pedicel and ovary, or, if equalling them, then more than 10 cm long 17
2. Stem very short; leaves in a fan 3
2. Leaves borne on a long stem 6
3. Leaves 50–60 cm long, 7–9 cm broad; lip angled on each side .... **43.1. J. amplifolia**
3. Leaves up to 45 cm long; 5 cm or less broad; lip not as above 4
4. Spur 12 mm or less long; sepals 25–27 mm long; lip 25 mm long, somewhat pandurate .... **43.2. J. brachycentra**
4. Spur 40–60 mm long; sepals 35–50 mm or more long; lip 40 mm or more long 5
5. Peduncle 10 mm long; lip oblong, subacute, 40 mm long .... **43.3. J. maxillarioides**
5. Peduncle 50–90 mm long; lip lanceolate, acuminate, 50–60 mm long .... **43.4. J. sagittata**
6. Stem more than 15 cm long, often much longer 7
6. Stem less than 15 cm long 14
7. Spur more than 2.5 cm long 8

26. Leaves 2.5–4 cm long; stem up to 50 cm long; peduncle 10 mm long; spur 12–16 cm long . . . . **43.26. J. majalis**

26. Leaves 5–8 cm long; stem up to 30 cm long; peduncle 2.5 cm long; spur 12 cm or less long 26

27. Sepals 20–25 mm long; lip 20–25 mm long; spur 12 cm long . . . . **43.24. J. pandurata**

27. Sepals 16 mm long; lip 18 mm long; spur 10.5–11 cm long . . . . **43.25. J. porrigens**

28. Leaves 16–25 mm broad; sepals 30 mm or more long 29

28. Leaves 5–15 mm broad; sepals usually 30 mm or less long 31

29. Sepals 23–24 mm long; lip 23–24 × 5.5 mm . . . . **43.27. J. arborescens**

29. Sepals 30 mm or more long; lip 30–32 × 20 mm 30

30. Leaves 1.7–1.9 cm broad; peduncle 25–35 mm long; lip trullate, 10–14 mm broad; spur 8.5–9 cm long . . . . **43.41. J. alionae**

30. Peduncle 10 mm long; sepals 30 mm or more long; lip ovate-lanceolate, 30–32 × 20 mm; spur 1–11 cm long . . . . **43.28 J. lignosa**

31. Plant creeping; dorsal sepal reflexed . . . . **43.40 J. stenoglossa**

31. Plant not creeping, erect, spreading or pendent; dorsal sepal erect 32

32. Stem short, 12 cm or less long; leaves clustered together 33

32. Stem elongate, 20 or more cm long; leaves well-spaced on stem 34

33. Leaves succulent, 5–7 × 1–2 cm; sepals 15–19 mm long; lip ovate, apiculate, 16–18 × 10–12 mm . . . . **43.31. J. densefoliata**

33. Leaves not succulent but pleated, 1–1.6 × 0.5–0.6 cm; sepals 13 mm long; lip elliptic-obovate, subobtuse, 13 × 7 mm . . . . **43.35. J. jumelleana**

34. Leaf basal sheaths black-spotted . . . . **43.37. J. punctata**

34. Leaf basal sheaths not spotted 35

35. Leaves 3–4 cm long . . . . **43.29. J. brevifolia**

35. Leaves 4–16 cm long 36

36. Inflorescences clustered at nodes; spur 7.5–8 cm long . . . . **43.33. J. gregariiflora**

36. Inflorescences produced one at a time from nodes; spur 9 cm or more long 37

37. Lip oblanceolate, acute . . . . **43.32. J. flavescens**

37. Lip ovate, ligulate or trullate, acute to acuminate 38

38. Sepals 17–20 mm long; lip ligulate . . . . **43.34. J. intricata**

38. Sepals 20 mm or more long; lip elliptic, ovate to trullate 39

39. Pedicel and ovary 30 mm long; lip ovate-trullate . . . . **43.39. J. similis**

39. Pedicel and ovary 40–60 mm long; lip rhombic or elliptic 40

40. Lip rhombic, acute; pedicel and ovary 50–60 mm long; spur 12–13 cm long . . . . **43.30. J. confusa**

40. Lip elliptic, acuminate; pedicel and ovary 40–50 mm long; spur 9–10 cm long . . . . **43.38 J. rigida**

43.1. J. amplifolia Schltr.

Leaves 50–60 cm long, 7–9 cm broad; sepals 3.5 cm long; lip elliptic-oblong, 3.3 × 1.3 cm; spur 4.3 cm long; related to *J. maxillarioides* (p. 310), which is smaller and has a different lip. Antsiranana. Mossy, montane forest; c. 2000 m. Fl. January.

43.2

43.3

43.4

43.5

43.2. **J. brachycentra** Schltr.

(p. 309, top left)

(syn. *Jumellea floribunda* Schltr.)

Plant large; leaves 15–37 × 1.4–2.5 cm; sepals 2.5–2.7 cm long; lip oblong, acute, 25 mm long, a little contracted between the lower third and the centre; spur 12 mm or less long, shorter than the lip. Antananarivo, Antsiranana, Fianarantsoa, Toamasina. Moss- and lichen-rich forest; 1800–2400 m. Fl. January–March.

43.3. **J. maxillarioides** (Ridl.) Schltr.

(p. 309, top right)

(syn. *Angraecum maxillarioides* Ridl.)

Epiphyte or lithophyte; plant robust; stem covered by fibrous, disintegrating sheaths; leaves 20–25 × 4–5 cm; flowers large and waxy; sepals 3.6 cm long; spur c. 4 cm long. Antananarivo, Fianarantsoa, Toamasina. On rocks and in humid, evergreen forest; 1200–2000 m. Fl. January–March.

43.4. **J. sagittata** H.Perrier

(p. 309, bottom left)

(syn. *Angraecum gracilipes* Rolfe)

Epiphyte; leaves 25–30 × 3–3.5 cm; sepals 3–4 cm long; lip lanceolate, 3.6–4 × 1.8–2 cm; spur 5–6 cm long. Antananarivo, Toamasina. Mossy, montane forest; c. 1400 m. Fl. January–February.

43.5. **J. major** Schltr.

(p. 309, bottom right)

Epiphyte; plant very large; stem thick, up to 60 cm long; leaves up to 45 × 7.3 cm; flowers large; lip 3.5 mm long; spur 6–7 cm long. Antsiranana. Mossy, montane forest and river margins on large trees; c. 1500 m. Fl. March, May.

43.6. **J. bathiei** Schltr.

Pendent epiphyte; stem 30 cm or more long; leaves 6–8 × 0.7–1.1 cm; sepals 19–21 mm long; lip oblanceolate, 22 × 7 mm; spur 35–45 mm long. Antananarivo. Mossy, upper montane forest on moss- and lichen-covered trees; c. 2400 m. Fl. March.

43.7. J. ibityana Schltr.

(p. 313, top left)

Erect, rigid lithophyte, 15–30 cm tall; leaves rigid and leathery, 3–4.5 × 0.7–1 cm; sepals 20 mm long; lip lanceolate, acute, 10 × 6 mm; spur c. 3.5 mm long. Antananarivo. Rocky quartzite outcrops; 2000–2100 m. Fl. February–March.

43.8. J. spathulata (Ridl.) Schltr.

(p. 313, top right)

(syn. *Angraecum spathulatum* Ridl.)

Epiphyte; plant up to 30 cm tall; leaves ligulate, 2.5–3 × 0.6 cm; sepals lanceolate; lip ovate-spathulate, c. 20 mm long; spur 2.5 cm long. Fianarantsoa. Rocky outcrops. Fl. March.

43.9. J. angustifolia H.Perrier

Plant up to 40 cm long; leaves numerous, 7–8 × 0.8 cm; sepals 13 14 mm long; lip very concave, 15 mm long, almost geniculate above the narrow base; spur 12 mm long. Antsiranana. Lichen-rich forest; 1400–2000 m. Fl. April.

43.10. J. dendrobioides Schltr.

Epiphyte to 40 cm or more long; leaves dense, 8–10 × 0.9–1.2 cm; sepals 14–16 mm long; lip 16 × 8–8.5 mm; spur c. 12 mm long. Antsiranana. In lichen-rich, montane forest; c. 2000 m. Fl. January.

43.11. J. françoisii Schltr.

(p. 313, bottom left)

Branching, ascending, many-stemmed epiphyte, 30–60 cm tall; leaves 8–11 × 1–1.4 cm; sepals 18 mm long; lip obovate, apiculate, 17 × 7 mm; spur 12–13 mm long. Antananarivo, Fianarantsoa, Toamasina. Humid, mossy, evergreen forest; 1200–1500 m. Fl. January–May.

43.12. J. pachyra (Kraenzl.) H.Perrier

(syn. *Angraecum pachyrum* Kraenzl.)

Stem up to 30 cm long; leaves 2.5 × 0.6–0.7 cm; sepals 10 mm long; lip close to *J. spathulata* (p. 311); spur 12 mm long. Fianarantsoa. Fl. unknown.

43.13. J. cyrtoceras Schltr.

(*opposite, bottom right*)

Plant c. 15 cm tall, densely leaved; leaves 7–10 × 0.8–1.1 cm; sepals 15 mm long; lip ligulate, 15 mm long; spur short, slightly longer than the sepals. Madagascar. In forest. Fl. unknown.

43.14. J. hyalina H.Perrier

(*p. 315, top left*)

Plant small, 10–15 cm tall; leaves linear, 1.8–2.5 × 0.5–0.6 cm; sepals 10–11 mm long; lip transversely fan-shaped, rounded and apiculate at tip, 10–11 × 8 mm, with a small callus at the base; spur 10–11 mm long. Antananarivo. Mossy, montane forest; c. 1500 m. Fl. January.

43.15. J. pachyceras Schltr.

(*p. 315, bottom*)

Epiphyte; close to *J. hyalina*, but 10–30 cm tall, with larger leaves, 3.5–9 × 0.6–0.9 cm, longer floral bracts, larger flowers, sepals up to 16 mm long, a 17–20 mm long lip, and a 15 mm long spur. Antananarivo, Fianarantsoa. Mossy, montane forest; c. 2000 m. Fl. December–February.

43.16. J. teretifolia Schltr.

(*p. 315, top right*)

Epiphyte; leaves 5–7, cylindrical, 10–20 × 0.2–0.3 cm; pedicel and ovary 4–5 cm long; sepals linear-lanceolate, acuminate, 35 mm long; lip narrow and sub-unguiculate at the base, rhombic above, aristate-acuminate, 30 × 7 mm; spur filiform, geniculate in basal part, 13 cm long. Antananarivo, Fianarantsoa, Toamasina. Humid, evergreen forest; 1100–1500 m. Fl. September–December.

43.7

43.8

43.11

43.13

43.17. J. peyrotii Bosser

Epiphyte; vegetatively similar to *J. teretifolia* (p. 312), leaves 3–5, semi-cylindrical, linear, 15–35 × 0.25–0.4 cm; pedicel and ovary 5–6 cm long; sepals and petals lanceolate, acute, 15–23 × 4–5 mm; lip narrowly clawed then ovate, apiculate, 20–24 × 10 mm; spur filiform, geniculate in basal quarter, 11–12 cm long. Toamasina. Humid, mossy, evergreen forest; 500–1000 m. Fl. February–March.

43.18. J. linearipetala H.Perrier

Epiphytic plant almost stemless, less than 10 cm tall; leaves 4.5–4.7 × 1 cm; pedicel and ovary 5 mm long; sepals 16–18 cm long; petals wide, short and narrow; lip oblanceolate, 17–20 × 4–6 mm; spur 10–12 cm long. Antsiranana, Mahajanga. Humid forest; 500–1000 m. Fl. February.

43.19. J. gracilipes Schltr.   *(p. 317, top left)*

(syn. *Jumellea ambongensis* Schltr.; *J. exilipes* Schltr.; *J. imerinensis* Schltr.; *J. unguicularis* Schltr.)

Epiphytic stemless plant; leaves 5–7, 8–40 × 0.4–1.5 cm; pedicel and ovary 3.5–7 cm long; sepals 22–23 mm long; lip rhombic-elliptic above, acute, 22–23 × 7–8 mm; spur 11–15 cm long. Antananarivo, Antsiranana, Fianarantsoa, Toamasina. Mossy, montane forest; 1400–2000 m. Fl. November–May.

43.20. J. zaratananae Schltr.   *(p. 317, top right)*

Robust epiphytic plant; leaves 5–6, up to 35 × 3.5 cm; sepals up to 22 mm long; lip rhombic-lanceolate, shortly clawed, 23 × 6 mm; spur 9 cm long. Antsiranana. Mossy evergreen forest; 600–1700 m. Fl. December–January.

43.21. J. ambrensis H.Perrier

Leaves 4–6, fairly little rigid, thickened, semi-cylindrical, narrowly linear, 15–23 cm × 5 mm; sepals 45–50 mm long; lip 20 × 8 mm; spur 12 cm long. Antsiranana. Mossy, montane forest; c. 1000 m. Fl. February.

43.14

43.16

43.15

43.22. J. longivaginans H.Perrier   (*opposite, bottom left*)
(syn. *Jumellea longivaginans* var. *grandis* H.Perrier)

Epiphyte; plant medium-sized; pedicel and ovary 8 cm long; sepals 18–20 mm long; lip lanceolate, 20 mm long; spur 9–10 cm long. *J. longivaginans* var. *grandis* is a larger form of the species and has a short ridge at the apex of the throat. Antsiranana, Toamasina. Lichen-rich, humid, evergreen forest; 950–2000 m. Fl. January, October.

43.23. J. papangensis H.Perrier

Epiphyte; plant 15–25 cm tall; leaves ligulate, 5–8 × 0.5–0.8 cm; sepals 14–15 mm long; lipobovate, 15–19 × 9 mm; spur 13 cm long. Fianarantsoa. Montane ericaceous scrub; 1300–1500 m. Fl. December.

43.24. J. pandurata Schltr.

Epiphyte or lithophyte; plant more than 20 cm tall; leaves ligulate, 6–8 × 1–1.8 cm; sepals 20–25 mm long; lip pandurate-lanceolate; spur c. 12 cm long. Antananarivo. Rocky outcrops; c. 1200 m. Fl. October.

43.25. J. porrigens Schltr.

Epiphyte, 12–25 cm tall; stem up to 20 cm long; leaves 5–8 × 0.6–0.9 cm; sepals 16 mm long; lip oblanceolate-spathulate; spur 10–11 cm long. Antsiranana. Lichen-rich, evergreen forest; c. 2000 m. Fl. January.

43.26. J. majalis (Schltr.) Schltr.
(syn. *Angraecum majale* Schltr.)

Epiphyte; stem long and pendent; leaves 2.5–4 × 0.5–0.7 cm; sepals 12–18 mm long; lip pandurate-ligulate, 16 × 5 mm; spur 12–16 cm long. Antsiranana, Toamasina. Lichen-rich, evergreen forest; 1500–2000 m. Fl. January–May.

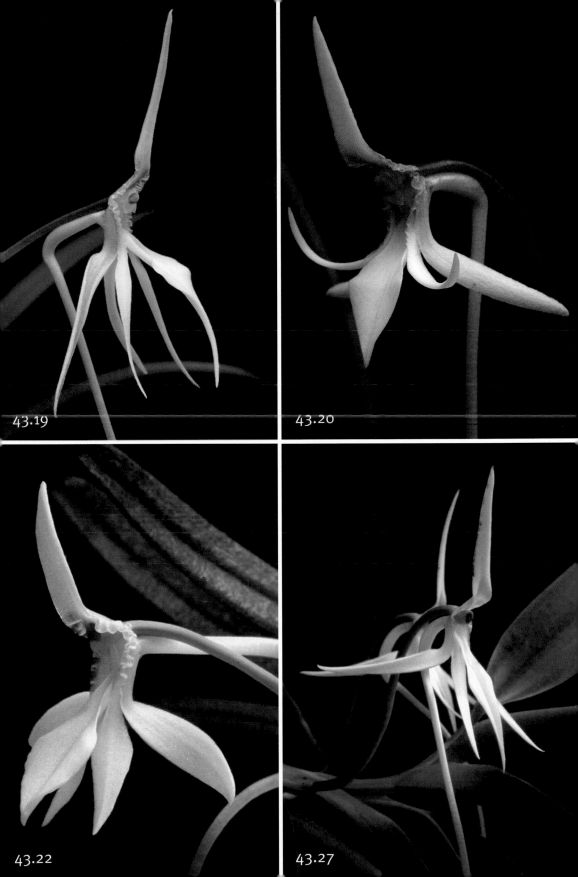

43.19

43.20

43.22

43.27

43.27. J. arborescens H.Perrier (p. 317, bottom right)

Epiphyte or lithophyte; plant up to 1 m tall; leaves up to 20, 11–12 × 1.6–2.2 cm; flowers fleshy; sepals 23–24 mm long; lip 24 mm long, expanded towards the middle; spur 11–12 cm long. Antananarivo, Toamasina. Mossy montane forest on trunks; 1100–1400 m. Fl. October, February.

43.28. J. lignosa (Schltr.) Schltr. (opposite, top left)

(syn. *Angraecum lignosum* Schltr.; *Jumellea ferkoana* Schltr.; *J. lignosa* subsp. *typica* H.Perrier; *J. lignosa* subsp. *ferkoana* (Schltr.) H.Perrier)

Epiphyte or lithophyte; plant very tall with a woody stem; leaves 6–25 × 2–2.5 cm; sepals 30–40 mm long; lip ovate-lanceolate, shortly clawed, c. 30 mm long, widest (c. 11 mm) in the middle; spur 10–11 cm long. Antananarivo, Antsiranana, Fianarantsoa. Lichen-rich forest; rocky outcrops of granite and gneiss; 700–2000 m. Fl. December–March.

43.28a. J. lignosa subsp. acutissima H.Perrier ex Hermans

(syn. *Jumellea lignosa* var. *acutissima* H.Perrier)

Differs from all other subspecies by its much shorter, 6 cm long spur; anther broadly indented at the front. Antananarivo. Mossy forest; c. 1400 m. Fl. March.

43.28b. J. lignosa subsp. latilabia H.Perrier ex Hermans

(syn. *Jumellea lignosa* var. *latilabia* H.Perrier)

Differs from the other subspecies in its broadly oval, acute lip with 17 obvious veins and a central keel all along the length of the lip, 8 cm long spur, and column in which the central tooth of the rostellum is almost as long as the wings that are obtuse at the back. Antananarivo. Humid, evergreen forest; c. 1000 m. Fl. February.

43.28c. J. lignosa subsp. tenuibracteata H.Perrier ex Hermans

(syn. *Jumellea lignosa* var. *tenuibracteata* H.Perrier)

Differs from the typical subspecies by the sheaths on the peduncle and its thin, tubular floral bracts that are shorter and neither keeled nor flattened. Antsiranana. Mossy montane forest; c. 2000 m. Fl. January.

43.28

43.30

43.31

43.35

43.29. J. brevifolia H.Perrier

Lithophytic plant robust, 25–35 cm tall; leaves leathery, 3–4 × 1.2–1.3 cm; sepals 18–23 mm long; lip 22 mm long, abruptly expanded into a wide triangular, acute lamina; spur filiform, 11 cm long. Toliara. Rocky outcrops of granite and gneiss; 1500–1650 m. Fl. November.

43.30. J. confusa (Schltr.) Schltr.   *(p. 319, top right)*

(syn. *Angraecum confusum* Schltr.; *Jumellea ankaratrana* Schltr.)

Epiphyte or lithophyte up to 60 cm tall; leaves 7–12 × 1–2.4 cm; sepals 22–30 mm long; lip c. 6 mm long, wide in the middle; spur filiform, 12–13 cm long. Antananarivo, Antsiranana, Toamasina. Humid, evergreen, montane forest; 600–1500 m. Fl. September, December–May.

LOCAL NAME *Fontilahyjanahary madiniky.*

43.31. J. densefoliata Senghas   *(p. 319, bottom left)*

Epiphyte or lithophyte; stem 8–10 cm tall; leaves clustered in tiers on a short stem, succulent, 5–7 × 1–2 cm; sepals 15–19 mm long; lip ovate, acute, 16–18 × 10–12 mm, unguiculate; spur 11–13 cm long. Antananarivo, Fianarantsoa. Humid, evergreen forest; shaded and humid rocks; in *Uapaca* forest; 1200–1600 m. Fl. September–November.

43.32. J. flavescens H.Perrier

Plant branching, 20–60 cm tall; leaves 4–11 × 0.5–1.4 cm; sepals 17–20 mm long; lip oblanceolate, acute, 2 × 0.7–0.8 cm wide; spur 10–12 cm long. Antananarivo, Antsiranana. Mossy, montane forest; c. 1200 m. Fl. January–February.

43.33. J. gregariiflora H.Perrier

Lithophyte; plant branching, up to 40 cm tall; leaves 5–10, 8–12 × 0.8–1 cm; flowers several, grouped at the inter-nodes; sepals 19–22 mm long; lip 20–22 mm long; spur 7.5–8 cm long. Toliara. Rocky outcrops of granite and gneiss; 1000–1150 m. Fl. November.

43.34. J. intricata H.Perrier

Plant spreading; stems branched and tangled, 20–60 cm long; leaves 4–11 × 0.5–1.4 cm; sepals 17–20 mm long; lip ligulate, 20 × 6 mm; spur 10 cm long. Antsiranana, Fianarantsoa, Toamasina. Lichen-rich montane forest; 1600–2000 m. Fl. January.

43.35. J. jumelleana (Schltr.) Summerh.   *(p. 319, botom right)*

(syn. *Angraecum jumelleanum* Schltr.; *Jumellea henryi* Schltr.)

Plant small, 7–11 cm tall; leaves pleated, oblong-ligulate, 10–16 × 5–6 mm; sepals 13 mm long; lip obovate-elliptical, subobtuse, 13 × 7 mm; spur c. 11 cm long. Antananarivo, Toamasina. On shrubs and small trees in humid, evergreen forest; 800–1500 m. Fl. September–November, February.

43.36. J. marojejiensis H.Perrier

Epiphyte; related to *J. cyrtoceras* (p. 312) but differs by the shorter leaves, the more acute, narrower sepals, petals and lip, and the filiform spur, which is slightly shorter than the sepals and lip. Antsiranana. Mossy, montane forest; 1400–2000 m. Fl. March–April.

43.37. J. punctata H.Perrier   *(p. 323, top left)*

Epiphyte; plant 40–50 cm tall; stem sheaths finely flecked with black dots; leaves narrowly linear, 10–13 × 0.9–1.3 cm; sepals 27–32 mm long; lip 30 mm long, oval-lanceolate; spur 10–11 cm long. Toamasina. Humid, evergreen forest on trunks; 600–1100 m. Fl. September–October.

43.38. J. rigida Schltr.   *(p. 323, top right)*

Epiphyte or lithophyte up to 60 cm tall; leaves 8–12 × 1.2–1.6 cm; sepals 20 mm long; lip elliptic, acuminate in front, 20 × 7 mm; spur 9–10 cm long; similar to *J. porrigens* (p. 316) but with shorter, wider leaves, bigger flowers, a different lip, and a shorter spur. Antananarivo. Granite outcrops; 1500–1800 m. Fl. December–March.

43.38a. J. rigida var. **altigena** Schltr.

Lithophyte; plant more compact, inflorescence shorter and flowers fleshy. Fianarantsoa. Rocky outcrops; c. 2400 m. Fl. January.

43.39. J. similis Schltr. *(opposite, bottom left)*

Epiphyte up to 55 cm tall; similar to *J. confusa* (p. 320) but the leaves narrower, 7–12 × 0.8–1 cm, and the flowers smaller; sepals 20 mm long; lip ovate-trulliform in front, 22 mm long; spur 11 cm long. Antananarivo. Humid, evergreen forest; c. 1500 m. Fl. November.

43.40. J. stenoglossa H.Perrier

Epiphyte; plant creeping, unbranched; leaves oblong-lanceolate, 4–6 × 0.9–1 cm; peduncle 15–45 mm long; pedicel and ovary 30–50 mm long; sepals 25 mm long, the dorsal reflexed; lip 30 × 7 mm, broadened at the base; spur 11–12 mm long. Toliara. Coastal forest; 20–100 m. Fl. March.

43.41. J. alionae P.J.Cribb *(opposite, bottom right)*

Clump-forming, lithophyte or epiphyte; stems branching at or near base, 50–100 cm tall; leaves narrowly oblong-ligulate, 9–10.5 × 1.7–1.9 cm, dark glossy green; inflorescences solitary from upper nodes; peduncle 2.5–3.5 cm long; flowers white, the spur pale green in the upper part, strongly sweetly fragrant; pedicel and ovary sinuous, 5–6 cm long; sepals 3–3.7 cm long; lip hastate, acute, 3.2–3.3 × 1–1.4 cm; spur curved and parallel to column then pendent, 8.5–9 cm long. Toliara. In evergreen forest on ridge-top, on low trees and on mossy boulders near stream; 500–1000 m. Fl. Feb.

43·37

43·38

43·39

43·41

44. ANGRAECUM

A large genus of about 210 species, mainly in Africa, Madagascar and the adjacent islands of the Indian Ocean. In Madagascar, 128 species. Small to large epiphytic or less commonly lithophytic plants. Stems erect to pendent, short to elongate, monopodial, leafy. Leaves thin-textured, fleshy or leathery, distichous or arranged in a fan, dorsiventrally flattened, V-shaped in cross-section, rarely bilaterally compressed or terete, sometimes unequally bilobed at tip. Inflorescences axillary, 1–many-flowered, usually unbranched. Sepals and petals free, subsimilar or petals smaller than sepals. Lip entire, with or without a basal linear callus, with a spur; spur globose, clavate or filiform. Column short, fleshy, lacking a foot. Pollinia 2, each attached to its own viscidium; rostellum deeply cleft, sometimes with a short median tooth.

Key to sections of *Angraecum*

1. Inflorescence 1-flowered, rarely 2-flowered, flowers produced simultaneously or in succession 2
1. Inflorescence several–many-flowered, simple or branching 8
2. Plants sessile with a very short stem; leaves few, basal . . . . **I. Sect. Acaulia**
2. Plants erect with a distinct stem; leaves several to many, all along the stem 3
3. Ovary lacking a stalk; inflorescence stalk very short, scarcely developed . . . . **II. Sect. Pectinaria**
3. Ovary with a distinct stalk; inflorescence stalk more or less distinctly developed 4
4. Inflorescence stalk covered entirely with 3–4, laterally compressed sheathing bracts; flower large; ovary triangular in cross-section . . . . **III. Sect. Perrierangraecum**
4. Inflorescence stalk covered at base only with 1–2 adpressed sheathing bracts; flowers usually smaller, ovary circular in cross-section 5
5. Inflorescence stalk short, shorter than the internodes, very rarely as long . . . . **IV. Sect. Filangis**
5. Inflorescence stalk prominent, slender, longer than the internodes 6
6. Flowers small; spur cylindrical or club-shaped . . . . **V. Sect. Angraecoides**
6. Flowers medium-sized to larger; spur filiform, elongate 7
7. Lip narrow . . . . **VI. Sect. Pseudojumellea**
7. Lip broad, flowers often produced in succession . . . . **VII. Sect. Arachnangraecum**
8. Flowers thin-textured, diaphanous 9
8. Flowers fleshy in texture 13
9. Stem more or less distinct 10
9. Stem very short, not obvious 11
10. Stem short or rather rarely elongate and leafy above; inflorescences borne in lower leaf axils; inflorescence stalk setiform, elongate, loosely few-flowered . . . . **VIII. Sect. Gomphocentrum**
10. Stem elongate, leafy; inflorescences in axils of upper leaves 12
11. Inflorescences many-flowered, as long as or longer than the leaves; rachis somewhat flexuous . . . . **IX. Sect. Lepervenchea**
11. Inflorescences few-flowered, rarely 1–3-flowered, shorter than the leaves; rachis hardly flexuous . . . . **X. Sect. Lemurangis**
12. Flowers minute, often forming a dense raceme, normally 1-sided . . . . **XI. Sect. Nana**
12. Flowers small to medium-sized, more or less laxly racemose, all round the rachis . . . . **XII. Sect. Boryangraecum**

I. Section Acaulia

Stems very short; leaves few, basal; inflorescences usually 1–2-flowered.

Key to species of Section Acaulia

1. Spur clavate, 7.5 mm long .... **44.1. A. brachyrhopalon**
1. Spur not clavate, 20 mm or more long 2
2. Leaves 11–15 mm long .... **44.3. A. rhynchoglossum**
2. Leaves 30 mm or more long 3
3. Flowers whitish yellow; spur not dilated at tip; lip 6 × 4 mm .... **44.2. A. chaetopodum**
3. Flowers yellow-green; spur slightly dilated at tip; lip 7 × 7 mm .... **44.4. A. setipes**

44.1. **A. brachyrhopalon** Schltr.

Plant 5–7 cm tall; leaves linear, 50 × 5 mm; inflorescences up to 4.5 cm long; flowers 1–2, yellow-green; sepals 4.5–5 mm long; lip ovate, obtuse; 5–6 × 3.5–4 mm; spur clavate, 7.5 mm long. Antsiranana. Seasonally dry, deciduous forest and woodland. Fl. January.

44.2. **A. chaetopodum** Schltr.   *(p. 327, top left)*

Stemless epiphyte close to *A. pergracile* (p. 372), 4–10 cm tall, but differs by its narrower, more pointed and fleshier leaves, 30–80 × 2.5–4 mm; inflorescences 1-flowered, shorter than the leaves; flowers whitish yellow; sepals 6 mm long; lip ovate-cucullate, shortly acuminate-rostrate, 6 × 4 mm; spur 20–30 mm long; also related to *A. rhynchoglossum* but differs by the narrower sepals and petals and the more extended lip. Antsiranana, Mahajanga(?). Mossy, montane forest; 1200–1500 m. Fl. January.

44.3. **A. rhynchoglossum** Schltr.   *(p. 327, top right)*

(syn. *Mystacidium viride* Ridl.; *Angraecum viride* (Ridl.) Schltr.; *A. foxii* Summerh.)

Small stemless epiphyte; leaves 11–15 × 4–5 mm; inflorescence 15–30 cm long, 1-flowered; flower with lip uppermost, yellow-green, 6 × 3 mm, expanded and very rounded at the base; spur funnel-shaped and narrowed below the entrance for 6 mm, then a little contracted, finally a little expanded towards the upper third, 25 mm long. Antananarivo, Toamasina. Moss- and lichen-rich forest; 1000–1400 m. Fl. September–April.

44.4. **A. setipes** Schltr.   *(opposite, bottom left)*

Slender, stemless epiphyte, 7–13 cm tall; leaves 5–6, linear-lanceolate, 35–110 × 3–8 mm; inflorescences numerous, 5–9 cm long, 1–2-flowered; flowers greenish yellow; sepals 6–7 mm long; lip suborbicular broadly rhombic, 7 × 7 mm, apiculate; spur horizontal, subfiliform-cylindrical, very slightly thickened at the tip, 22 mm long. Antananarivo, Fianarantsoa, Toamasina, Toliara. Mossy forest; 600–2000 m. Fl. September–February.

II. Section Pectinaria

Stems elongate; leaves narrow, fleshy, flattened in a vertical plane; flowers subsessile, solitary, white.

Key to species of Section Pectinaria

1. Leaves very narrow, more than 10 times as long as broad 2
1. Leaves less than 5 times as long as broad 3
2. Leaves fleshy, subcylindrical, 10–16 mm long; lip obtuse; ovary hispid ....
   .... **44.6. A. humblotianum**
2. Leaves flat, 30–40 mm long; lip acute; ovary glabrous .... **44.7. A. panicifolium**
3. Bract 1.5 mm long; sepals acute, 3–3.5 mm long; leaves dorsally keeled or
   winged; ovary glabrous; lip acute .... **44.9. A. pterophyllum**
3. Bract 2 mm long or more; sepals 6-6.5 mm long 4
4. Lip acute; ovary glabrous; spur less than half the length of the ovary, 2 mm
   long.. **44.8. A. pectinatum**
4. Lip obtuse; ovary densely shortly hairy; spur 5.5 mm long, longer than the
   ovary .... **44.5. A. dasycarpum**

44.5. **A. dasycarpum** Schltr.   *(opposite, bottom right)*

Small semi-succulent epiphyte or lithophyte with small leaves; stem 10 cm or more long; leaves oblong, 7–10 × 4–7 mm; flowers white; sepals 6.5 mm long; lip concave, oblong-ligulate, obtuse, 5.5–6 mm long, slightly narrowed above the middle; spur oblong-cylindrical, obtuse, 4 mm long, slightly narrowed below the middle, straight, pressed against the ovary; pedicellate ovary and roots hairy. A form without spur has been recorded. Toamasina. Evergreen forest; *Philippia* scrub; coastal forest; on rocks covered in moss; sea level–800 m. Fl. January–April.

44.6. **A. humblotianum** (Finet) Schltr.

(syn. *Macroplectrum humblotii* Finet; *Angraecum finetianum* Schltr.; *A. humblotii* (Finet) Summerh.; *A. abietianum* Schltr.)

Small epiphyte; stem up to 15 cm long; leaves long and narrow, almost cylindrical, 10–16 × 1–1.5 mm; sepals 5–5.5 mm long; lip ovate-oblong, obtuse, 5 × 1 mm; spur cylindrical, 2.2–3.5 mm long, slightly inflated, partly hidden by the floral bracts; the ovary bears dark tufts of hair. Fianarantsoa, Toamasina. Humid, evergreen forest; sea level–900 m. Fl. January–March.

44.2

44.3

44.4

44.5

44.7. **A. panicifolium** H.Perrier

(opposite, top)

Epiphyte; stem 10–15 cm long, slender; leaves linear, 35–38 × 1.3–2.2 mm; inflorescence 1-flowered; flowers small, white; sepals 7–8 mm long; lip concave, ovate, 7–8 mm long, 4 mm wide above the base, a little contracted towards the middle; spur 5–6 mm, at first narrow, then cylindrical. Toamasina. Moss- and lichen-rich, evergreen forest; 1000–1200 m. Fl. February.

44.8. **A. pectinatum** Thouars

(opposite, bottom)

(syn. *Pectinaria thouarsii* Cordem.)

Medium-sized epiphyte; stem 10–20 cm long, simple or branching; leaves elongate-linear, thick, 12–16 mm long; flowers small, fleshy, white; sepals 6 mm long; lip similar to the sepals, almost flat; spur almost straight, a little inflated at the tip, just over 2 mm long; ovary a little rough. Antananarivo, Fianarantsoa, Toamasina; also in the Mascarenes and Comoros. Humid, evergreen forest; in tapia (*Uapaca*) woodland; 200–1750 m. Fl. December–May.

44.9. **A. pterophyllum** H.Perrier

Small epiphyte, 3 cm tall; leaves elliptic, 8–12 × 2.5–3.5 mm; flowers small, white; ovary covered by small mossy hairs; sepals 3–3.5 mm long; lip ovate, acute, 2.5 × 1.5 mm, very concave; spur obtusely sac-like, 2 × 1.2 mm. Antananarivo, Fianarantsoa, Toamasina. Mossy forest; 1100–1200 m. Fl. February.

III. Section Perrierangraecum

Small to medium-sized, usually very short stemmed plants with 1- or rarely 2-flowered inflorescences; flower on short peduncle covered by 3–4 compressed sheaths; lip white; ovary triquetrous.

Key to species of Section Perrierangraecum

1. Spur 5 cm long or more 2
1. Spur 4 cm long or less 3
2. Spur 18 mm long, hook-shaped at tip .... **44.12. A. drouhardii**
2. Spur 30–40 mm long, not hooked at tip 4
3. Stem elongate, up to 30 cm long; leaves linear, 3–3.5 mm broad; lip pandurate; spur 12 mm long .... **44.10. A. ambrense**
3. Stem short, less than 14 cm long; leaves 4 mm or more broad; lip not pandurate; spur 18 mm or more long 5
4. Leaves 1.6–2 cm long; lip cordate, 15 × 6 mm; spur 3–3.3 cm long .... **44.13. A. rigidifolium**
4. Leaves 3.5–6 cm long; lip broadly rhombic-suborbicular, 13 × 12 mm; spur 4 cm long .... **44.11. A. curvicalcar**

44.7

44.8

44.10. **A. ambrense** H.Perrier

Pendent epiphyte up to 30 cm long; leaves linear, 5 × 0.3–0.35 cm; flowers white; sepals 12 mm long; lip pandurate with a narrow base, 12 × 6 mm; spur 12 mm long, its entrance expanded into a thick lobule. Antsiranana (Mt Ambre). Moss- and lichen-rich forest; c. 1200 m. Fl. September.

44.11. **A. curvicalcar** Schltr.

Small epiphyte, 4–7 cm tall; leaves 4–5, 35–60 × 4–6 mm; flowers small, white; sepals 13 mm long; lip broadly rhombic-suborbicular, 13 × 12 mm, the central vein thickened into a keel at the base; spur 4 cm long, thread-like and gradually expanding towards the opening. Antsiranana. Lichen-rich forest; c. 2000 m. Fl. January.

44.12. **A. drouhardii** H.Perrier

Plant 2–7 cm tall; leaves oblong, 13–16 × 5–6 mm; flowers white; sepals 15 mm long; lip subrhombic, 15–17 × 8–12 mm; spur 18 mm long. Resembles *A. curvicalcar*, but differs in its short and wide leaves, the inflorescence longer than the leaves, the ample floral bracts that are not contracted, the wide lip folded around the column at the base, and the short spur with a hook-shaped apex, which is slightly longer than the lip. Antsiranana (Mt Ambre). Mossy, montane forest; c. 1200 m. Fl. April, August.

44.13. **A. rigidifolium** H.Perrier

Epiphyte up to 14 cm tall; leaves 16–20 × 4–5 mm; flowers white; sepals 14–16 mm long; lip 15 × 6 mm cordate towards the base; spur almost filiform, 3–3.3 cm long. Antananarivo. Moss- and lichen-rich forest; c. 1300 m. Fl. February.

44.14. A. aloifolium Hermans & P.J.Cribb   (*opposite, top*)

Epiphyte; stem 6–10.5 cm long; leaves narrowly ensiform, thick, leathery, 42–50 × 10–12 mm, with a pitted surface; flowers with white-yellow sepals, white petals, lip and spur and a green column; sepals 22–25 mm long; lip concave, elliptic, acute, 49–52 × 15–20 mm; spur 8–10 cm long, funnel-shaped at base. Mahajanga. Dry, deciduous forest. Fl. September.

44.15. A. clareae Hermans, la Croix & P.J.Cribb   (*opposite, bottom left*)

Epiphyte 3–13 cm tall; leaves 10–20, glaucous, glabrous, succulent, elliptic-oblong, 25–38 × 9–12 mm; flowers 1–2, white, lemon-scented; sepals 28–32 mm; lip ovate, acute, 25–30 × 15–19 mm; spur broad-mouthed, erect then pendent, 9–12 cm long. Its closest ally is *A. kraenzlinianum* (p. 334) but the stem is shorter, the leaves longer and closer together on the stem, and the lip and spur are a different shape. Antananarivo. Evergreen forest; 1350 m. Fl. September–October.

44.16. A. compactum Schltr.   (*opposite, bottom right*)

Widespread and variable epiphyte; stem robust with 5–6 leathery leaves, 5–10 × 1.3–2.8 cm; flowers large; tepals somewhat reflexed, 2 cm long; lip oblong, concave, 25 × 12–20 mm, very rounded at the tip with a short terminal apicule; spur 12–13 cm long, wide at the base, curved and then filiform and descending. Antananarivo, Antsiranana, Fianarantsoa, Toamasina, Toliara. Humid, evergreen forest on trees and shrubs including *Philippia*; 700–2000 m. Fl. August–December.

44.17. A. dollii Senghas   (*p. 335, top left*)

Epiphyte with stem up to 40 cm long; leaves canaliculate, up to 14 × 1.5 cm long; inflorescence 1-flowered, 5 cm long; flower large, slightly glossy; sepals and petals 4.5–5 cm long, pale green; lip ovate, apiculate, 35 × 22 mm, apex acute, white; spur funnel-shaped and erect and base, then decurved and slender, up to 14 cm long. Antananarivo, Fianarantsoa. Seasonally dry forest; c. 1000 m. Fl. unknown.

44.18. A. equitans Schltr.

Erect, compact epiphyte, 9–10 cm tall; leaves conduplicate, overlapping, 30–40 × 11–14 mm; flower medium-sized and resembles *A. clareae*; sepals 17–20 mm long; lip concave, elliptic, acuminate, 17 × 9 mm; spur funnel-shaped at the base and then filiform, c. 8 cm long. Antsiranana. Moss- and lichen-rich forest; c. 2000 m. Fl. December.

44.14

44.15

44.16

44.19. A. kraenzlinianum H.Perrier

(syn. *Aeranthes englerianus* Kraenzl.; *Angraecum englerianum* (Kraenzl.) Schltr.; *A. robustum* Kraenzl. non Schltr.)

Epiphyte related to *A. curnowianum*, stem short; leaves 11–25, fleshy, recurved, 2.5–4 × 0.5–0.7 cm; flower white; sepals 25–30 mm long; lip broadly roundly dilated on sides, cuspidate-subacute at the front; spur 6–8 cm long. Antsiranana, Fianarantsoa. Mossy and lichen-rich forest; 1500–2000 m. Fl. January.

44.20. A. ankeranense H.Perrier

Epiphyte up to 15 cm tall; leaves 5–6, fleshy, ovate-oblong, 2–3 × 1.1–1.4 cm; sheaths with strong longitudinal ridges; flower white; ovary very broadly winged; sepals 3.5–4 cm long; lip broadly ovate, 35–37 × 22 mm; spur cylindrical, 8 cm long. Antananarivo. On tree trunks in mossy forest; 700–2000 m. Fl. February.

44.21. A. bicallosum H.Perrier

Epiphyte reportedly close to *A. obesum* (p. 338) and *A. didieri* (p. 336); stem up to 14 cm long; roots glabrous; leaves broadly elliptic, 15–25 × 8–11 mm; flower white with sepals and spur reddish yellow; sepals 30 mm long; lip broadly ovate, subacuminate-acute, 30–15 mm, with 2 pronounced calli at the base; spur 12 cm long. Antsiranana. Moss- and lichen-rich forest; c. 1000 m. Fl. November.

44.22. A. breve Schltr.   *(opposite, top right)*

Epiphyte related to *A. rutenbergianum* (p. 340); roots warty; leaves thicker, 20–55 × 5–8 mm, grey-green; flower smaller; sepals and petals 15–17 mm long; lip rhombic-elliptic, 18 × 14 mm; spur 11 cm long. Antsiranana (Mt Tsaratanana). Lichen-rich, montane forest. c. 1800 m. Fl. January.

44.23. A. curnowianum (Rchb.f.) T.Durand & Schinz

(opposite, bottom left)

(syn. *Aeranthus curnowianus* Rchb.f.; *A. suarezense* Toill.-Gen. & Bosser; ?*A. subcordatum* (H.Perrier) Bosser)

Robust epiphyte, up to 40 cm tall; leaves 13–18 × 1.5–1.9 cm; inflorescence 1-flowered; flower large, white; sepals 25–33 mm long; lip broadly ovate, acute-subapiculate, 30–32 × 20–21 mm; spur filiform, 8.5–9 cm long. *A. subcordatum* may be referable to this species, further work is needed. Antananarivo, Antsiranana, Toamasina. Mossy, evergreen forest; 800–1500 m. Fl. February–May.

44.17

44.22

44.23

44.24

44.24. A. didieri (Baill. ex Finet) Schltr.   *(p. 335, bottom right)*
(syn. *Macroplectrum didieri* Baill. ex Finet)
Epiphyte with a prominent stem 12–15 cm long; roots warty; leaves ligulate,
17–42 × 6–10 mm; flowers large for the plant; sepals 22–35 mm long; lip
elliptic or elliptic-oblong, shortly acute at the tip, 23–32 × 12–15 mm; spur
narrowing to an acute tip, 8–15 cm long. Antananarivo, Toamasina, Toliara.
Humid, evergreen forest on plateau; on *Philippia*; 600–1500 m. Fl.
October–January.

44.25. A. dryadum Schltr.   *(opposite, top left)*
Widespread epiphyte; 7–10 leaves, 20–40 × 6–10 mm; flowers very variable
in size; sepals 22–30 mm long; lip broadly ovate, acute, 22–33 × 11–13 mm;
spur 10–13 cm long. The closest relative is *A. rutenbergianum* (p. 340), but it
has a much longer stem and wider and thicker leaves, as well as narrower
sepals and a differently shaped lip. Antananarivo, Antsiranana,
Fianarantsoa, Toamasina, Toliara. Humid, lichen-rich, evergreen forest on
plateau; 1000–2100 m. Fl. October–March.

44.26. A. elephantinum Schltr.
Epiphyte related to *A. didieri*; plant short, stout, 10–15 cm tall; roots
glabrous; leaves leathery, ligulate, 7.5–10 × 2–2.5 cm; flower white; sepals
35 mm long; lip ovate, obtuse, 30 × 20 mm; spur filiform, 10–11 cm long.
Antananarivo. Fl. unknown.

44.27. A. imerinense Schltr.
Slender and long epiphyte, up to 25 cm long; leaves linear, thin-textured,
25–35 × 2.5–3.5 mm; flowers medium-sized, white; sepals 19 mm long; lip
ovate-elliptic, slightly obtuse and apiculate, 19 × 8–9 mm, concave at the
base, the middle vein thickened into a ridge; spur pendent, filiform, slightly
narrowed from the entrance to the apex, 9.5 cm long. Antananarivo. In woods
on west-facing slopes; c. 1700 m. Fl. September.

44.28. A. lecomtei H.Perrier   *(opposite, right)*
Small epiphyte up to 5 cm tall, with narrow linear leaves, 1.5–3.5 × 0.1–0.25
cm; flowers small, white; sepals 10 mm long; lip with a narrow base, the
blade narrowed at the spur entrance, broadly ovate, 12 × 8 mm; spur very
slender, 10–12 cm long, shortly narrowed below the wide opening.
Toamasina. Mossy, evergreen forest; 1000–1200 m. Fl. February–March.

44.25

44.32

44.28

44.29. A. letouzeyi Bosser

Small epiphyte up to 4 cm tall, with verrucose roots; leaves 4–5, erect, cylindrical, 6–10 × 0.1–0.2 cm; flowers small, white; sepals 14–17 mm long; lip concave, ovate-lanceolate, acute, 15–16 × 7 mm; spur very slender, 6–7 cm long, shortly narrowed below the wide opening. Antananarivo, Toamasina. Mossy, evergreen forest; 900–1200 m. Fl. October–February.

44.30. A. littorale Schltr.

Small epiphyte; roots not warty; leaves 4.5–5 × 0.5–0.8 cm; flower relatively large; sepals 27 mm long; lip elliptic, rather longly acuminate, 28 × 12 mm, the central vein thickened into a ridge at the base; spur filiform, S-shaped, 9 cm long. Fianarantsoa, Toamasina. Coastal forest; up to 200 m. Fl. October.

44.31. A. longicaule H.Perrier

Slender epiphyte or lithophyte; stem elongate, 4–30 cm long; leaves small, 14 × 2 mm; sepals 12–15 mm long; lip suborbicular, apiculate, 15–20 × 14–17 mm, 17–19-nerved; spur filiform, 45–50 mm long. Antsiranana. Montane, ericaceous scrub and lichen-rich forest; terrestrial in moss; 500–2100 m. Fl. March–April.

44.32. A. obesum H.Perrier   (p. 337, bottom left)

Epiphyte or lithophyte; stem up to 50 cm long; leaves tapering, 10–12 × 1.3–1.5 cm; flower thick-textured; sepals 30–35 mm long; lip with the base contracted but then immediately expanded into a very concave, broadly ovate-boat-shaped, 20–34 × 14–20 mm; spur cylindrical, 5–10 cm long, gradually narrowing from the base to the apex. Antananarivo. Moss- and lichen-rich forest; 1200–1500 m. Fl. January–March.

44.33. A. palmicolum Bosser

Small epiphyte easily distinguished by the bluish-green, glaucous leaves and the smooth roots; lip broadly ovate to suborbicular, acute and apiculate at the tip, 11–16 × 8–10 mm; spur filiform, 10–11 cm long. Antananarivo. Highland forest on trunks of the palm *Dypsis decipiens*, amongst lichens; 1300–1400 m. Fl. September.

44.36

44.34. **A. peyrotii** Bosser *(opposite, top left)*

Epiphyte; stem 5–6 cm long; leaves 6–8, fleshy, semicylindrical, 10–12 × 0.7–0.8 cm; flower white; sepals 35–45 mm long; lip broadly elliptic, 35–40 × 20–25 mm; spur filiform, 10–15 cm long; related to *A. rutenbergianum* but differs by its longer, very fleshy, semicylindrical leaves. Antananarivo. Evergreen forest; 1300–1500 m. Fl. September.

44.35. **A. pseudodidieri** H.Perrier *(opposite, top right)*

Erect epiphyte, up to 20 cm tall; leaves fleshy, linear, 6.5–8 × 0.8–1 cm; flowers white; sepals 15 mm long; lip almost flat, broadly obovate-cuspidate, 25 × 18 mm; spur 10–11 cm long, narrowed from the entrance to the tip. Antsiranana. Moss- and lichen-rich forest; c. 1000 m. Fl. November.

44.36. **A. rutenbergianum** Kraenzl. *(p. 339 and opposite, bottom left)*
(syn. *Angraecum catati* Baill.)

A variable epiphyte or lithophyte; plant 4–12 cm tall, almost stemless; roots glabrous, warty; leaves 4–12, rigid, thick, 22–55 × 5–7 mm; flowers large; sepals 30–37 mm long; lip rhombic-elliptic, 30–35 × 17–18 mm; spur filiform, variable in length, 6–14 cm long. Antananarivo, Fianarantsoa. Humid, evergreen forest on trunks and branches of *Agauria salicifolia* on plateau and on shaded wet rocks; 1500–2600 m. Fl. November–February.

44.37. **A. sambiranoense** Schltr.

Pendent epiphyte, up to 35 cm long; leaves oblong-ligulate, 50–60 × 10–13 mm; sepals 30 mm long; lip lanceolate, acuminate, 30 × 10 mm; spur 10–11 cm long. Antananarivo, Antsiranana. Humid, evergreen forest; 700–800 m. Fl. February.

44.38. **A. urschianum** Toill.-Gen. & Bosser *(opposite, bottom right)*

A tiny epiphyte, 2–3 cm tall; stem 1.5 cm long, with characteristic darkly pitted leaves, 0.8–2.3 × 0.5–0.8 cm; flowers comparatively large, yellow-green with a white lip; sepals 10–11 mm long; lip broadly triangular-ovate, flat, 10–13 × 7.5–9 mm; spur slender, thread-like, 11–12 cm long. Toamasina. Humid, evergreen, mossy forest; c. 1100 m. Fl. September.

44.34

44.35

44.36

44.38

IV. Section Filangis

Stems very long; leaves in 2 rows (distichous); inflorescence 1-flowered; peduncle short, covered by 1–2 sheaths; flowers medium-sized, white or greenish white; spur elongate, filiform.

Key to species of Section Filangis

1. Sepals 16–17 mm long; lip broadly triangular, acute, 15 × 9 mm .... **44.39. A. amplexicaule**
1. Sepals 12 mm long; lip lanceolate-subpandurate, 11–12 mm long .... **44.40. A. filicornu**

44.39. A. amplexicaule Toill.-Gen. & Bosser

Epiphyte; stem pendent, to 30 cm long with thin, narrowly oblong leaves, 40–55 × 10 mm; sepals 16–17 mm long; lip uppermost, broadly acute-triangular, 15 × 9 mm, deeply concave with a broad base; spur 10–11 cm long, slender, curved in the middle. Toamasina. Humid, lowland and coastal forest; sea level–400 m. Fl. February, July.

44.40. A. filicornu Thouars   (p. 345, top left)
(syn. *Aeranthes thouarsii* S.Moore)

Epiphyte; stems long; roots verrucose; leaves linear, grass-like, 45–90 × 1.5–4 mm; inflorescences 1-flowered; flowers medium-sized: sepals 12 mm long; lip lanceolate, somewhat pandurate, 11–12 mm long; spur filiform, 9–11 cm long. Antananarivo, Toamasina, Toliara; also in the Mascarenes. Coastal forest; low elevation, humid, evergreen forest; on *Philippia*; sea level–800 m. Fl. September–July.

V. Section Angraecoides

Stems long, slender, leafy; leaves in 2 rows (distichous); inflorescences 1–2-flowered, much longer than the internodes; peduncle slender, with 1–2 sheaths at base; flowers small, green or yellow-green; spur cylindrical or clavate.

Key to species of Section Angraecoides

1. Spur shorter than or equalling the sepals 2
1. Spur much longer than the sepals 6
2. Inflorescence longer than the leaves 3
2. Inflorescence shorter than or equalling the leaves 4
3. Lip broadly triangular, acute; spur slightly hooked 8 mm long; leaves 3–5 mm
   broad .... **44.41. A. chermezoni**
3. Lip broadly ovate, acuminate; spur cylindrical, 7 mm long; leaves 5–6.5 mm
   broad .... **44.43. A. curvicaule**
4. Leaves 8–14 mm long; flowers white; spur 4 mm long .... **44.47. A. scalariforme**
4. Leaves 15–25 mm long; flowers yellowish or greenish; spur 8–9 mm long 5

44.41. **A. chermezoni** H.Perrier

(p. 345, top right)

Epiphyte; stems 20–50 cm long, branched; leaves many, short, lanceolate or linear, 20–60 × 3–5 mm; inflorescences 1-flowered, shorter than the leaves; flowers produced in abundance, small, yellowish; sepals 8 mm long; lip broadly triangular, acute, 7 × 4.5 mm, expanded abruptly above the narrow base, acuminate towards the thick apex; spur slightly expanded towards the hook-shaped tip, 8 mm long. Toamasina. Moss- and lichen-rich montane forest on trunks; 1000–1200 m. Fl. February.

44.42. **A. clavigerum** Ridl.

(p. 345, bottom left)

Small epiphyte; stem 5–10 cm long; leaves 6–12, ligulate, 15–25 × 3–6 mm; flowers relatively large; sepals 13–16 mm long; lip ovate-navicular, acuminate, 11–12 × 4 mm, the area around the column bright green; the spur 6–10 mm long, inflated at the end. Antananarivo, Antsiranana, Fianarantsoa, Toamasina. Moss- and lichen-rich forest; 1100–2000 m. Fl. January–March.

44.43. **A. curvicaule** Schltr.

(p. 345, bottom right)

Epiphyte; stem 35–40 cm long with many linear-ligulate leaves, 40–50 × 5–6.5 mm; inflorescence equalling the leaves, 1–2-flowered; flowers greenish yellow; sepals 7 mm long; lip broadly ovate, apiculate, at its base, 7 mm long, with a central ridge at the base; spur 7 mm long, cylindrical, obtuse. Fianarantsoa, Toamasina. Humid, evergreen forest; 900–1050 m. Fl. January.

44.44. **A. pingue** Frapp. *(p. 347, top left)*
(syn. *A. nasutum* Schltr.)

Erect epiphyte 20 cm tall; leaves narrowly ligulate, thick, 25–30 × 5–7 mm; flowers fleshy, yellow; sepals 14 mm long; lip concave, folding around the column, broadly ovate, acuminate, 15 mm long; spur slender, cylindrical, thickened and almost club-shaped at the apex, 15 mm. Antsiranana; also in the Mascarenes. Moss- and lichen-rich forest; 2000–2200 m. Fl. January.

44.45. **A. rhizomaniacum** Schltr.

Similar to *A. zaratananae* (p. 346), but larger in all parts; epiphyte up to 20 cm long; leaves arcuate-recurved, ligulate, 20–50 × 4–6 mm; flowers small, fleshy, greenish white; sepals 5 mm long; lip very widely rhomboid-suborbicular, 4.5 × 5 mm, concave and shortly apiculate-acuminate; spur slender, cylindrical, a little thickened at the obtuse tip, 6.5 mm long. Antsiranana. Moss- and lichen-rich forest; c. 2000 m. Fl. October, January.

44.46. **A. rostratum** Ridl.

Epiphyte with a 30 cm long stem; leaves lanceolate, obtuse, 18–25 mm long; inflorescences 1-flowered; flowers greenish; sepals 7–8 mm long; lip 12 mm long, tapering to a point; spur 8–9 mm long, a little expanded into a club and recurved at the end. Fianarantsoa. Moss- and lichen-rich forest; c. 1500 m. Fl. March.

44.47. **A. scalariforme** H.Perrier

Elongate epiphyte or lithophyte, 15–40 cm long, unbranched; leaves lanceolate, 8–14 × 1.7–2.5 mm; inflorescence 1-flowered, 16–18 mm long; flower small, white; sepals 9 mm long, acute; lip oblong, acuminate at the tip, 9 mm long; spur 4 mm long, narrowed at the base. Antsiranana. Lichen-rich forest; 1000–1400 m. Fl. March.

44.48. **A. sedifolium** Schltr. *(p. 347, top right)*

Small epiphyte; stem up to 15 cm long, many-leaved; leaves resemble those of *Sedum*, narrowly linear, subcylindrical, 15–25 × 1.5–2.5 mm; inflorescences shorter than the leaves; flowers yellow-green, similar to *A. pingue*, but much smaller plant and flowers; sepals 8 mm long; lip broadly rhombic, 7.5 × 5 mm; spur 13 mm long. Antananarivo, Toamasina. Moss- and lichen-rich, evergreen forest, 900–2000 m. Fl. December–May.

44.40

44.41

44.42

44.43

44.49. **A. triangulifolium** Senghas

Small, tufted, erect or pendent epiphyte; stem simple or branched near base, up to 30 cm long, many-leaved; leaves narrowly triangular, 17 × 7 mm; inflorescences 2–3 per axil, much longer than the leaves; flowers yellowish, diaphanous; sepals 10–12 mm long; lip triangular, 10 × 3 mm; spur 25 mm long, slender from a broad mouth. Antananarivo, Toamasina. Seasonally dry montane woods, 1200 m. Fl. December–February.

44.50. **A. zaratananae** Schltr.

Epiphyte; stem simple, elongate; leaves ligulate, sometimes slightly recurved, 20–50 × 4–6 mm; inflorescences 1-flowered; flowers whitish yellow; sepals 12 mm long; lip very broadly rhombic, 12 × 9 mm; spur a little expanded at the opening, then slender and cylindrical and finally recurved-inflated into a club-shape at the apex, 4 cm long. Related to *A. rhizomaniacum* (p. 344) which has smaller leaves and much smaller flowers with a shorter spur. Antsiranana. Moss- and lichen-rich forest; 500–2200 m. Fl. January–February.

VI. Section Pseudojumellea

Stems elongate, slender; leaves distichous, widely spaced; inflorescence 1-flowered, with a long slender peduncle; flowers white, medium-sized; lip narrow; spur slender, elongate.

Key to species of Section Pseudojumellea

1. Leaves semicylindrical, 1–1.2 mm in diameter . . . . **44.53. A. danguyanum**
1. Leaves dorsiventrally flattened, 5 mm or more in breadth 2
2. Spur 1.2–1.6 cm long, subclavate; sepals 8–10 mm long . . . . **44.55. A. elliotii**
2. Spur 3.5 cm long or more; sepals 11 mm or more long 3
3. Spur 9–13 cm long 4
3. Spur 8 cm or less long 7
4. Inflorescence 2–4-flowered; spur 9–10 cm long . . . . **44.56. A. florulentum**
4. Inflorescence 1–2-flowered; spur 11–13 cm long 5
5. Stem short, up to 10 cm long; leaves 2.2–4 × 0.3–0.6 cm; lip ovate-lanceolate,
   40 mm long . . . . **44.58. A. meirax**
5. Stem up to 60 cm long; leaves 4–10 × 1.5–2.5 cm; lip triangular-ovate,
   15–26 mm long 6
6. Sepals 38–40 mm long; lip 25–26 × 18 mm . . . . **44.51. A. ampullaceum**
6. Sepals 20–22 mm long; lip 15–18 mm long . . . . **44.60. A. moratii**
7. Stem sheaths black-spotted; sepals 11 mm long; lip 9 mm long . . . **44.59. A. melanostictum**
7. Stem sheaths not spotted; sepals 14 mm or more long; lip 12 mm or more long 8
8. Flowers resupinate; lip lanceolate, acute; leaves oblong-lanceolate, acute . . . .
   . . . . **44.57. A. mauritianum**
8. Flowers non-resupinate; lip ovate, broadly ovate or rhombic; leaves not acute 9
9. Leaves 5–14 cm long; inflorescence 2–5-flowered; sepals 22 mm or more long;
   lip ovate-rhombic or broadly ovate 10
9. Leaves 1.3–3 cm long; inflorescence 1–2-flowered; sepals 19–21 mm; lip ovate
   or triangular 11

44.44

44.48

44.53

44.51. **A. ampullaceum** Bosser

(syn. *Jumellea humbertii* H.Perrier)

Epiphyte up to 60 cm tall; leaves numerous, 8–10 × 2.2–2.5 cm; inflorescence 9–10 cm long, 1–2-flowered; peculiar for its spur-shaped growths at the base of the peduncle surrounded by tissue that seems to contain a resin; flowers white; sepals narrowly lanceolate, 3.8–4 cm long; lip triangular, acute, 2.5–2.6 × 1.8 cm; spur 11–13 cm long; and anther clearly 3-toothed at the front. Toamasina, Toliara. Humid, evergreen forest, on trunks; 500–1400 m. Fl. March, November.

44.52. **A. coutrixii** Bosser

Tall lithophyte with small, oblong leaves, 1.5–3 × 0.5–1 cm; inflorescence 1–2-flowered; flowers white; sepals 19–21 mm long; lip ovate, apex subapiculate, slightly concave, 17–22 × 9–12 mm; spur filiform, 5–5.5 cm long. Fianarantsoa. Rocky outcrops, confined to the quartzite mountain ranges of Itremo where it grows with *A. protensum* (p. 382); 1600–2000 m. Fl. January, May.

44.53. **A. danguyanum** H.Perrier   *(p. 347, bottom)*

Epiphyte; stem 15–20 cm long, slender; leaves slightly rigid, semi-cylindrical, linear, 30–35 × 1–1.2 mm; inflorescence 1-flowered; flowers white; sepals 15 mm long; lip large, very broadly elliptic, obtuse, 15 × 12 mm; spur 6–7 cm long, becoming thinner towards the tip. Antananarivo, Fianarantosoa, Toamasina. Humid, evergreen forest on trunks; 700–1000 m. Fl. February.

44.54. **A. dendrobiopsis** Schltr.   *(opposite, top left)*

Pendent epiphyte, 25–60 cm long; leaves lanceolate-linear, 5–12 × 0.5–0.8 cm; inflorescences 4–12 cm long, 2–4-flowered; flowers white; sepals 22–27 mm long; lip rhombic-ovate, 18–22 × 8–13 mm; spur cylindrical, 35 mm long, with a broad mouth. Antsiranana, Mahajanga, Toamasina. Mossy forest; 1000–2000 m. Fl. January–April.

44.54

44.56

44.57

44.59

44.55. **A. elliotii** Rolfe

Epiphyte; stem elongate; leaves linear-lanceolate, acute, distant, 37–50 × 6–8 mm; flowers medium-sized; sepals 8–10 mm long; lip broadly ovate, concave, longly acuminate, 6–8 × 4 mm; spur subclavate, 12–16 mm long. Antananarivo, Fianarantsoa, Toliara. Humid, lowland forest. Fl. unknown.

44.56. **A. florulentum** Rchb.f. *(p. 349, top right)*

Epiphyte; stem elongate and often pendent; flowers similar to *A. dendrobiopsis* (p. 348); leaves short, loriform, 4.5–7 × 1–1.5 cm; inflorescences 2–4-flowered; flowers white; sepals 20–24 mm long; lip ovate, embracing the column at the base; spur 9–10 cm long, 3 times longer than the pedicel and ovary. Possibly endemic to the Comores. On trunks in evergreen forest; 900–1900 m. Fl. November–December.

44.57. **A. mauritianum** (Poir.) Frapp. *(p. 349, bottom left)*

(syn. *Orchis mauritiana* Poir.; *Angraecum gladiifolium* Thouars)

Epiphyte or terrestrial; stem elongate and slender, 22–40 cm long; leaves oblong-lanceolate, acute, 4.5–6 × 1–1.4 cm, evenly spaced; flowers white; sepals 15–20 mm long; lip lanceolate, acute, 15–20 × 7 mm; spur filiform, 7–8 cm long. Antananarivo, Antsiranana, Fianarantsoa, Toamasina, Toliara; also in the Comoros and Mascarenes. Humid, evergreen forest, edge of forests, in humus; 200–1450 m. Fl. February–March.

44.58. **A. meirax** (Rchb.f.) H.Perrier

(syn. *Aeranthus meirax* Rchb.f.)

Small stemless epiphyte, up to 10 cm tall; leaves 3–4, broadly linear, 2.2–4 × 0.3–0.6 cm; inflorescence 1-flowered; flowers relatively large, white; sepals 32–40 mm long; lip ovate-lanceolate, very acutely narrowed above the middle, 40 × 12 mm, with a central ridged callus; spur filiform, 12–13 cm long. Antananarivo, Antsiranana, Toamasina; also in the Comoros. Humid, lowland forest; up to 200 m. Fl. October.

44.59. **A. melanostictum** Schltr. *(p. 349, bottom right)*

Epiphyte; stem up to more than 15 cm long, the stem sheaths densely spotted black; leaves linear-ligulate, 50 × 6–7 mm; inflorescence 1-flowered; flower white; sepals 11 mm long; lip ovate-lanceolate, obtuse, almost flat, 9 mm long; spur filiform, c. 7 cm long. Toamasina, Toliara. Evergreen forest; 100–1450 m. Fl. March.

44.60. A. moratii Bosser   *(p. 353, top left)*

Epiphyte similar to *A. florulentum* (p. 350), but more robust; leaves 4–7.5 × 1.5–2 cm; inflorescences 1–2-flowered; flowers white; sepals 20–22 mm long; lip oblong, boat-shaped-concave with an apiculate apex, 15–18 × 8–9 mm; spur pendent, filiform, pale green, 12–13 cm long. Antsiranana, Toamasina. Humid, evergreen and mossy forest; 500–1000 m. Fl. September.

44.61. A. oblongifolium Toill.-Gen. & Bosser   *(p. 353, top right)*

Epiphyte; stem branching, up to 20 cm long, densely leaved; leaves short, 1.5–2 × 0.5–1 cm, twisted at base to lie in one plane; flowers fleshy, white; sepals 14–17 mm long; lip concave, broadly triangular, 12–13 mm long; spur narrowly cylindrical, narrowing towards the tip, 6–7 cm long. Fianarantsoa. Humid, highland forest. Fl. May.

44.62. A. penzigianum Schltr.

Pendent epiphyte, up to 50 cm long, related to *A. florulentum* (p. 350); leaves ligulate, 8–14 × 1.2–2.3 cm; inflorescences 3–5-flowered, slightly longer than the leaves; flowers white; sepals 23–32 mm long; lip concave, broadly ovate, long acuminate, with a basal keel; spur filiform from a broader mouth, 5 cm long. Fianarantsoa, Toamasina. Mossy forest; 1500–1800 m. Fl. February–March.

VII. Section Aranchnangraecum

Plants with long stems; leaves in 2 rows (distichous); inflorescences arise above the leaf axil on a long, slender peduncle which persists after the flower has died; sepals white or more usually pale yellow or pale pink-buff; lip usually uppermost in the flower, conch-like, apiculate; spur filiform from a broader mouth, elongate.

Key to species of Section Aranchnangraecum

5. Leaves flat or ensiform; flowers white or greenish white; sepals 15 mm or
    more long; lip as broad as long 6
6. Lip cornet-shaped, with a very broad mouth .... **44.71. A. platycornu**
6. Lip filiform, sometimes with a broad mouth but not cornet-shaped 7
7. Leaves ensiform; spur 7–8 cm long; sepals 15 mm long .... **44.72. A. sterrophyllum**
7. Leaves not ensiform; spur 10 cm or more long; sepals 35 mm or more long 8
8. Lip with a 3 cm long acumen; sepals 5–7 cm long; leaves 12–16 cm long ....
    .... **44.70. A. humbertii**
8. Lip lacking a long acumen; sepals 4.5 cm or less long; leaves 10 cm or less long 9
9. Leaves 8–10 cm long, 2–2.5 cm broad; lip triangular-cordate, 25–30 × 18 mm
    .... **44.67. A. ampullaceum**
9. Leaves 25 mm, or less long, 7 mm or less broad; lip 17 mm or less long 10
10. Lip subcircular, 17 × 17 mm .... **44.69. A. conchoglossum**
10. Lip ovate-elliptic, 15–16 × 6–7 mm .... **44.68. A. arachnites**

44.63. **A. linearifolium** Garay   *(opposite, bottom left)*

(syn. *Angraecum palmiforme* H.Perrier non Thouars)

Epiphyte often confused in cultivation with *A. teretifolium*. Plant slender, pendent, with 6–12 filiform leaves, 5–11 cm long, 0.6–0.9 mm in diameter; flowers spidery, with salmon pink, acuminate tepals; lip very concave, semi-circular, edged rounded, with a strong central ridge at the base; spur conical at the base, then filiform, 9–10 cm long; anther truncate. Antananarivo, Antsiranana, Fianarantsoa. Toamasina, Toliara. Humid, mossy, evergreen forest; 500–2500 m. Fl. February, July, October–December.

44.64. **A. teretifolium** Ridl.   *(opposite, bottom right)*

Epiphyte often confused with *A. linearifolium* but with 4–6 leaves, 4.5–8 × 0.1–0.2 cm, a little narrower, the sepals and petals 1.2–2 cm long, acuminate, the lip very concave and with a central ridge at the base, ovate, acute, up to 3.2 × 1.4 cm; spur filiform, 6–10 cm long, and the anther truncate and a little emarginate at the front. Antananarivo, Fianarantsoa, Toamasina. Shady, evergreen, highland forest, lichen-rich forest; 1200–1500 m. Fl. October–November.

44.65. **A. popowii** Braem.   *(p. 355, top left)*

Erect epiphyte similar to *A. teretifolium* but with semiterete, channelled leaves, 6–8.5 × 0.2 cm, the sepals and petals 3 × 0.4–0.5 cm long, acuminate, whitish green, the lip very concave and with a central ridge at the base, ovate, acute, 3.2 × 1.4 cm, white; spur filiform, about 14 cm long, and the anther truncate and a little emarginate at the front. Antananarivo, Fianarantsoa, Toamasina. Shady, highland *Uapaca bojeri* (tapia) forest; 1200 m. Fl. October–November.

44.60

44.61

44.63

44.64

44.66. A. pseudofilicornu H.Perrier *(opposite, top right)*

Epiphyte, very similar to *A. scottianum* from the Comores; stem 10–20 cm long; roots warty; leaves cylindrical, thick-textured, 30–80 × 3–5 mm; flowers with lip uppermost, with salmon-pink sepals and petals and a white lip; sepals 18 mm long; lip very concave, 25 × 18 mm, cuspidate, with a ridge at the base; spur 15 cm long, compressed and narrowed below the 6 mm wide opening, then filiform. Antsiranana, Toamasina. Mossy forest on moss- and lichen-covered trees; c. 1000 m. Fl. November.

44.67. A. ampullaceum Bosser

Epiphyte; stem erect, up to 60 cm long; leaves linear, roundly bilobed, 8–10 × 2–2.5 cm, articulated near base; flowers white; sepals 38–40 mm long; lip triangular, cordate, shortly acuminate, 25–30 × 18 mm; spur 11–13 cm long, slender, tapering from a 5 mm wide mouth. Antananarivo, Toamasina, Toliara. Humid, evergreen forest; 700–1400 m. Fl. November.

44.68. A. arachnites Schltr. *(opposite, bottom left)*

(syn. *Angraecum ramosum* subsp. *typicum* var. *arachnites* (Schltr.) H.Perrier)

Small epiphyte with a long stem carrying very small, somewhat thickened ovate leaves with the margins curved downwards, 6–20 × 2–3.5 mm; flowers white; sepals and petals brownish-yellow, up to 45 mm long; lip white rounded and narrow, 15–16 × 6–7 mm; spur 10–11 cm long. Antananarivo, Antsiranana, Fianarantsoa, Toamasina. Mainly in eastern evergreen forest; 1380–1500 m. Fl. November–March.

44.69. A. conchoglossum Schltr. *(opposite, bottom right)*

(syn. *Macroplectrum ramosum* Finet; *Angraecum bathiei* Schltr.; *A. ramosum* subsp. *typicum* H.Perrier; *A. mirabile* Schltr.)

Variable epiphyte; roots rough to verrucose; leaves oblong-ligulate, unequally obtusely bilobed, 1.8–2.5 × 0.6–0.7 cm, more or less flat, thick- or thin-textured; flowers are similar to *A. germinyanum* from the Comoros, but have 4 cm long, thin, yellowish tepals; lip rounded, 17 × 17 mm, white; spur 11.5 cm long, filiform. Antananarivo, Antsiranana, Fianarantsoa, Toamasina; also in the Mascarenes. Mainly in mossy, evergreen forest; on *Philippia*; 1000–1600 m. Fl. September–May.

44.70. A. humbertii H.Perrier

Lithophyte or epiphyte; stem short; leaves long and leathery, 12–16 × 1–1.3 cm; flower greenish white with a pure white lip; sepals 5–7 cm long; lip very concave, conch-shaped at the spur opening, 5 cm long including the 3 cm long acumen; spur at first broad, then acuminate-filiform, up to 14 cm long. Toliara. Dry, deciduous scrub and rocky outcrops; 800–900 m. Fl. January–February.

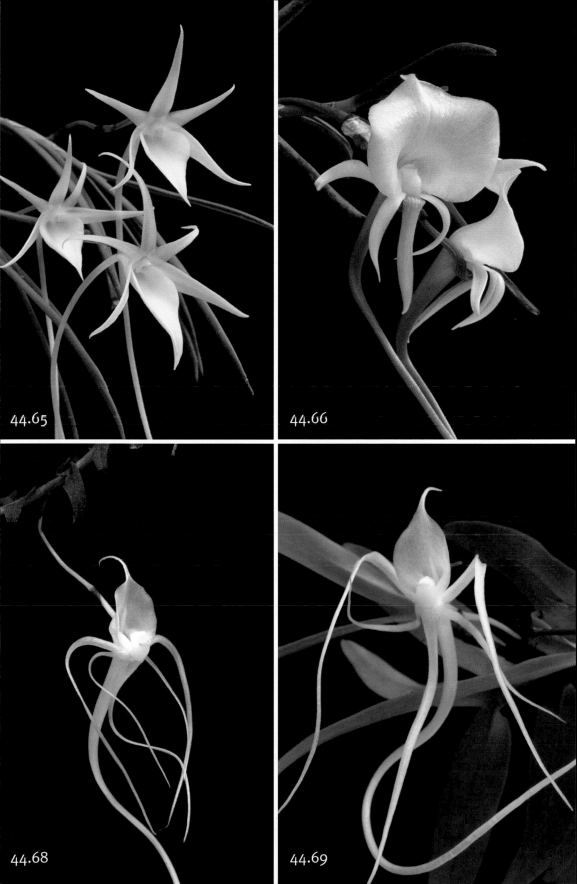

44.65

44.66

44.68

44.69

44.71. **A. platycornu** Hermans, P.J.Cribb & Bosser   (*opposite, left*)

Epiphyte close to *A. conchoglossum* (p. 354) but the flower is distinguished by the lip which is much broader than long, very obscurely 3-lobed, and with a broad-mouthed, horn-like spur at its base. Its rostellum is also distinctive in having a distinctive short tooth between the large lateral lobes. Toamasina. Humid, lowland forest; up to 100 m. Fl. July.

44.72. **A. sterrophyllum** Schltr.

Epiphyte; stem 8–15 cm long; roots warty; leaves ensiform, 40–50 × 2.5–3 mm; flowers 1–3, white; sepals 15 mm long; lip suborbicular, concave, apiculate, wider than long, 12 × 15 mm, with a central ridge at the base; spur arched-pendent expanded at the entrance, then filiform, 7–8 cm long. Antananarivo. On *Uapaca bojeri* (tapia), in drier places on small trees, in humid evergreen forest on plateau; c. 1400 m. Fl. October–January.

44.73. **A. triangulifolium** Senghas   (*opposite, right*)

A pendent epiphyte up to 30 cm long; leaves narrowly triangular, 17 × 7 mm; inflorescences 1-flowered; flower diaphanous, yellow; sepals 10–12 mm long; lip obovate, 8–9 × 1.8 mm, with a narrow callus in the middle extending into the spur; spur 22–27 mm long, fairly wide at the opening and narrowing towards the base. Toamasina. Lower-montane and humid, plateau, evergreen forest; c. 1200 m. Fl. August.

VIII. Section Gomphocentrum

Stem short but distinct; leaves ligulate; inflorescences from the axils of the lower leaves, few–many-flowered; flowers small, yellow-green, diaphanous.

Key to species of Section Gomphocentrum

1. Spur 12–25 mm long 2
1. Spur 10 mm or less long 4
2. Sepals 8–11 mm long; spur twisted at the base, 12 mm long .... **44.82. A. guillauminii**
2. Sepals 6–8 mm long; spur not twisted at base, 12–24 mm long 3
3. Stem short; inflorescence shorter than the leaves, simple, 2–4-flowered;
   spur filiform, 22–24 mm long .... **44.75. A. andringitranum**
3. Stem elongate, up to 40 cm long; inflorescence as long as or longer than the
   leaves, branched, many-flowered; spur cylindrical, 12–15 mm long .... **44.76. A. calceolus**
4. Stem branching; leaves 1.5–2 cm long, falciform, conduplicate ...... **44.81. A. falcifolium**
4. Stem not branched; leaves 3 cm or more long, not falciform, conduplicate 5
5. Inflorescences long-lived, flowering on a new branch in a parallel series in
   successive years; spur 2–4 mm long .... **44.84. A. multiflorum**
5. Inflorescences annual; spur 4 mm or more long 6

6. Leaves 1.5–2 cm broad; inflorescences 8–12 cm long . . . . **44.77. A. caulescens**
6. Leaves 1.4 cm or less broad 7
7. Stem very short 8
7. Stem elongate 10
8. Flowers yellow; sepals 8 mm long; spur 8 mm long . . . . **44.89. A. vesiculatum**
8. Flowers yellow-green, sometimes with a whitish lip; spur 5 mm long 9
9. Leaves ligulate, 4–9 cm long; lip 5 mm long . . . . **44.74. A. acutipetalum**
9. Leaves grass-like, very narrowly lanceolate, acute, 5–14 cm long; lip 7 mm
   long . . . . **44.78. A. cornucopiae**
10. Leaves 12.5–18 cm long; 5–7-flowered . . . . **44.80. A. dauphinense**
10. Leaves 12 cm or less long; inflorescence 1–4(–5) flowered 11
11. Sepals 8 mm long, recurved clavate at tip; lip 7.5 × 4 mm . . . . **44.79. A. corynoceras**
11. Sepals 7 mm or less long, not curved at tip; lip 6 mm or less long 12
12. Flowers green or greenish white 13
12. Flowers yellow 15
13. Leaves 10–14 × 0.8–1.4 cm long; spur 7.5–10 mm long . . . . **44.88. A. verecundum**
13. Leaves 5–8.5 × 0.3–0.6 cm; spur 5–6 mm long 14
14. Spur and lip 6 mm long . . . . **44.83. A. ischnopus**
14. Spur and lip 5 mm long . . . . **44.87. A. tenuipes**
15. Leaves 5–9 × 0.7–0.8 cm; spur cylindrical . . . . **44.85. A. rhizanthium**
15. leaves 3.7–6 × 0.4–0.7 cm; spur saccate . . . . **44.86. A. sacculatum**

44.71   44.73

44.74. **A. acutipetalum** Schltr. *(opposite, top left)*

Epiphyte; stem 10–12 cm long; leaves 5–9, ligulate, 4–12 × 0.5–1 cm; inflorescences shorter than the leaves, 2–5-flowered; flowers pale yellow-green with a whitish lip; sepals 5.5 mm long; lip broadly ovate-cucullate, acuminate, 5 mm long; spur cylindrical, 5 mm long, dilated abruptly at apex. Antananarivo, Antsiranana. Mossy and lichen-rich forest; 1000–2000 m. Fl. September–January.

44.74a. **A. acutipetalum** var. **analabeensis** H.Perrier ex Hermans

Differs from the typical variety in its 3 mm long sepals, shorter spur and anther which is broadly indented at the front, the margins clearly acutely angular, and the 2 linear stipes being united at the base by a large viscous gland. Its flowers are white, with the sepals and petals slightly yellowish. Antananarivo. Humid, highland forest. Fl. unknown.

44.74b. **A. acutipetalum** var. **ankeranae** H.Perrier ex Hermans

Differs from the typical variety in having a shorter spur, slightly downy disk of the lip and an anther with a rounded anterior indentation. Its flower is also generally smaller. Antananarivo. Humid, highland forest. Fl. unknown.

44.75. **A. andringitranum** Schltr.

Epiphyte; stem up to 40 cm long; leaves ligulate, 11–13 × 1.1–1.3 cm; inflorescences shorter than the leaves, 2–4-flowered; flowers small, greenish white; sepals 7 mm long; lip uppermost, ovate, acute, 7 × 4 mm; spur 22–24 mm long and filiform. Fianarantsoa (Andringitra). Moss- and lichen-rich forest; c. 1600 m. Fl. March.

44.76. **A. calceolus** Thouars *(opposite, top right)*

(syn. *Angraecum carpophorum* Thouars; *Aerobion calceolus* Spreng.; *A. paniculatum* Frapp.; *A. rhopaloceras* Schltr.; *A. patens* Frapp.; *A. laggiarae* Schltr.)

Variable and widespread epiphyte, terrestrial or lithophyte; leaves 3–10, narrowly lanceolate, 6–20 × 0.8–1.8 cm; inflorescence 10–15 cm long, simple or up to 3-branched; flowers small, greenish-yellow; sepals 6–8 mm long; lip cucullate, ovate-lanceolate, 6–10 × 4–5 mm; spur cylindrical sometimes a little thickened at the end, 12–15 mm long. Antananarivo, Antsiranana, Fianarantsoa, Mahajanga, Toliara, Tomasina; also in the Comoros, Mascarenes and Seychelles. Humid, evergreen forest, on rocks and bases of tree trunks amongst moss; at the base of ericaceous scrub; dry, semi-deciduous, coastal forest; tapia forest; sea level–2000 m. Fl. October–May.

44.74

44.76

44.77

44.77. **A. caulescens** Thouars   *(p. 359, bottom)*

Stem up to 10 cm long; leaves up to 10, ligulate, 15–20 × 1.5–2 cm; inflorescences 8–12 cm long, unbranched; flowers 3–5, pale green; sepals 6 mm long; lip ovate, acute, 5 mm long; spur 8–10 mm long, curved and dilated at tip; differing from *A. multiflorum* (p. 361) by the stem being shorter than the leaves, and by its simple and few-flowered raceme, which is as long as the leaves. Antananarivo; also in the Comoros and Mascarenes. In forest; intermediate altitudes. Fl. March.

44.78. **A. cornucopiae** H.Perrier

Large, short-stemmed epiphyte, 15–25 cm tall; leaves grass-like, linear-lanceolate, acute, 5–14 × 0.4–0.7 cm; inflorescences 2–4, 3–5 cm long, similar to *A. multiflorum* (p. 361), 2–7-flowered; flowers small, yellow-green; sepals 6 mm long; lip widely boat-shaped-ovate, apiculate, 7 × 4 mm; spur cylindrical-tapering, 5 mm long. Antsiranana. Lichen-rich, evergreen forest; 1000–2000 m. Fl. November–December.

44.79. **A. corynoceras** Schltr.

Epiphyte; stem up to 25 cm long; leaves slender, 60–100 × 5–7 mm; inflorescences 1–2-flowered, slender, equalling the leaves; sheaths short, brown; flowers yellowish; sepals 8 mm long; lip 7.5 × 4 mm; spur 8 mm long, cylindrical then recurved and clavately inflated towards the tip. Antsiranana. Moss- and lichen-rich forest; c. 2000 m. Fl. January.

44.80. **A. dauphinense** (Rolfe) Schltr.

(syn. *Mystacidium dauphinense* Rolfe)

Epiphyte; stem elongate; leaves linear, 12.5–18 × 0.6–1 cm; inflorescences 5–7-flowered; sepals 6 mm long; lip broadly ovate, 6 mm long; spur cylindrical, straight, 6 mm long. Allied to *A. caulescens*, but with much longer leaves and slightly different flowers. Toliara. Coastal forest. Fl. unknown.

44.81. **A. falcifolium** Bosser

Long and branching epiphyte; stems leafy, up to 50 cm long; leaves conduplicate, leathery, falciform, 1.5–2 × 0.8–1 cm; inflorescences shorter than the leaves; flowers 3–5, small, fleshy, green; sepals 6–6.5 mm long; lip ovate-lanceolate, acute, concave, 6 × 2 mm, downy-papillose on the inside; spur oblong at the back and thickened at the front, apex rounded, 3 mm long. Fianarantsoa. Mossy forest. Fl. February.

44.82. A. guillauminii H.Perrier

Plant c. 15 cm long; leaves ligulate, 6.5–15 × 1.5–2.5 cm; inflorescence 12 cm long, 5–6-flowered; flowers small; sepals 8–11 mm long; lip with the base broadly ovate, concave, longly acuminate; spur 12 mm long, twisted at base. Madagascar. Fl. September.

44.83. A. ischnopus Schltr.

Epiphyte; stem up to 30 cm long; leaves linear, 5–8.5 × 0.3–0.6 cm; inflorescences many, equalling the leaves, 1–3-flowered; flowers small, greenish; sepals 5–7 mm long; lip ovate, 6 × 3 mm, shortly acuminate; spur reflexed in a dilated apex, 6 mm long. Antsiranana. Lichen-rich montane forest; 2500 m. Fl. October.

44.84. A. multiflorum Thouars   (p. 363, top left)

(syn. *Aerobion multiflorum* (Thouars) Spreng.; *Angraecum caulescens* var. *multiflorum* (Thouars) S.Moore)

Medium-sized epiphyte, up to 20 cm tall; leaves oblong-lanceolate, attenuate at the bilobed tip, 3–5 × 0.4–0.6 cm; inflorescences grouped in a unilateral linear series, some in flower, the others having flowered in previous years, very shortly branched; flowers 1–3 per branch, yellowish; sepals 3.5–5.5 mm; lip 3–5 mm long, deeply bowl-shaped; spur with a wide opening, 2–4 mm long, more or less expanded and curved at the apex. Antananarivo, Antsiranana; also in the Mascarenes and Seychelles. Humid, evergreen forest on plateau; 1200–1500 m. Fl. November–February.

44.85. A. rhizanthium H.Perrier

Epiphyte, 10–15 cm tall; leaves ligulate-linear, 5–9 × 0.7–0.8 cm; inflorescence 5–6 cm long, 1–3-flowered; sepals 10 mm long; differs from *A. multiflorum* by its very narrow conch-shaped lip, the cylindrical and straight spur and the reflexed tip of the dorsal sepal. Antsiranana. Mossy, evergreen forest; 1000–1500 m. Fl. January–February.

44.86. A. sacculatum Schltr.

Epiphyte with a stem up to 17 cm long; leaves linear, 37–60 × 4–7 mm; inflorescences 1–2-flowered; flowers small, yellowish; sepals 4–4.5 mm long; lip ovate, acuminate, up to 5 mm long; spur oblong saccate, obtuse, scarcely 4 mm long. Antananarivo, Fianarantsoa. Mossy forest; 1300–1800 m. Fl. September–February.

44.87. A. tenuipes Summerh. *(opposite, top right)*

(syn. *Angraecum ischnopus* Schltr. non Schltr.)

Epiphyte; stem up to 30 cm long; leaves linear, 5–8.5 × 0.3–0.6 cm; related to *A. caulescens* (p. 360), but differs in its thin inflorescence and pointed leaves; flowers 1–3, small, greenish; sepals 5–7 mm long; lip ovate, concave, shortly acuminate-subacute, 5 × 3 mm; spur as long as the lip, curved, gradually inflated from the entrance to the apex into a broad purse-shape. Antananarivo, Antsiranana. Moss- and lichen-rich forest; 1400–2500 m. Fl. September–December.

44.88. A. verecundum Schltr.

Epiphyte; stem 15–20 cm long; leaves linear-ligulate, 10–14 × 0.8–1.4 cm; inflorescences shorter than the leaves, 1–2-flowered; flowers medium-sized, whitish green; sepals 7 mm long; lip broadly ovate, apiculate, 6.5 × 6 mm; spur subcylindrical, a little expanded towards the obtuse tip, 7.5–10 mm long. Antananarivo. Mossy forest; c. 2000 m. Fl. February.

44.89. A. vesiculatum Schltr.

Epiphyte up to 10 cm tall; leaves ligulate, up to 50 × 7 mm; flowers small, yellowish; sepals 8 mm long; lip 7 × 7 mm, ovate-elliptic and acuminate; spur sac-shaped, almost as long as the sepals, 8 mm long. Antsiranana. Lower-montane and plateau, humid, evergreen forest; 1500–2000 m. Fl. October.

IX. Section Lepervenchea

Stems elongate, leafy, often pendent and branched; leaves in 2 rows; inflorescences axillary, towards base of stem, longer than the leaves, 1–several-flowered; flowers small, greenish yellow, diaphanous.

Key to species of Section Lepervenchea

1. Inflorescences 2–4 times as long as the leaves . . . . **44.93. A. pauciracemosum**
1. Inflorescences more or less as long as or shorter than the leaves 2
2. Sepals 4 mm long; lip 4 mm long . . . . **44.92. A. musculiferum**
2. Sepals 3 mm or less long; lip 3 mm or less long 3
3. Leaves 1.3–1.6 cm long; flowers greenish yellow . . . . **44.90. A. appendiculoides**
3. Leaves 3–7 cm long; flowers yellow 4
4. Leaves 4 mm broad; sepals 3 mm long; lip broadly ovate, 3 mm long . . . **44.91. A. caricifolium**
4. Leaves 2 mm long; lip ovate-rhombic, 2.25 mm long . . . . **44.94. A. tenuispica**

44.84

44.87

44.90

44.91

44.90. A. appendiculoides Schltr.   *(p. 363, bottom left)*

Pendent epiphyte; stem very long, pendent, densely leafy; leaves ovate-lanceolate, 13–16 × 7 mm; inflorescences 4–7-flowered; floral bracts longer than the pedicellate ovary; flowers small, greenish yellow; sepals 2.5 mm long; lip broadly ovate, 2 mm long; spur almost straight, 1.75 mm long. Antananarivo. Mossy, evergreen forest; 1700–2000 m. Fl. September–December.

44.91. A. caricifolium H.Perrier   *(p. 363, bottom right)*

Long, slender epiphyte; stem 10–25 cm long; leaves linear, thin-textured, 5–7 × 0.4 cm; flowers very small; inflorescence 2–6 cm long, 5–10-flowered; sepals 3 mm long; lip concave, ovate-lanceolate, 3 × 2.5 mm, with tiny hairs inside; spur thick, cylindrical, obtuse, 2.5 mm long. Antananarivo, Antsiranana, Toamasina. Moss- and lichen-rich forest; 900–1400 m. Fl. January–May, July, October.
LOCAL NAME *Velomitongoa*.

44.92. A. musculiferum H.Perrier

Epiphyte with 6–30 cm long stem; leaves linear, 2.5–5 × 0.7–0.8 cm; inflorescences many, as long as or longer than the leaves, 4–10-flowered; flowers small, yellowish; sepals 4 mm long; lip very concave, boat-shaped, 4 × 2.6 mm, acuminate-acute; spur cylindrical, 2 mm long. Antananarivo, Antsiranana, Toamasina. Mossy, montane forest; 1200–1400 m. Fl. September–April.

44.93. A. pauciramosum Schltr.   *(opposite, top left)*

(syn. *Mystacidium graminifolium* Ridl.; *Angraecum graminifolium* (Ridl.) Schltr. non (Kraenzl.) Engl.; *A. poöphyllum* Summerh.)

Pendent epiphyte; stem 15–20 cm long, branching; leaves ligulate or almost linear, 10–20 × 3–4 mm; inflorescence 2– 4 times as long as the leaves, 4–7-flowered; flowers small, yellowish; sepals 1.5 mm long; lip very concave, ovate, acuminate, 2 mm long; spur almost cylindrical, about 2 mm long. A variable species close to *A. caricifolium*. Antananarivo, Fianarantsoa, Toamasina. Humid, evergreen forest; 800–1200 m. Fl. September–July.
LOCAL NAME *Hazomiavona*.

44.94. A. tenuispica Schltr.

Related to *A. pauciramosum*; stem to 25 cm long; leaves linear-ligulate, 3–7 × 0.6–0.8 cm; inflorescences equalling the leaves, 5–10-flowered; flowers yellowish; sepals 2 mm long; lip ovate-rhombic, obtuse, 2.25 × 1.5 mm; spur cylindrical, 2.25 mm long. Madagascar. Fl. unknown.

44.93

44.96

44.97

44.99

X. Section Lemurangis

Stems long, leafy; leaves in 2 rows (distichous); inflorescences shorter than the leaves; flowers 1–5, small.

Key to species of Section Lemurangis

1. Sepals and petals 11–13 mm long; lip 9–11 mm long; spur 8 mm long, spherical at the tip ....**44.98. A. floribundum**
1. Sepals and petals 2–3.5 mm long; lip 2–3.5 mm long; spur 1–2 mm long, mostly not spherical at tip 2
2. Leaves ovate or broadly ovate, obtuse 3
2. Leaves linear or lanceolate 4
3. Inflorescence as long as or longer than the leaves ....**44.95. A. alleizettei**
3. Inflorescence shorter than the leaves ....**44.99. A. madagascariense**
4. Leaves not recurved; flowers pale green; spur curved and dilated but not globular at tip ....**44.96. A. baronii**
4. Leaves recurved; flowers white; spur 2 mm long, globular at tip ....**44.97. A. decaryanum**

44.95. **A. alleizettei** Schltr.

Epiphyte; stem 8–20 long, branched; leaves ovate, obtuse, 6–12 × 3–6 mm; inflorescences equalling or longer than the leaves, 2–5-flowered; flowers very small; sepals 1.8–2 mm long; lip deeply boat-shaped, 2 mm long; spur straight, 1.5 mm long. Antananarivo, Fianarantsoa, Toamasina. Mossy forest; 1400–1700 m. Fl. January, July, November.

44.96. **A. baronii** (Finet) Schltr.   *(p. 365, top right)*

(syn. *Macroplectrum baronii* Finet; *Angraecum dichaeoides* Schltr.)

Pendent epiphyte; stem long; leaves lanceolate, up to 20 × 4 mm, distichous; inflorescences 1–2-flowered; flowers small, pale greenish; sepals 3 mm long; lip cucullate, concave, 3 mm long; spur curved and a little thickened towards the tip, 1–1.5 mm long. Antananarivo, Antsiranana, Fianarantsoa. Moist, moss- and lichen-rich, evergreen forest; 1400–2500 m. Fl. September–April, July.

44.97. **A. decaryanum** H.Perrier   *(p. 365, bottom left)*

Epiphyte; stem long; leaves thick, recurved-linear, acute, 10–20 × 1–5 mm; inflorescences 1–4-flowered; flowers small, white; sepals 3–3.5 mm long; lip elliptic, apiculate-obtuse, 3–3.5 × 2 mm; spur short, c. 2 mm long, tip globular. Toliara. Dry, deciduous scrubland on *Didiereaceae* and coastal forest; up to 200 m. Fl. June.

44.98. A. floribundum Bosser

Epiphyte; stem 40–60 cm long; leaves distichous, ligulate, 8–13 × 1.5–2 cm; inflorescences short, 1–2-flowered; flowers green; sepals 11–13 mm long; lip broad, elliptic, boat-shaped, 9–11 × 4–5 mm, papillose; spur 8 mm long, clavate, apex dilated into a sphere, laterally compressed. Antananarivo. Mossy forest. Fl. August.

44.99. A. madagascariense (Finet) Schltr.   *(p. 365, bottom right)*
(syn. *Macroplectrum madagascariense* Finet)

Medium-sized epiphyte; stem 6–10 cm long; leaves broadly ovate, 10–13 × 4.5–5 mm; inflorescences short, 3–4-flowered; flowers very small; sepals 2 mm long; lip very concave, broadly ovate, shortly acuminate-acute at the front, 1.5 × 1 mm; spur cylindrical, 1 mm, very obtuse at the apex. Antsiranana. Humid, highland forest; 1500 m. Fl. unknown.

XI. Section Nana

Dwarf plants with short stems; inflorescences densely several-flowered, often subsecund; flowers tiny.

Key to species of Section Nana

1. Flowers covered by red hairs on outer surface of sepals; leaves reddish . . . . **44. 107. A. rubellum**
1. Flowers glabrous, white or yellow; leaves green 2
2. Inflorescences 5- or more-flowered 3
2. Inflorescences 1–3-flowered 6
3. Inflorescences up to 33 cm long, 7–12-flowered . . . . **44.101. A. bemarivoense**
3. Inflorescence 20 cm or less long, 5- to many-flowered 4
4. Leaves 7–8.5 cm long; inflorescences 5–9-flowered . . . . **44.104. A. onivense**
4. Leaves 1.2–2.8 cm long; inflorescences many-flowered 5
5. Leaves 4, ligulate, 2.8 cm long; lip 1.1 mm long . . . . **44.102. A. microcharis**
5. Leaves 2–3, lanceolate, 1.2 cm long; lip 1.5 mm long . . . . **44.108. A. tenellum**
6. Sepals 4 mm long; lip 4–5.5 mm long 7
6. Sepals 1.25–2.8 mm long; lip 2.3–2.5 mm long 8
7. Inflorescences 7–15 cm long, twice as long as the leaves or more . . . **44.100. A. andasibeense**
7. Inflorescences shorter than or as long as the leaves, up to 7.5 cm long . . . . **44.103. A. muscicolum**
8. Leaves 0.8–2.5 × 0.1–0.3 cm; sepals 2 mm long; lip broadly ovate, 2.3 × 3 mm, broader than long . . . . **44.105. A. perhumile**
8. Leaves 0.45–0.8 × 0.2–0.25 cm; lip pandurate, apiculate, 2.4 × 1.6 cm . . . . **44.106. A. perparvulum**

44.100. A. andasibeense H.Perrier

Stemless epiphyte; leaves 5–7, narrowly linear, 30–60 × 3–4 mm; inflorescences numerous, 7–15 cm long, 1–2-flowered; flowers small, yellowish; sepals 4 mm long; lip ovate-funnel shaped, 5.5 × 4 mm; spur short and slender, 1.3 mm long. Antananarivo, Antsiranana, Toamasina. Mossy and lichen-rich forest; 1000–1700 m. Fl. February.

44.101. A. bemarivoense Schltr.

Small epiphyte similar to *A. tenellum* (p. 370) and *A. microcharis*, 4–5 cm tall; leaves 3–5, obliquely oblong, up to 15 × 5 mm; inflorescences up to 33 mm long, 7–12-flowered; flowers white or whitish; sepals 1.25 mm long; petals oblong, somewhat obtuse, a little narrower than the sepals; lip ovate, 1.2 mm long; spur a little longer than the ovary, 1.35 mm long. Antananarivo, Antsiranana. Humid, evergreen forest on branches; 800–1600 m. Fl. March, September.

44.102. A. microcharis Schltr. *(opposite, top left)*

Tiny epiphyte, 3–3.5 cm tall; leaves 4, ligulate, 28 × 3.5–4 mm; inflorescences numerous, many-flowered; flowers whitish; sepals 1.25 mm long; lip ovate, obtuse 1.1 × 1.25 mm; spur thick, oblong, obtuse, 1.25 mm long, a little shorter than the ovary. Antsiranana. Seasonally dry, deciduous forest and woodland, on branches of shrubs. c. 600 m. Fl. January.

44.103. A. muscicolum H.Perrier *(opposite, top right)*

Stemless epiphyte; leaves 3–4, linear, 2.5–7.5 × 0.1–0.3 cm; inflorescences 1–3-flowered, shorter or as long as the leaves; flowers yellow; sepals 4 mm long; lip very concave, very broadly boat-shaped, 4 × 4 mm, rounded-subcordate at the base, shortly acuminate or acute at the tip; spur 3 mm, narrowing from the wide entrance to the middle, then expanded in an oval club-shape. Antananarivo. Mossy, montane forest; c. 1400 m. Fl. February.

44.104. A. onivense H.Perrier

Small stemless epiphyte up to 10 cm tall; leaves 7–8, lanceolate-linear, 7–8.5 × 0.3–0.4 cm; flowers small, white; sepals 1.8 mm long; lip very concave, ovate, 1.8 × 1.3 mm; spur saccate, 1.4 mm long, contracted in the middle. Antananarivo, Toamasina. Mossy forest; c. 1000 m. Fl. February.

44.102

44.103

44.105

44.107

44.105. A. perhumile H.Perrier   *(p. 369, bottom left)*

Small epiphyte, up to 5 cm tall; leaves narrow, 8–25 × 1–3 mm; inflorescences 1–3-flowered, as long as the leaves; sepals 2 mm long; lip very concave, wider than high, 2.3 × 3 mm; spur cylindrical, 1.5 mm long, cylindrical. Antananarivo. Mossy forest on branches and lichen-covered trees; c. 1400 m. Fl. February.

44.106. A. perparvulum H.Perrier

Tiny epiphyte; leaves 4.5–8 × 2.2–2.5 mm; flowers small; sepals 2.5–2.8 mm long; lip almost pandurate, shortly acuminate, 2.4 × 1.6 mm; spur scrotiform, 1.4 × 1.2 mm. Antananarivo, Antsiranana. Humid, evergreen forest on plateau on branches; 1500–2000 m. Fl. February–April.

44.107. A. rubellum Bosser   *(p. 369, bottom right)*

Small epiphyte with reddish leaves; flowers tiny, fleshy, pink and carrying red hairs on the outer surface; lip 2.5 × 2.5 mm, acute at the tip, with rounded edges, raised and narrowly surrounding the column; spur 2.5–3 mm long, slightly flattened dorsi-ventrally, expanded at the tip. Toamasina. Humid, evergreen forest; on *Pandanus*; 900 m. Fl. March.

44.108. A. tenellum (Ridl.) Schltr.   *(p. 373, top left)*

(syn. *Mystacidium tenellum* Ridl.; *Angraecum waterlotii* H.Perrier; *Saccolabium microphyton* Frapp.; *A. microphyton* (Frapp.) Schltr.)

Tiny epiphyte, 3–3.5 cm tall; leaves 2–3, lanceolate, 12 × 4 mm; inflorescences 25 or more mm long, longer than the leaves, many-flowered; flowers white; sepals 1.5 mm long; lip ovate, obtuse, 1.8 × 1.25 mm; spur thick, cylindrical, obtuse, a little shorter than the ovary. Antananarivo, Fianarantsoa. Humid, evergreen forest; 1000–1500 m. Fl. October, February–June.

XII. Section Boryangraecum

Plants with short stems; inflorescences several-flowered; flowers small to medium-sized.

Key to species of Section Boryangraecum

44.109. **A. aviceps** Schltr.

Small, stemless epiphyte, 6–7 cm tall; leaves ligulate, 5–6 × 1–1.2 cm; inflorescence slightly shorter than the leaves, 3–5-flowered; flowers small, yellowish white; lip with its pointed end, resembling a bird's head, with the sepals and petals as wings; sepals 7 mm long; lip ovate, rostrate, 6.5 × 2.5 mm; spur horizontal, subfiliform, 9–10 mm long. Antsiranana. Humid, evergreen forest on plateau; c. 1000 m. Fl. July.

44.110. **A. flavidum** Bosser

Small stem-less epiphyte; leaves 3–4, distichous, ligulate, 3–5 × 0.3–0.45 cm; inflorescences 2–4-flowered; flowers yellowish, larger than any other species of the section; its sepals and petals are narrow and tapering, 15–16 mm long; lip ovate, acuminate, 14–15 × 3–4 mm; spur 8–10 mm long. Antsiranana. Humid, evergreen forest; trees on limestone formations; sea level–500 m. Fl. January.

44.111. **A. myrianthum** Schltr.

Almost stemless epiphyte or lithophyte close to *A. pusillum* and *A. burchellii* from South Africa; leaves 7–9, linear, 12–20 × 0.7–0.9 cm; inflorescence very densely several-flowered, longer than the leaves; sepals 2 mm long; lip ovate-lanceolate, acuminate, 2 mm long; spur conical, subacute, a little contracted at the base, 0.75 mm long. Toliara. Dry, deciduous scrub, in Didieraceae scrub on branches; 250–1400 m. Fl. March–June.

44.112. **A. ochraceum** (Ridl.) Schltr.   *(opposite, top right)*

(syn. *Mystacidium ochraceum* Ridl.)

Epiphyte; stem short; leaves narrowly linear-lanceolate, 10 x 0.6 cm; inflorescences shorter than the leaves, 1-flowered; flowers ochre-coloured; sepals lip concave, ovate, acute at the tip, a little longer than the petals; spur slender, filiform, with the apex a little expanded. Antananarivo, Fianarantsoa, Toamasina. Fl. April.

44.113. **A. pergracile** Schltr.

Stemless epiphyte; leaves 3–4, oblong or oblong-ligulate, 4.5 × 0.6–0.8 cm; inflorescence up to 8 cm long, 1-flowered; peduncle setiform; flower whitish yellow; sepals 7–8 mm long; lip ovate, 6.5 × 4.5 mm; spur cylindrical, 25 mm long; related to *A. chaetopodum* (p. 325) but differs in its shorter and wider leaves and narrower flowers. It differs from *A. rhynchoglossum* (p. 325)which has narrower sepals and petals, a longer lip and a thinner spur. Antsiranana. Seasonally dry, deciduous forest and woodland, in dry woods on trunks; c. 200 m. Fl. January.

44.114. **A. pinifolium** Bosser   *(opposite, bottom left)*

Epiphyte; leaves narrowly linear, up to 10 cm long, 2 mm wide; flowers yellow-green; sepals 7–7.5 mm long; lip ovate-acuminate, 4.5–5.5 mm long, with a prominent ridge at the base; spur 6–6.5 mm long, 2 mm in diameter, gradually inflated at the tip, with a few thin hairs at the spur entrance. Toamasina. Humid, evergreen forest, on plateau in mossy forest; 900 m. Fl. January.

44.115. **A. pumilio** Schltr.

Small epiphyte, 2–10 cm tall; differing from *A. acutipetalum* (p. 358) by the wider leaves and petals and the longer, ascending spur; leaves linear-ligulate, 12–75 × 2.5–7 mm; inflorescences 1–2-flowered, equalling or shorter than the leaves; flowers small, greenish yellow; sepals 5 mm long; lip thick, fleshy, suborbicular, cochlear-concave, 4.5 × 5 mm; spur narrowed from the wide entrance to the middle, then inflated and clavate, 6–7 mm long. Antsiranana, Fianarantsoa. Montane, ericaceous scrubland and lichen-rich forest; 1000–2400 m. Fl. February–April.

44.108

44.112

44.114

44.119

44.116. A. sinuatiflorum H.Perrier

Medium-sized epiphyte, without a stem; leaves grass-like, 8–15 × 0.25–0.4 cm; inflorescence in a series of 2–4, as in *A. multiflorum* (p. 361), 2–5-flowered; sepals 4 mm long; lip slipper-shaped, 4 × 2 mm, obtusely apiculate; spur pendent, 13–14 mm long. Fianarantsoa. Humid, evergreen forest on plateau. Fl. September.

44.117. A. tamarindicolum Schltr.

Small epiphyte with typically 3–5, very thin, papery leaves, 20–50 × 3.5–5 mm, reminiscent of species in *Aeranthes*; inflorescences shorter than the leaves, 1–2-flowered; flowers small, whitish yellow; sepals 3 mm long; lip broadly rhombic-ovate, 3 × 3 mm; spur 7.5 mm long, narrowed in the basal part, then inflated into a somewhat obtuse club. Antsiranana. Humid, evergreen forest at low elevation, on Tamarind trees; below 100 m. Fl. February.

44.118. A. vesiculiferum Schltr.

Epiphyte differing from *A. setipes* (p. 326) in its thicker and pointed leaves; leaves 3, lanceolate-ligulate, up to 50 × 8 mm; inflorescences half length of leaves, 2–3-flowered; sepals 5 mm long; lip broadly ovate, 4.5 × 3 mm, concave, acuminate; spur horizontally recurved, inflated at the tip, obtuse, 2.5 mm long. Antananarivo. Mossy forest; 1500–2000 m. Fl. September.

XIII. Section Chlorangraecum

Stems short; leaves long, in a fan; inflorescences 1–3-flowered; flowers yellow to pale green.

Key to species of Section Chlorangraecum

1. Spur 20–24 mm long; sepals 15 mm long; lip entire, 10 × 10 mm . . . . . **44.119. A. ferkoanum**
1. Spur 12–20 mm long; sepals 12–13 mm long; lip 9–12 × 5–8 mm . . . **44.120. A. huntleyoides**

44.119. A. ferkoanum Schltr.   *(p. 373, bottom right)*

Stemless epiphyte close to *A. huntleyoides* (p. 376); leaves ligulate, 30–37 × 2–2.7 cm; flowers medium-sized, yellowish green; sepals 15 mm long; lip conical, very broadly ovate, almost hollowed into a sac in the middle, 10 × 10 mm; spur curved, then straight, subfiliform, 20–24 mm long. Antananarivo. Montane forest. Fl. unknown.

44.120

44.120

44.121

44.121

44.120. A. huntleyoides Schltr.   *(p. 375, top left and right)*
(syn. *Angraecum chloranthum* Schltr.)

Large epiphyte; stem up to 30 cm long; leaves linear-ligulate, up to 30 × 2–2.5 cm; inflorescences 1–3-flowered; flowers yellow, medium-sized, on short peduncles; sepals 12–13 mm long; lip very broadly elliptic, acuminate, concave, 9–12 × 5–8 mm, with numerous long hairs inside; spur subfiliform, arched-descendant, 12–20 mm long. Toamasina. Humid, evergreen forest; in the crowns of *Dracaena*; 580–1100 m. Fl. February, July.

XIV. Section Humblotiangraecum

Stems short to medium-sized; leaves fleshy or leathery, sometimes iridiform; inflorescences few-flowered; flowers white or tinged with green or orange, large, fleshy; pedicel and ovary often winged; spur emerging from a broad mouth, elongate, slender.

Key to species of Section Humblotiangraecum

1. Leaves bilaterally compressed, iris-like, borne in a fan .... **44.121. A. leonis**
1. Leaves dorsiventrally compressed 2
2. Stems short, covered by leaf bases; flowers white .... **44.122. A. magdalenae**
2. Stems elongate; flowers with orange-brown sepals, petals and spur .... **44.123. A. viguieri**

44.121. A. leonis (Rchb.f.) André   *(p. 375, bottom left and right)*
(syn. *Aeranthes leonis* Rchb.f.; *Angraecum humblotii* Rchb.f. ex Rolfe non (Finet) Summerh.)

Epiphyte or lithophyte; leaves 4, narrowly ensiform, fleshy, 5–20 cm long; inflorescences 1–4-flowered, shorter than the leaves; flowers large, white; sepals 2 cm long; lip concave, broadly ovate-orbicular; disc with a short, central ridge; spur filiform below a funnel-shaped base, twisted-undulate, 7–9 cm long. The Comorean form generally has larger flowers. Antananarivo, Antsiranana, Mahajanga, Fianarantsoa, Toliara. Comoros: Anjouan, Grande Comore, Mayotte. Edge of streams, coastal forest, seasonally dry, deciduous forest and woodland, humid evergreen forest on plateau, on baobab on limestone, on trunks; sea level–1500 m. Fl. November–March.

44.122. A. magdalenae Schltr. & H.Perrier   *(opposite, top and bottom left)*
Large lithophyte; leaves 6–8, leathery, oblong-ligulate, 13–16 × 2.5–3.5 cm, finely veined across on both sides; inflorescences 1–2-flowered; flowers large, white; sepals 4–5 cm long; lip concave, suborbicular-ovate, shortly acuminate; spur with a wide mouth, attenuate-filiform, 10–11 cm long. Antananarivo, Fianarantsoa. Rocky quartz and granite outcrops, *Uapaca* woodland; 800–2000 m. Fl. November–January.

44.122

44.122

44.123

44.122a. A. magdalenae var. latilabellum Bosser

Epiphyte much bigger than the typical variety with larger wider leaves, reaching 30 × 7 cm, and larger flowers, with the lip more spreading and blunter. Antsiranana. Humid, high elevation forest. Fl. November.

44.123. A. viguieri Schltr.   *(p. 377, bottom right)*

Robust epiphyte with warty roots and thin ligulate-linear leaves, 6–14 × 0.8–1.2 cm; flowers large, orange-brown with a white lip; sepals and petals 5.5–6 cm long; lip obovate, apiculate, c 5.5 cm × 3.5 cm; spur strongly funnel-shaped at the base, then filiform, up to 10 cm long. Antsiranana, Toamasina. Humid, evergreen forest on plateau, mossy ridge-top forest; 900–1100 m. Fl. October–November.

XV. Section Angraecum

Plant large; stem elongate. Leaves distichous, leathery. Inflorescence few- to many-flowered, never secund. Flowers fleshy; spur elongate, filiform.

Key to species of Section *Angraecum*

1. Spur 20 cm or more long 2
1. Spur 15 cm or less long 3
2. Sepals 70–90 mm long; lip ovate-subpandurate, 65–80 × 35–40 mm ....
   .... **44. 132. A. sesquipedale**
2. Sepals 50–60 mm long; lip suborbicular-ovate, 50–60 × 30–33 mm .... **44.133. A. sororium**
3. Lip uppermost in the flower; sepals and petals green, lip white 4
3. Lip lowermost in flower; sepals, petals and lip white 5
4. Spur 6–8 cm long .... **44. 125. A. eburneum**
4. Spur 40 cm long .... **44. 126. A. longicalcar**
5. Inflorescence 5–12-flowered .... **44. 124. A. crassum**
5. Inflorescence 1–5-flowered 6
6. Sepals 22 mm or less long 7
6. Sepals 25–40 mm long 8
7. Spur 8 cm long, filiform; sepals 17–18 mm long; leaves 5–8 mm broad ....
   .... **44.127. A. mahavavense**
7. Spur 5–6 cm long, tapering to the tip; sepals 14–15 mm long; leaves 10–12 mm
   broad .... **44.128. A. potamophilum**
8. Inflorescence up to 4-flowered; leaves 25–30 × 2.2–3 cm .... **44.129. A. praestans**
8. Inflorescence 1-flowered; leaves 8–15 × 0.7–1.5 cm 9
9. Leaves grey-green; sepals 30–40 mm long; lip obovate-rhombic, acute ....
   .... **44.130. A. protensum**
9. Leaves green; sepals 25–30 mm long; lip obovate, with a 12 mm long
   apicule .... **44.131. A. serpens**

44.125

44.124. **A. crassum** Thouars

(syn. *Angraecum crassiflorum* H.Perrier; *A. sarcodanthum* Schltr.)

Epiphyte, up to 40 cm tall; stem thick; leaves linear, 8–20 × 1–2.5 cm; inflorescences 5–12-flowered, secund; flowers white, turning yellow; sepals 15–20 mm long; lip cucullate, broadly ovate, very shortly acuminate-apiculate at the front, 13 × 10 mm, with 2 crests on the upper surface; spur short, 25–30 mm long, with a wide opening, thick, cylindrical and straight. Toamasina. Humid, evergreen, lowland and coastal forest on trunks; up to 400 m. Fl. September–March, July.

44.125. **A. eburneum** Bory   *(p. 379)*

(syn. *Limodorum eburneum* (Bory) Willd.; *Angraecum eburneum* var. *virens* Hook.)

Large, robust epiphyte, lithophyte or semi-terrestrial; stem up to 100 cm long; leaves 10–15, loriform, up to more than 30 × 3 cm; inflorescences 15–30-flowered, as long as the leaves; flowers showy on a long inflorescence, green with a white lip; sepals 4–5 cm long; lip uppermost, broadly ovate, concave, 35 × 30 mm; spur narrowing, 6–7 cm long. Antsiranana, Toamasina, Toliara; also in the Comoros, Mascarenes and Seychelles. On rocks or sand in wet littoral forest; sea level–1500 m. Fl. May–August.

LOCAL NAMES *Fontsylahyjamahavy* and *Ahaka*.

44.125a. **A. eburneum** subsp. **superbum** (Thouars) H.Perrier

(syn. *Angraecum superbum* Thouars; *A. brongniartianum* Rchb.f. ex Linden; *A. comorense* Kraenzl.; *A. voeltzkowianum* Kraenzl.; *A. eburneum* var. *brongniartianum* (Rchb.f. ex Linden) Finet)

Large epiphyte or lithophyte with much larger flowers than the typical subspecies with the subrectangular lip, reaching 4.7 cm in length and 4–7 cm in breath, and a spur up to 8 cm long. Antananarivo, Toamasina, Toliara; also in the Comoros and Seychelles. On rocky outcrops and in coastal forest; sea level–500 m. Fl. September–June.

44.125b. **A. eburneum** subsp. **xerophilum** H.Perrier

Large epiphyte or terrestrial, differing from the type in its smaller flowers, lip which is wider than long and the relatively longer 7–8 cm long spur. Toliara. Dry, deciduous scrubland and xerophytic bush; sea level–200 m. Fl. February.

44.127

44.126

44.129

44.126. A. longicalcar (Bosser) Senghas   *(p. 381, bottom left)*

(syn. *Angraecum eburneum* subsp. *superbum* var. *longicalcar* Bosser)

Large epiphyte or lithophyte, close to *A. eburneum* subsp. *superbum* (p. 380), but with a much longer spur reaching up to 40 cm long. Antananarivo, Fianarantsoa, Mahajanga. On trachyte rock; in xerophytic vegetation; in gallery forest; 1000–2000 m. Fl. February.

44.127. A. mahavavense H.Perrier   *(p. 381, top)*

Stem-less epiphyte, up to 15 cm tall; leaves broadly linear, 7–14 × 1–1.2 cm; inflorescences 3–5 cm long, 3–5-flowered; flowers medium-sized, white; sepals 14–15 mm long; lip acutely ovate, 13–16 × 7–10 mm; spur with a narrow opening, linear or slightly narrowing towards the tip, 5–6 cm long. Antsiranana. Humid, highland forest. Fl. October–November.

44.128. A. potamophilum Schltr.

Short-stemmed epiphyte, 12–18 cm tall; leaves thick, linear, 12–15 × 0.5–0.8 cm; inflorescence 5 cm long, 3–5-flowered; flowers white; sepals 17–18 mm long; lip ovate-lanceolate, 17–18 × 8 mm, narrowed-acute towards the tip, a little contracted at the wide base; spur filiform, acute, c. 8 cm long. Mahajanga. By rivers on trunks of *Eugenia*. Fl. November.

44.129. A. praestans Schltr.   *(p. 381, bottom right)*

Robust epiphyte or lithophyte with loriform-ligulate, leathery leaves, 25–30 × 2.2–3 cm; flowers large; sepals 4 cm long; lip obovate-rhombic, 4 × 2.2 cm; spur funnel-shaped at the base, then attenuate-filiform, curved and 9 cm long. Antsiranana, Mahajanga, Toliara. Dry coastal forest on tamarind trees; on rock formations; at the base of large trees; on limestone cliffs; sea level–100 m. Fl. November–June.

44.130. A. protensum Schltr.   *(opposite and p. 385, top left)*

Erect lithophyte, up to 35 cm; leaves greyish-green, linear-ligulate, 8–11 × 0.7–1 cm; inflorescences usually 1-flowered; flowers large, white; sepals 3–4 cm long; lip ovate-concave, acute or apiculate, 4 × 2.3 cm; spur pendent, 14–15 cm long. Antananarivo, Fianarantsoa (Itremo massif). On quartz outcrops; 1600–2000 m. Fl. January–March.

44.130

44.131. **A. serpens** (H.Perrier) Bosser
(syn. *Jumellea serpens* H.Perrier)

Stem up to 50 cm long, branching at base; leaves linear, 10–15 × 0.8–1.5 cm; inflorescence 4–5 cm long, 1-flowered; flower white; sepals 25–30 mm long; lip 30 × 8 mm, narrow at base, dilated above and with a 12 mm long cylindrical point; spur 11–12 mm long; epiphyte resembling *A. penzigianum* (p. 351) but differs in its 1-flowered inflorescence, the very large floral sheath, which reaches 2.5 cm in length, the flower with a totally different lip, and the spur which is heavier, twisted and shorter. Toamasina. Creeping over rocks in humid, evergreen forest; c. 600 m. Fl. July.

44.132. **A. sesquipedale** Thouars   *(opposite, top right)*

Large epiphyte, lithophyte or semi-terrestrial with ligulate, coriaceous leaves, 22–30 cm long; flowers waxy, large, showy, white, star-shaped; sepals 7–9 cm long; lip concave, ovate-subpandurate, obtusely acuminate at apex, 6.5–8 cm long, 3.5–4 cm broad; spur pendent, 30–35 cm long, gradually narrowing from the base to the apex. Antsiranana, Fianarantsoa, Toamasina, Toliara. Coastal forest; dry, semi-deciduous forest; sea level–700 m. Fl. May–November.

44.132a. **A. sesquipedale** var. **angustifolium** Bosser & Morat
(opposite, bottom left)
(syn. *Angraecum bosseri* Senghas)

Terrestrial; a smaller xerophytic variant of the above. Toliara. Seasonally dry, deciduous forest, woodland and scrub, on exposed scree. Fl. May–June.

44.133. **A. sororium** Schltr.   *(opposite, bottom right and p. 387, top)*

Erect lithophyte, terrestrial or epiphyte up to 1 m tall; leaves ligulate, 18–30 × 3–4.5 cm; inflorescences 1–4-flowered, shorter than the leaves; flowers large and fleshy, white; sepals 5–6 cm long; lip suborbicular-ovate, acute, 50–60 × 30–33 mm; spur 25–32 cm long, cylindrical. Antananarivo, Antsiranana, Fianarantsoa, Toamasina. On inselbergs and in *Tapia* woodland in mountainous regions of the island; 1600–2200 m. Fl. December–March.

44.130

44.132

44.132a

44.133

45. AMBRELLA

A monotypic genus endemic to Madagascar. Vegetatively resembling *Aerangis*. The lip completely envelops the column and is rolled into a narrow tube.

45.1. **A. longituba** H.Perrier   *(opposite, bottom left)*

Plant stemless, up to 10 cm tall; leaves 5–6, elliptic or obovate-oblong, 5–9 × 1.6–2.2 cm; inflorescence equalling leaves; flower tubular; sepals 4 cm long; lip 3-lobed, 4.5 cm long; spur recurved in middle, 3.5 cm long. Antsiranana. Humid, evergreen forest, on branches of *Calliandra*; 500–1000 m. Fl. November.

46. LEMURELLA

A small genus of 4 species endemic to Madagascar. Short-stemmed epiphytes. Leaves distichous, twisted at base to lie more or less in one plane, leathery. Inflorescences 1–several-flowered, unbranched, axillary. Flowers small, yellow or yellowish green, pedicel and ovary much longer than the bract. Sepals free, spreading. Petals similar but smaller. Lip 3-lobed, spurred at the base, base of lip sometimes papillose; side lobes rounded, embracing the column; midlobe much larger than the side lobes; spur slender, elongate, with a broad mouth. Column short, fleshy, lacking a foot; pollinia 2, each attached to its own elliptic viscidium.

Key to species of *Lemurella*

1. Floral axis pubescent-papillose; inflorescence axis prolonged into a filiform appendage lacking a flower .... **46.1. L. papillosa**
1. Floral axis glabrous; inflorescence axis not so prolonged 2
2. Lip fleshy and truncate at the tip, shortly apiculate in the apical sinus ... **46.2. L. culicifera**
2. Lip midlobe deltoid or lanceolate and acute, lacking an apicule 3
3. Inflorescence 1–2-flowered on a 8–15 mm long peduncle; lateral sepals 10–17 mm long; spur 15–20 mm long .... **46.3. L. pallidiflora**
3. Inflorescence many-flowered; peduncle 25–35 cm long; lateral sepals 4.5–5.5 mm long; spur less than 10 mm long .... **46.4. L. virescens**

46.1. **L. papillosa** Bosser   *(opposite, bottom right)*

Flowers spaced along the stem but axis of inflorescence lacking a flower; sepals and petals 6–8 mm long; spur slender, 25 mm long. Antsiranana, Toamasina. Humid, evergreen forest; 800–1000 m. Fl. January–May.

44.133

45.1

46.1

46.2. L. culicifera (Rchb.f.) H.Perrier   *(opposite, top left)*

(syn. *Angraecum culiciferum* Rchb.f.; *A. ambongense* Schltr.; *Lemurella ambongensis* (Schltr.) Schltr.; *Beclardia humbertii* H.Perrier)

Plant loses its leaves during the dry season; leaves blunt; lip with the terminal lobe truncate, with a short apicule in the sinus; spur ascending over ovary. Mahajanga, Toliara. Also in the Comoros. Seasonally dry, deciduous forest and woodland, on tree trunks; up to 700 m. Fl. October–November.

46.3. L. pallidiflora Bosser   *(opposite, top right)*

Differs from *L. culicifera* and *L. virescens* in its habit, its 1–2-flowered inflorescence which is shorter than the leaves, the short peduncle and the larger flowers, with 8–17 mm long sepals; spur 15–20 mm long, slightly S-shaped in side view. Antsiranana, Toamasina. Mossy, montane forest; 800–1200 m. Fl. January–March.

46.4. L. virescens H.Perrier

Inflorescence several-flowered, on a 2.5–3.5 cm long peduncle; flowers small; sepals 4.5–5.5 mm long; spur less than 1 cm long. Toamasina. In mossy forest; 1000–1200 m. Fl. February, June.

47. OEONIELLA

A genus of 2 species endemic to Madagascar and the Mascarenes. A single species in Madagascar. Epiphytes with long stems and many adventitious roots scatted along stem. Leaves distichous, ligulate. Inflorescences lateral, 7–12-flowered. Flowers longer than the leaves, distichously arranged, white. Sepals and petals subsimilar, lanceolate. Lip trumpet-shaped with a long apicule, shortly spurred at the base. Column short; pollinia 2, each attached by a thin stalk to a saddle-shaped viscidium.

47.1. O. polystachys (Thouars) Schltr.   *(opposite, bottom left)*

(syn. *Epidendrum polystachys* Thouars; *Oeonia polystachae* var. *longifoliae* Rchb.f.)

A variable and locally common species; stems 15–30 cm long; leaves oblong, 5–11 × 1.5–2 cm; flowers 7–15, 12–16 mm long, white; sepals and petals 1.2–14 cm long; lip trumpet-shaped, longly apiculate; spur 4 mm long. Antsiranana, Fianarantsoa, Toamasina, Toliara; also in the Comoros, Mascarenes and Seychelles. Coastal, lowland and lower montane forest, on trunks; up to 100 m. Fl. August–May.

LOCAL NAME *Foutsilang zanahar*.

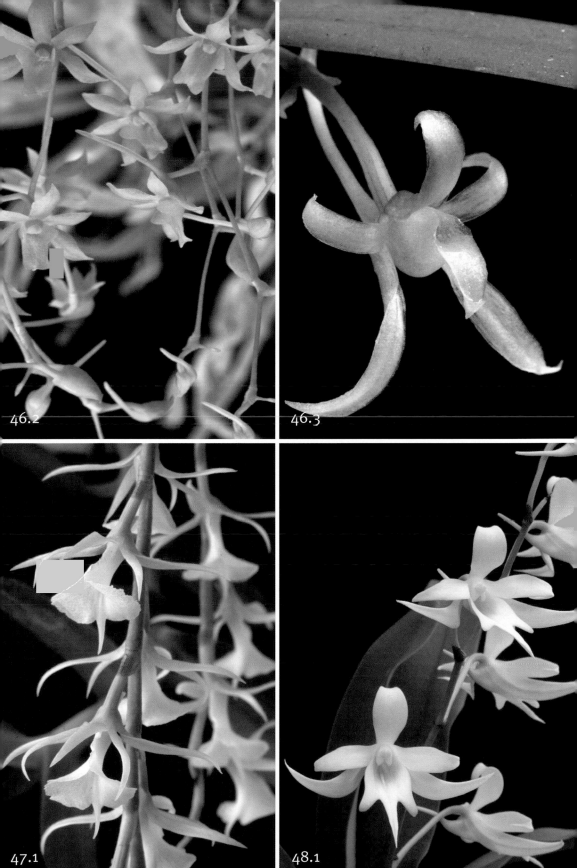

46.2

46.3

47.1

48.1

48. SOBENNIKOFFIA

A Madagascan endemic genus of 3 species. Plants similar to *Angraecum eburneum* but flowers, white with a green mark in the throat of the spur, with a prominent 3-lobed lip and upcurved spur.

Key to species of *Sobennikoffia*

1. Lip 30–32 mm long . . . . **48.1. S. humbertiana**
1. Lip 20–24 mm long 2
2. Spur 15–17 mm long; leaves up to 10 cm long . . . . **48.2. S. poissoniana**
2. Spur 28 mm long; leaves up to 35 cm long . . . . **48.3. S. robusta**

48.1. **S. humbertiana** H.Perrier
(p. 389, bottom right)

Lithophyte; leaves less than 20 cm long; sepals 2.5 cm long; lip 3–3.2 cm long, the midlobe elongate, 10–12 mm long, the lateral lobules obtuse; spur a third shorter than the lip, 1.8 cm long. Fianarantsoa, Toliara. Evergreen forest on plateau; dry, deciduous scrubland; 400–1200 m. Fl. October–November.

48.2. **S. poissoniana** H.Perrier
(opposite, top)

Epiphyte or lithophyte; plant small; leaves up to 10 cm long; lip 3-lobed, 2–2.3 mm long; spur cylindrical 15–17 mm long. Antsiranana, Mahajanga. On rocks and in coastal vegetation; sea level. Fl. November–December.

48.3. **S. robusta** (Schltr.) Schltr.
(opposite, bottom)

Epiphyte or lithophyte; leaves up to 35 cm long; sepals 2.5 cm long; lip 2.4 cm long, the midlobe shorter and the lateral lobules subacute; spur 2.8 cm long. Mahajanga, Toliara. Seasonally dry deciduous woods and scrub; 1500–2000 m. Fl. November–January.

48.2

48.3

49. CRYPTOPUS

A small genus of 3 species, endemic to Madagascar and the Mascarenes. Scrambling epiphytes with long, branching, leafy stems. Leaves distichous. Inflorescences axillary, simple or branched, many-flowered. Flowers flat, white. Sepals and petals free, the latter often lobed. Lip 3–4-lobed. Column very short; pollinia 2, each attached to its own viscidium, spreading.

Key to species of *Cryptopus*

1. Inflorescence paniculate; lip 3-lobed, the midlobe narrowly triangular, the side lobes anchoriform; petals anchoriform .... **49.3. C. paniculatus**
1. Inflorescence simple; lip 4-lobed, the midlobe with 2 long arms at the tip; spur 20 mm or more long 2
2. Inflorescence 30–60 cm long; petals clawed and irregularly dilated or 4-lobed at the top; spur 25 mm long .... **49.2. C. dissectus**
2. Inflorescence 7–16 cm long; petals T-shaped, the apical lobes linear; spur 20 mm long .... **49.1. C. brachiatus**

49.1. C. brachiatus H.Perrier

Stem 5–10 cm long, branching, sinuous; leaves 4–5, elliptic-oblong, 1.5–5.6 × 1–2 cm; flowers 5–10; sepals 13–14 mm long; petals T-shaped, similar to the midlobe of the lip; lip with basal lobes oblong, rounded at the tip and contracted at the base into a short tooth; midlobe T-shaped, truncate at the tip extended laterally into 2 elongated filiform appendices; spur 2 cm long. Fianarantsoa, Toamasina. In humid, evergreen forest; 600–1200 m. Fl. February–April.

49.2. C. dissectus (Bosser) Bosser

(syn. *Cryptopus elatus* subsp. *dissectus* Bosser)

Plant 30–40 cm tall; flowers yellow-green; sepals and petals 13–15 mm long; terminal lobes of the lip and the petals deeply divided into narrow lobules. Fianarantsoa; also in Mascarenes. In humid, mossy, evergreen forest; 500–1000 m. Fl. December.

49.3. C. paniculatus H. Perrier   (*opposite, top*)

Epiphytic with a stem up to 40 cm long; leaves elliptic, 3.5–5 x 1.3–1.8 cm; inflorescence paniculate, 20–40 cm long, the branches 12–18 cm long; flowers 3–10 per branch; flowers white; sepals 6.5–8 mm long; petals anchor-shaped, 9 x 12 mm; lip deeply 3-lobed, concave an papilose at the base, the side lobes incurved, subanchoriform, the midlobe narrowly triangular, emarginated; spur 2 mm long. Toamasina. In montane forest, 500–1000 m. Fl. Feb.

49.3

50.2a

50. OEONIA

A small genus of 5 species endemic to Madagascar and the Mascarenes. Scramblers or epiphytes with long leafy stems and elongate adventitious roots. Leaves distichous, leathery, oblong-elliptic. Inflorescences axillary, few-flowered. Flowers showy, predominantly white or with yellowish or yellow-green obovate sepals and petals. Lip 3–5-lobed, spurred but lacking a callus. Column short, enclosed by infolded side lobes of lip; pollinia 2, waxy.

Key to species of *Oeonia*

1. Flower greenish . . . . **50.1. O. madagascariensis**
1. Flowers white or white with yellowish sepals and petals 2
2. Stem 6–8 cm long; lip 5–6-lobed, broader than long, 8 x 10 mm . . . . **50.2. O. brauniana**
2. Stem 20 cm or more long, climbing or scrambling; lip 3–4-lobed, as long as or longer than broad 3
3. Sepals and petals greenish or yellow-green; lip white with red base; spur 5–7 mm long . . . . **50.3. O. rosea**
3. Sepals and petals white; lip white, yellowish in throat 4
4. Sepals 10–12 mm long; lip 25 × 25 mm; spur incurved, 13–15 mm long . . . . **50.4. O. curvata**
4. Sepals 12–25 mm long; lip 17–30 cm long; spur cylindrical, 3.5–6 mm long . . . **50.5. O. volucris**

50.1. **O. madagascariensis** (Schltr.) Bosser
(syn. *Perrieriella madagascariensis* Schltr.)

Stem pendent, 10–25 cm long; leaves fleshy, linear or narrowly oblong, 1.5–3.5 × 0.5 cm; inflorescence 1-flowered; flower greenish; lip 3-lobed, 7–10 mm long and wide, margins erose; spur cylindrical, 1.2–1.5 cm long; anther with a 2-lobed, elongated lobule at the front. Antsiranana. Humid evergreen forest on plateau and in lichen-rich forest; 1500–2000 m. Fl. January–March.

50.2. **O. brauniana** H.Wendl. & Kraenzl.

Stem up to 30 cm long; leaves ovate-oblong, 3.5–7.5 × 1.7–2.5 cm; inflorescence 2–4.5 cm long; flowers 1–3, the sepals and petals orange with green bases, the lip white with 2 red spots at base; sepals 12–14 mm long; lip 12–15 mm long; lip 6-lobed, the terminal lobe incised. Toamasina. Humid lowland forest; up to 100 m. Fl. October.

50.2a. **O. brauniana** var. **sarcanthoides** (Schltr.) Bosser
(p. 393, bottom)

(syn. *Oeoniella sarcanthoides* Schltr.; *Oeonia subacaulis* H.Perrier; *Lemurella tricalcariformis* H.Perrier; *L. sarcanthoides* (Schltr.) Senghas)

Sepals and petals 7–9 mm long; lip 5-lobed, the terminal lobe retuse or only slightly indented. Antsiranana, Fianarantsoa, Toamasina. Humid, evergreen forest; 500–1000 m. Fl. January–February.

50.3. **O. rosea** Ridl. *(p. 397, top left)*

(syn. *Oeonia oncidiiflora* Kraenzl.; *O. forsythiana* Kraenzl.)

Stems elongate, simple or branched; leaves leathery, ovate or narrowly ovate, 2–2.5 × 0.5 cm; sepals and petals yellowish green, 10–12 mm long; lip 25 mm long, white with a red throat, 4-lobed, the terminal lobes large, spathulate, divergent; spur 5–7 mm long, slightly inflated. Antananarivo, Fianarantsoa, Toamasina, Toliara; also in the Mascarenes. Humid, mossy, evergreen forest, on branches; 500–2000 m. Fl. September–May.

50.4. **O. curvata** Bosser

Stem branched, elongate; leaves ovate, 2 × 1 cm; inflorescence 5–7 cm long; flowers 1–2, white; sepals and petals 10–12 mm long; lip 25 × 25 mm, 4-lobed, the 2 lateral lobes erect, broad, rounded, the edges undulate, and the terminal lobes smaller and narrower; spur incurved, 13–15 mm long. Fianarantsoa. Humid forest; 1200–1500 m. Fl. April.

50.5. **O. volucris** (Thouars) Spreng. *(p. 397, top right)*

(syn. *Epidendrum volucre* Thouars; *Oeonia auberti* Lindl.; *O. elliotii* Rolfe; *O. humblotii* Kraenzl.)

Stem erect or pendent, simple or branched; leaves ovate-oblong or elliptic, 2–2.6 × 0.6–0.9 cm; flowers completely white; sepals 12–25 mm long; lip 3-lobed, 17–30 mm long, the midlobe more or less indented; spur cylindrical and slender, 3.5–6 mm long. Antsiranana, Fianarantsoa, Toamasina, Toliara; also in the Mascarenes. Coastal forest; humid, evergreen forest, on branches; up to 1500 m. Fl. March–July.

51. EULOPHIA

A large genus of some 230 species, most diverse in tropical and southern Africa, Madagascar and tropical Asia, extending into temperate C Asia, NE Australia, the SW Pacific Islands and with 2 species in the tropical Americas. Small to large terrestrial or less commonly lithophytic herbs, rarely saprophytic and leafless. Roots with a well-defined white velamen. Pseudobulbous if above ground or rhizomatous if subterranean; pseudobulbs cylindrical, fusiform, conical or ovoid, few–several-noded, 1–many-leaved. Leaves pleated or less commonly not and leathery, linear to elliptic-lanceolate. Inflorescence basal, laxly to subdensely few–many-flowered, simple or branching. Flowers small to large, occasionally bicoloured. Dorsal sepal free; lateral sepals oblique at base, otherwise similar to dorsal sepal. Petals free, similar or dissimilar to sepals, often broader than the sepals. Lip free to base or fused to base of column, 3-lobed, more or less spurred at the base, callose; callus 2–3-ridged or papillose. Column short to elongate, usually with a column-foot; pollinia 2, subglobose attached by a short stipes to an oblong, elliptic to lunate viscidium.

Key to species of *Eulophia*

1. Plants saprophytic, lacking green leaves 2
1. Plants autotrophic, with green leaves 3
2. Flowers white with a rose pink lip; lip with 3 denticulate keels; spur
   conical, 3.5 mm long ....**51.1. E. hologlossa**
2. Flowers yellowish-chestnut brown with purple striations and a white lip
   marked with purple; lip with a single glabrous callus ridge; spur cylindrical,
   3 mm long ....**51.2. E. mangenotiana**
3. Leaves linear-filiform, subterete ....**51.3. E. filifolia**
3. Leaves flat, neither linear-filiform nor terete 4
4. Inflorescence branched 5
4. Inflorescence simple, unbranched 8
5. Plant with elongated 15–50 cm long cylindrical stems, 3–4-leaved at the
   top; lip cornet-shaped, bilobed ....**51.7 E. beravensis**
5. Plant with ovoid to conical pseudobulbs; 3- or more-leaved at top; lip
   entire or 3-lobed 6
6. Pseudobulbs subterranean; spur inflated-clavate ....**51.6. E. perrieri**
6. Pseudobulbs above ground; spur cylindrical to conical-tapering 7
7. Sepals 8–10 mm long; lip 3-lobed, midlobe crispate-undulate; spur 2 mm
   long ....**51.5. E. ramosa**
7. Sepals 5–6 mm long; lip entire, obovate, fimbriate on margins; spur 5 mm
   long ....**51.4. E. macra**
8. Flowers or lip bright rose-purple 9
8. Flowers not bright rose-purple 10
9. Petals 18–20 mm long; lip midlobe conduplicate, larger than the side lobes,
   bearing 5 low ridges in basal half ....**51.8. E. livingstoniana**
9. Petals 11 mm long; lip midlobe not conduplicate, ovate, much smaller than
   the side lobes, the callus comprising 2 conical calli at the base tapering
   into 2 rows of verrucose lines to the tip of the midlobe ....**51.9. E. grandidieri**
10. Flowers predominantly yellow, yellow-brown or orange-brown 11
10. Flowers predominantly greenish or purplish green with white petals and lip 14
11. Flowers not opening widely, bright yellow with 2 maroon marks at the base
    of the lip ....**51.13. E. rutenbergiana**
11. Flowers opening widely, predominantly yellow-, orange- or red-brown, marked
    with purple 12
12. Pseudobulbs cylindrical-fusiform, up to 6 cm long, above the ground ...**51.10. E. ibityensis**
12. Pseudobulbs subterranean 13
13. Sepals spatulate, 9.5–12 mm long, with recurved tips; petals porrect, oblong,
    9–12 × 2–4.5 mm; lip midlobe subquadrate, decurved ....**51.11. E. pileata**
13. Sepals ovate, 7–8 mm long; petals elliptic, 9 × 8 mm; lip midlobe convex,
    obcuneate, obtuse, with crispate margins; callus of several verrucose keels ....
    ....**51.12. E. reticulata**
14. Spur pendent, conical, 5–7 mm long; lip midlobe convex; leaves narrowly
    lanceolate ....**51.16. E. plantaginea**
14. Spur not pendent, cylindrical, 2–3 mm long; lip midlobe not convex; leaves
    grass-like, linear 15
15. Lip callus 3–5-ridged, papillose; lip oblong, 3-lobed, 10 × 9 mm, midlobe
    entire ....**51.14. E. hians**
15. Lip callus 2-ridged, not papillose; lip obovate, obscurely 3-lobed, 11 × 8 mm;
    midlobe trilobulate ....**51.15. E. nervosa**

50.3

50.5

51.2

51.3

51.1. E. hologlossa Schltr.

Plant 13–40 cm tall; inflorescence laxly 6–18-flowered; flowers white with a rose-pink lip; sepals 8 mm long; lip oblong, 8 × 4 mm, with 3 toothed keels; spur conical-attenuate, 3.5 mm long. Fianarantsoa, Toliara. In marshes and sand dunes; up to 200 m. Fl. October–November.

51.2. E. mangenotiana Bosser & Veyret   (p. 397, bottom left)

Plant entirely brownish yellow, 15–25 cm tall; flowers 7–12; sepals yellowish chestnut brown and purple-striped; sepals 6.5 mm long; lip white with purple markings, 5.5–6 mm long, ovate-ligulate, obscurely 3-lobed, the basal lobes not prominent, the midlobe oblong-ovate, with a central crest in the centre; spur cylindrical, 3–4 mm long. Antsiranana, Fianarantsoa. In humid, evergreen forest at low elevation. Fl. unknown.

51.3. E. filifolia Bosser & Morat   (p.397, bottom right)

Pseudobulbs ovoid; leaves 3–4, linear-filiform; flowers close to those of *E. ramosa*, but a little larger and the lip has 3 well-developed ridges on the upper surface. Toliara. In sandy soil and chalk; up to 200 m. Fl. February, July–October.

51.4. E. macra Ridl.   (opposite, top left)

(syn. *Eulophia ambositrana* Schltr.; *Lissochilus macer* H.Perrier)

Plant terrestrial or lithophytic, up to 60 cm tall, forming clumps; pseudobulbs ovoid-subconical, 3–5 cm tall; leaves 3–4, narrowly linear, 40–60 cm long; inflorescence paniculate, laxly many-flowered; flowers dull in colour; sepals and petals 5–6 mm long; lip entire, obovate, 8 × 5 mm, fimbriate in the front, the disk with a few small keels; spur short, 3 mm long. Fianarantsoa, Toliara. Open forest and exposed areas on rock; up to 1500 m. Fl. April–June.

51.5. E. ramosa Ridl.

(syn. *Eulophia leucorhiza* Schltr.; *E. pseudoramosa* Schltr.)

Terrestrial or lithophyte, 50–80 cm tall; pseudobulbs narrowly cylindrical or stem-like, less than 6 cm long; leaves developing after flowering, grass-like; inflorescence branching, very laxly several-flowered; flowers yellowish and lined with red; sepals 8–10 mm long; lip oblong, 3-lobed, side lobes small and obtuse, midlobe larger and crispate-undulate at the front, the disk with 5 verrucose keels; spur cylindrical-tapering, 2 mm long. Antananarivo, Fianarantsoa, Mahajanga, Toamasina, Toliara. Open ground; dry forest; limestone outcrops; up to 1000 m. Fl. February, July–November.

51.4

51.7

51.8

51.10

51.6. E. perrieri Schltr.

Lithophyte; pseudobulbs fusiform, underground; leaves fleshy, linear, acute, up to 26 × 1.2 cm, with an inflated sheathing base; inflorescence branched, overtopping leaves; flowers similar to those of *E. ramosa* (p. 398) but lip side lobes scarcely broader than the midlobe and spur more inflated-clavate. Mahajanga. On shaded and wet rocks; up to 500 m. Fl. July.

51.7. E. beravensis Rchb.f.   *(p. 399, top right)*

Terrestrial up to 90 cm tall. Stem cylindrical, 15–50 cm long, 2–4-leaved; leaves leathery, linear, 25–35 × 0.6–1 cm, with a serrate margin; inflorescence up to 1 m tall, branched. Flowers small; sepals 7–10 mm long; lip bilobed, 7 × 10 mm, callus 3-lobed; spur short. Toliara, Mahanjanga. Deciduous woodland on sand and dunes; sea level–800 m. Fl. April.

51.8. E. livingstoniana (Rchb.f.) Summerh.   *(p. 399, bottom left)*

(syn. *Lissochilus rutenbergianus* Kraenzl.; *Eulophia jumelleana* Schltr.; *Lissochilus jumelleanus* (Schltr.) Schltr.; *L. laggiarae* Schltr.)

Rhizome subterranean, corm-like; leaves 3–5, linear, 40–60 cm long, appearing after flowering; flowers pale to deep mauve-pink, the lip with darker callus ridges and green or yellowish side lobes flushed and edges with pink; sepals reflexed, lanceolate 2–2.5 cm long; petals elliptic, 1.8–2 cm long; lip 3-lobed, conduplicate, with 5 low ridges; spur conical, short. Antananarivo, Antsiranana, Fianarantsoa, Mahajanga, Toamasina, Toliara; also in the Comoros, E and S-C Africa. In grassland and marshes; up to 1300 m. Fl. October–March.
LOCAL NAME *Felatrandraka*. The tubers are eaten.

51.9. E. grandidieri H.Perrier

Plant 50–70 cm tall; leaves developing after the flowers, 4–5, grass-like, 10–30 × 1–1.2 cm; inflorescence 20–25-flowered; sepals linear-lanceolate, 15–16 mm long; petals elliptic, obtuse, 11 × 6 mm; lip deeply 3-lobed, 1.8–2 × 1.5 cm, rose-purple, the side lobes broadly rounded, the midlobe ovate, obtuse, the margins fimbriate, callus of 2 raised basal protuberances running into 2 verrucose ridges in front; spur 3 mm long. Toamasina. On a hill. Fl. April.

51.10. E. ibityensis Schltr.   *(p. 399, bottom right)*

Plant 30–40 cm tall; pseudobulbs fusiform-cylindrical, up to 6 cm tall; leaves produced after flowering; 25–30 cm long; inflorescence 25–30 cm tall, laxly 15–20-flowered; flowers yellowish-brown; sepals ligulate, 10–12 mm long; lip 10 × 4–6 mm, 3-lobed below the middle, the midlobe almost rectangular, a little indented at the front, the margins pleated-fringed, the disc with 5 longitudinal ridges; spur 3 mm long. Antananarivo, Fianarantsoa. On rocky outcrops or laterite and in tapia (*Uapaca*) forest; 1000–2000 m. Fl. November–April.

51.11

51.11. E. pileata Ridl. (p. 401)

Plant tall, slender; pseudobulbs fusiform-cylindrical, 6–7 cm long, 3–5-noded, green; leaves 5–7 in a fan, grass-like, 5–10 mm broad; inflorescence 40–50 cm tall, unbranched, 20–30-flowered; flowers orange-yellow, with brick-reddish streaks; sepals spathulate, 9.5–12 mm long; petals porrect; lip obovate, 10 × 8 mm, 3-lobed in the upper third, midlobe obovate; the disc with 2 very protruding keels towards the base, heavily spotted with purple on midlobe; spur 3 mm long. Antananarivo; also in the Comoros. In montane forest; 1000–1500 m. Fl. July–November.

51.12. E. reticulata Ridl. (opposite, top left)

(syn. *Lissochilus madagascariensis* Kraenzl.; *Eulophia camporum* Schltr.)

Corms underground; leaves 5–6, grass-like, appearing after flowering. Inflorescence 30–40 cm long; flowers with red-brown sepals, petals yellow marked with red veins, the lip with a large yellow callus; sepals ovate, obtuse, 7–8 mm long; petals elliptic, 9 × 8 mm; lip 3-lobed, the lateral lobes small and obtuse, the midlobe convex, obcuneate and obtuse, margins upcurved, crispate-undulate, callus of several verrucose ridges. Antananarivo, Antsiranana, Fianarantsoa, Toliara. Grassland and amongst rocks; 500–2000 m. Fl. October–December.

LOCAL NAMES *Kamasina, Tandrokondrylahy, Kitandrokondrilahy*; leaves and capsules are eaten and the roots used as an aphrodisiac.

51.13. E. rutenbergiana Kraenzl. (opposite, top right)

(syn. *Lissochilus kranzlinii* H.Perrier)

Rhizome underground; pseudobulbs corm-like, irregular; leaves suberect, 3–4, grass-like; inflorescence 40–60 cm tall; flowers sulphur-yellow with dark maroon marks on the base of the lip; sepals and petals 1.8–2 cm long; lip 20 mm long, 3-lobed; midlobe oblong-obovate; spur 3 mm long. Antananarivo, Antsiranana, Fianarantsoa, Toamasina, Toliara. Marshes; secondary grassland; open ground; rocky slopes; 500–2000 m. Fl. December–May.

51.14. E. hians Lindl. (opposite, bottom left)

(syn. *Eulophia clavicornis* Lindl.; *E. madagascariensis* Kraenzl.; *E. vaginata* Ridl.; *Graphorchis madagascariensis* (Kraenzl.) O.Kuntze; *E. bathiei* Schltr.)

Plant 25–93 cm tall; pseudobulbs underground; leaves developing after flowering, grass-like, 20–25 cm long; flowers with dark purplish-brown or purplish green sepals, white petals tinged pink or blue, a white lip flushed bluish; sepals and petals 9–11 mm long; lip 3-lobed, 10 × 9 mm, with a pale papillose 3-ridged callus, and a 4 mm long, unswollen spur. Antananarivo, Fianarantsoa, Toamasina, Toliara; also in E, S-C and S Africa. In grassland, dry deciduous scrubland and rocky outcrops; 1500–2500 m. Fl. December–March.

LOCAL NAME *Tongolomboalavo*. Used medicinally against boils.

51.12

51.13

51.14

51.16

51.15. E. nervosa H.Perrier

(syn. *Lissochilus nervosus* (H.Perrier) H.Perrier)

Plant up to 35 cm tall; pseudobulbs subterranean; leaves grass-like, up to 30 × 1 cm wide; sepals narrowly lanceolate, 12 mm long; petals oblong-lanceolate, 9 × 4 mm, with red veins; lip obovate, 3-lobed, 11 × 8 mm, the lateral lobes obtuse, the midlobe 3-lobulate, the margins undulate-fimbriate, the disk with 2 calli. Antananarivo. In peaty marshes; 1000–1500 m. Fl. January.

51.16. E. plantaginea (Thouars) Rolfe ex Hochr.

(opposite and p. 403, bottom right)

(syn. *Eulophia grandibracteata* Kraenzl.; *Lissochilus plantagineus* (Thouars) H.Perrier)

Pseudobulbs underground, lobed; leaves narrowly lanceolate; inflorescence robust, 30–100 cm tall, unbranched; flowers with yellowish green sepals, pure white petals and a white lip, spur and disk tinted with violet-red; sepals 13–22 mm long; lip 3-lobed, 12–15 mm long; keels 3; spur conical, 5–7 mm long. Antananarivo, Fianarantsoa, Mahajanga, Toamasina, Toliara; also in the Comoros. In grassland; by ditches in arable land; marshes; up to 1300 m. Fl. January–March.

LOCAL NAMES *Tenondahy, Tongolombato, Ovinakanga*.

52. OECEOCLADES

A genus of about 40 species in Africa, Madagascar and the adjacent islands but with one species, *O. maculata*, extending to the Americas, and another, *O. pulchra*, to the Pacific islands. Terrestrial plants with clustered, conical, ovoid or fusiform pseudobulbs. Leaves leathery and not pleated to thin-textured and pleated, 1–several at apex of pseudobulbs, linear, elliptic, lanceolate or ovate. Inflorescences basal, simple of branching, few–many-flowered. Flowers relatively small, white, green or yellow green, often marked with purple on the lip. Sepals free. Petals free, similar to sepals but smaller. Lip usually 4-lobed, the lobes spreading, with a basal callus and a scrotiform to shortly cylindrical spur. Column short to elongate-incurved; pollinia 4, waxy, attached to a lunate viscidium.

Key to species of *Oeceoclades*

1. Leaves plicate 2
1. Leaves not plicate 3
2. Lip midlobe bilobulate, the lobules obovate, rounded at tips, overlapping; lip callus 3-ridged; spur cylindrical, incurved .... **52.1. O. ambrensis**
2. Lip midlobe somewhat flabellate, very broadly emarginate; lip callus 2-ridged; spur scrotiform-clavate .... **52.2. O. pulchra**
3. Leaves linear, 10 times or more longer than broad 4
3. Leaves elliptic, ovate or lanceolate, rarely linear, lamina less than 10 times as long as broad 12

51.16

4. Inflorescence 120–150 cm tall, branched, many-flowered; leaves 3–4 5
4. Inflorescence 75 cm or less tall, simple; leaves 1–2 6
5. Pseudobulbs 8–12 cm long, 4–5-angled; leaves green; flowers green with a white callus and purple-veined lip .... **52.6. O. calcarata**
5. Pseudobulbs 2 cm long; leaves mottled; flowers yellow, flushed purple-brown on sepals and petals, the callus ridges pink edged .... **52.8. O. hebdingiana**
6. Pseudobulbs above ground, fusiform or conical and 5-angled 7
6. Pseudobulbs subterranean, conical to flattened ovoid 11
7. Pseudobulbs fusiform to subcylindrical; leaves green 8
7. Pseudobulbs conical, 4–5-angled; leaf mottled 9
8. Leaves 1–2, spreading, 15–23 × 1–1.2 cm, often flushed with red; sepals spatulate, 16–20 mm long; petals 10–12 mm long; lip recurved 11–16 × 10–12 mm; spur conical 3–4 mm long .... **52.7. O. decaryana**
8. Leaf 1, erect, 40–50 × 0.5–0.7 mm; sepals and petals oblong, 5 mm long; lip straight, 5 × 6 mm long, the side lobes much larger than the midlobes; spur clavate, 3 mm long .... **52.10. O. perrieri**
9. Leaf solitary, 30–32 × 1–1.3 cm; flowers white marked with purple ... **52.5. O. callmanderi**
9. Leaves 1–2, up to 11 × 0.7 cm; flowers yellow marked with purple veins and flushing 10
10. Leaves obscurely mottled; sepals 7–8 mm long; lip 10–12 × 14–16 mm; spur scrotiform, 3 mm long .... **52.3. O. angustifolia**
10. Leaves green; sepals 4 mm long; lip 6.3 × 5 mm; spur clavate, 5.5 mm long .... **52.4. O. quadriloba**
11. Leaves 1–2, erect, 30–45 × 0.35–0.5 cm, tessellated; bracts much longer than the flowers, purple-black-spotted; peduncle black-spotted; spur conical, 4–5 mm long .... **52.9. O. longibracteata**
11. Leaves 2–3, 10–15 × 0.6–1.2 cm; bracts shorter than the flowers, not spotted; peduncle not spotted; spur globose to clavate .... **52.11. O. spathulifera**
12. Leaves longly stalked 13
12. Leaves sessile or shortly petiolate 19
13. Leaf or leaves tessellated or mottled 14
13. Leaves green, not tessellated 15
14. Leaf 16–19 × 2–2.4 cm broad; inflorescence up to 10-flowered; lip disc glabrous .... **52.13. O. alismatophylla**
14. Leaf 6–8 × 1.2–1.3 cm; inflorescence simple or few-branched; lip disc hairy .... **52.14. O. analamerensis**
15. Leaf oblanceolate, 5–8 × 0.8–1.1 cm; spur bifid at tip .... **52.15. O. analavelensis**
15. Leaves elliptic, lanceolate or ovate, usually 1.5 cm or more broad 16
16. Plants 50–80 cm tall; inflorescence several branched; pseudobulbs conical, sometimes several-angled 17
16. Plants 40 cm or less tall; inflorescence simple or very few-branched; pseudobulbs fusiform, not angular 18
17. Leaf lamina broadly elliptic, 13–19 × 6–8 cm; sepals 7 mm long; spur conical, 7 mm long .... **52.16. O. cordylinophylla**
17. Leaf lamina lanceolate, 10–20 × 2–4.4 cm; sepals 5–6 mm long; spur cylindrical, 2.5–3 mm long .... **52.19. O. petiolata**
18. Leaf 1, erect, narrowly elliptic, 15–22 × 2–3 cm; flowers pale yellow with purple veins on the side lobes .... **52.17. O. flavescens**
18. Leaves 2–3, ovate-lanceolate, 3.5–8 × 1.5–2.5 cm; flowers not as above .... **52.18. O. pandurata**

52.1. **O. ambrensis** (H.Perrier) Bosser & Morat
(syn. *Lissochilus ambrensis* H.Perrier; *Eulophia ambrensis* (H.Perrier) Butzin)

Vegetatively similar to *O. pulchra* but distinct in having the lip with more developed, rounded terminal lobes; pseudobulbs fusiform; leaves 3, elliptic, longly stalked; sepals 9–12 mm long; lip 4-lobed; midlobes obovate, overlapping; spur cylindrical, incurved under lip, 2–3 mm long. Antsiranana. Humid forest at medium elevation; 1000–1500 m. Fl. December–January.

52.2. **O. pulchra** (Thouars) P.J.Cribb & M.A.Clem.   *(p. 408, top left)*
(syn. *Eulophia pulchra* (Thouars) Lindl.; *E. striata* Rolfe; *Eulophidium pulchrum* (Thouars) Summerh.)

Pseudobulbs fusiform-cylindrical; leaves plicate, 2–4, long-petiolate, narrowly elliptic, glossy green; inflorescence simple, up to 20-flowered; flowers yellow or pale green, marked with purple on the side lobes of lip and with an orange callus; sepals 11 mm long; petals oblong, 9 mm long; lip recurved, 4-lobed, 7 mm × 18 mm; callus of 2 short fleshy ridged at mouth of spur; spur scrotiform. Antananarivo, Antsiranana, Fianarantsoa, Toamasina, Toliara; also in the Mascarenes, and from tropical Africa across to the SW Pacific islands. In humid, evergreen forest; open forest; littoral forest; up to 1000 m. Fl. December–April.

52.3. O. angustifolia (Senghas) Garay & P.Taylor   *(opposite, top right)*

(syn. *Eulophidium angustifolium* Senghas; *E. angustifolium* subsp. *diphyllum* Senghas)

Leaves 1–2, shortly petiolate, linear, acuminate, up to 10 × 0.7 cm, obscurely variegated on upper side, paler underneath; petiole 1–1.5 cm long; flowers yellowish white, purple-striped and marked on the side lobes of the lip; sepals oblanceolate, 7–8 mm long; lip 4-lobed, 10–12 mm × 14–16 mm, the lobes distinctly convex, the lateral lobes semi-orbicular, the midlobe spreading, obovate; with a rounded 2-lobed callus at the base; spur scrotiform, 3 × 4 mm, bilobed. Antsiranana, Toliara. Dry forest; up to 500 m. Fl. July, November.

52.4. O. quadriloba (Schltr.) Garay & P.Taylor

(syn. *Eulophia quadriloba* Schltr.; *Eulophidium quadrilobum* (Schltr.) Schltr.)

Plant to 35 cm tall; leaves 1–2, suberect, oblanceolate-linear, 8–11 × 0.3–0.5 cm, green; inflorescence unbranched; flowers 8–12, small; sepals 4 mm long; lip 4-lobed, 6.3 × 5 mm, with 2 short triangular calli at the base; spur 5.5 mm long. Mahajanga, Toliara. Coastal forest; semi-deciduous and riverine forest. Fl. February. Terrestrial, sandstone and sand. Fl. January.

52.5. O. callmanderi Bosser

Plant 65–75 cm tall; pseudobulbs fusiform, 8–9 cm long. Leaf solitary, 30–32 × 1–1.3 cm, linear, acute. Inflorescence simple, 65–75 cm tall, 10–15-flowered; flowers white washed with purple; sepals narrowly obovate, 11–15 mm long; lip 4-lobed, 11–12 × 17–18 mm; callus bifid with apices outcurved; spur scrotiform, 1.5–2 mm in diameter. Antsiranana (Masoala). Littoral forest; 10 m. Fl. September.

52.6. O. calcarata (Schltr.) Garay & P.Taylor   *(opposite, bottom left)*

(syn. *Eulophia paniculata* Rolfe; *Lissochilus paniculatus* (Rolfe) H.Perrier; *Eulophidium paniculatum* (Rolfe) Summerh.)

Plant large, with conical, 4–5-angled pseudobulbs, 8–12 × 1.5–3 cm; leaves 3–4, linear, 20–50 × 1.5–3 cm, green; inflorescence 1.2–1.5 m tall, paniculate; flowers green with faint purple veins on lip and a white callus; sepals spatulate, 16–20 mm long, almost twice as long as the petals; lip recurved, 4-lobed, 10–16 × 10–12 mm, with 3 dentate keels; spur conical, 3–4 mm long. Antsiranana, Fianarantsoa, Mahajanga, Toliara. Rocks in highlands; semi-deciduous, western forest; up to 1100 m. Fl. November–April.

LOCAL NAMES *Bekapiaky, Tsikapiaky*.

52.2

52.3

52.6

52.7

52.7. **O. decaryana** (H.Perrier) Garay & P.Taylor   *(p. 409, bottom right)*

(syn. *Eulophia decaryana* H.Perrier ex Guillaumin & Manguin; *Lissochilus decaryanus* (H.Perrier) H.Perrier; *Eulophidium decaryanum* (H.Perrier) Summerh.)

Pseudobulbs obscurely 5-angled; leaves 1–2, sessile, 15–23 × 1–1.2 cm, dark green or brownish red with grey or pale brown mottling or marbling; inflorescence unbranched; flowers creamy yellow, heavily veined with purple; sepals spatulate, 2 cm long, twice as long as the petals; lip recurved, 4-lobed, 1.6 × 1.6 cm; callus at the base of the lip with 2 raised oblong lobes, V-shaped; spur cylindrical, 4 mm long. Fianarantsoa, Mahajanga, Toliara; also in Kenya, Zimbabwe, Mozambique and South Africa. Scrubland, amongst xerophytic vegetation; up to 300 m. Fl. January–April.

LOCAL NAME *Tsikapiaky*.

52.8. **O. hebdingiana** (Guillaumin) Garay & P.Taylor   *(opposite, top left)*

(syn. *Lissochilus hebdingianus* Guillaumin)

Pseudobulbs small, conical, 2 cm tall; leaves 2–4, erect, fleshy, develop after flowering, linear, up to 100 cm × 2 cm, mottled; inflorescence branching, up to 140 cm tall; flowers similar to those of *O. calcarata* (p. 408); sepals spatulate; lip recurved, 4-lobed; callus 2-ridged; spur cylindrical-conical, bifid at tip. Mahajanga. Scrubland and forest undergrowth; 80–100 m. Fl. January.

52.9. **O. longibracteata** Bosser & Morat   *(opposite, bottom left and right)*

Plant to 60 cm tall; pseudobulbs underground, conical, 2 cm tall; roots verrucose; leaves 1–2, erect, linear, 30–45 × 0.35–0.5 cm, tessellated light and dark green, spotted blackish brown; inflorescence unbranched, 20–25-flowered, the stalk spotted; bracts much longer than the flowers, dark spotted; flowers with green sepals and petals veined with purple, a yellow-green lip with purple veins on side lobes, and a white callus flushed with mauve; sepals 12–20 mm long; lip recurved, 4-lobed, 12–15 mm long; side lobes rounded; midlobe divergent, 5–6 mm long; callus 2-ridged; spur conical, 4–5 mm long. Fianarantsoa, Toliara. Dry forest undergrowth. Fl. February–March.

52.10. **O. perrieri** (Schltr.) Garay & P.Taylor

(syn. *Eulophia ambongensis* Schltr.; *Eulophidium perrieri* Schltr.)

Plant to 60 cm tall; pseudobulbs narrowly conical, 4–5-angled; leaf erect, linear, acute, 40–50 × 0.5–0.7 cm, tessellated; inflorescence simple or 1–2-branched; flowers brownish green with purple venation and red-spotting on the lip; sepals oblong, 5 mm long; petals similar; lip straight, 4-lobed, 5 × 6 mm, side lobes larger than the midlobes; callus bilobed, oblong; spur clavate-incurved, 3 mm long, half the length of the lip. Antsiranana, Mahajanga. Seasonally dry, deciduous woodland on sandstone and sand. Fl. December.

52.8

52.11

52.9

52.9

52.11. O. spathulifera (H.Perrier) Garay & P.Taylor   *(p. 411, top right)*
(syn. *Eulophia spathulifera* H.Perrier; *Eulophidium spathuliferum* (H.Perrier) Summerh.)

Pseudobulbs underground, flattened-ovoid; leaves 2–3, elliptic-linear, 10–15 × 0.6–1.2 cm, marked with green-blackish spots; inflorescence simple; basal bracts longer than the flowers; sepals 17–22 mm long, distinctly spatulate; petals 8 mm long, spatulate; lip recurved, 4-lobed, 12 × 12 mm; callus 2-ridged; spur globose. Antsiranana, Mahajanga, Toliara. Seasonally dry, deciduous woodland on sand and coastal dunes. Fl. November–January.

52.12. O. sclerophylla (Rchb.f.) Garay & P.Taylor
(syn. *Eulophia sclerophylla* Rchb.f.; *E. elliotii* Rolfe; *Eulophidium sclerophyllum* (Rchb.f.) Summerh.)

Leaves 2, linear, 2–3 × 0.4–0.6 cm; flowers 9–12; sepals 8 mm long; lip 8 × 4 mm, the front lobe deeply bifid; spur 4 mm long, a little wider at the obtuse tip than at the base. Toliara; possibly also in the Comoros. Seasonally dry, deciduous coastal forest on sand; sea level–200 m. Fl. April.

52.13. O. alismatophylla (Rchb.f.) Garay & P.Taylor
(syn. *Eulophia alismatophylla* Rchb.f.; *Eulophidium alismatophyllum* (Rchb.f.) Summerh.)

Psudobulbs small, 1-leafed; leaf erect, lanceolate, acute, up to 20 × 2.4 cm; petiole longer than the leaf lamina; inflorescence simple; up to 12-flowered; flowers yellow; petals oblong; lip 4-lobed; side lobes roundly subquadrate; midlobes smaller, oblong, rounded at tip; spur conical, 2–3 mm long. Antsiranana. Also possibly in the Comoros. Fl. unknown.

52.14. O. analamerensis (H.Perrier) Garay & P.Taylor
(syn. *Lissochilus analamerensis* H.Perrier; *Eulophidium analamerense* (H.Perrier) Summerh.)

Plant 35–45 cm tall; pseudobulbs small, ovoid; leaf 1, petiolate, lanceolate, 6–8 × 1.2–1.3 cm, tessellated; petiole up to 20 cm long; inflorescence simple or slightly branched; flowers 17–22, sepals 6–7 mm long; petals elliptic, 7 mm long; lip 4-lobed, 6–7 mm long, with a hirsute disc; callus of 2 small lamellae at the base; spur incurved-tapering, 2.2 mm long; vegetatively similar to *O. alismatophylla*. Antsiranana. Shaded and wet limestone rocks; up to 500 m. Fl. January.

52.15. **O. analavelensis** (H.Perrier) Garay & P.Taylor

(syn. *Eulophidium analavelense* (H.Perrier) Summerh.; *Lissochilus analavelensis* H.Perrier)

Plant 30–40 cm tall; pseudobulbs narrowly ovoid, 2–3 cm long; leaves 1–2, oblanceolate, 5–8 × 0.8–1.1 cm, longly petiolate, tesselated, petiole articulated in middle; sepals 6 mm long, slightly longer than the petals; lip recurved, 4-lobed, 11 × 11 mm, with 3 thickened ridges in front of the 2-ridged callus; spur clavate, 4 mm long, slightly bifid at tip. Toliara. Dry forest; 500–1000 m. Fl. March.

52.16. **O. cordylinophylla** (Rchb.f.) Garay & P.Taylor   (*p. 415, top left*)

(syn. *Eulophia cordylinophylla* Rchb.f.; *Eulophia lokobensis* H.Perrier; *Eulophidium cordylinophyllum* (Rchb.f.) Summerh.; *Eulophidium lokobense* (H.Perrier) Summerh.)

Plant large, up to 80 cm tall; pseudobulbs conical; leaf longly petiolate, broadly elliptic, 13–19 × 6–8 cm; petiole 16–30 cm long; inflorescence branched; flowers small; sepals spatulate, 7 mm long; lip 4-lobed, villous near the base and with 2 calli, spur conical, 6 mm long. Antsiranana; also in the Comoros. Shaded and humid rocks; up to 300 m. Fl. February.

52.17. **O. flavescens** Bosser & Morat

Plant up to 45 cm tall; pseudobulbs conical-cylindrical, 1.5–3 cm tall, on a distinct rhizome; leaf solitary, erect, thin-textured, narrowly elliptic, 15–22 × 2–3 cm, green; stalk 10–12 cm long; inflorescence simple, 10–12-flowered; flowers pale yellow with purple stripes on the lip side lobes; sepals narrowly oblong, 8–10 mm long; lip straight, 4-lobed, 8–9 × 2.5–3 mm; side lobes rounded; midlobes spreading, obovate, 3–5 mm long; callus 2-ridged; spur incurved-clavate, 1–1.2 mm in diameter. Toamasina. Coastal forest on sand; up to 100 m. Fl. November.

52.18. **O. pandurata** (Rolfe) Garay & P.Taylor   (*p. 415, top right*)

(syn. *Eulophia pandurata* Rolfe; *Eulophidium panduratum* (Rolfe) Summerh.)

Plant 20–30 cm tall; pseudobulbs fusiform, on a distinct rhizome; leaves 2–3, ovate-lanceolate, petiolate, 3.5–8 × 1.5–2.5 cm; petiole 2–3.5 cm long, articulated near apex; inflorescence simple or rarely 1-branched; flowers with similar sepals and petals, 8 mm long; lip 3-lobed, 8 × 6 mm, the lateral veins on the disc papillose-ciliolate; side lobes narrowly oblong; midlobe obovate; callus 2-ridged; spur 4 mm long, straight, subclavate. Antananarivo, Toliara. Seasonally dry, deciduous forest; up to 1500 m. Fl. December, May.

52.19. **O. petiolata** (Schltr.) Garay & P.Taylor *(opposite, bottom left)*

(syn. *Eulophia petiolata* Schltr.; *Eulophidium petiolatum* (Schltr.) Schltr.)

Close to *O. lonchophylla* from SE Africa and the Comoros. Pseudobulbs conical, 6–7-angled, 5 × 3 cm; leaf erect, narrowly lanceolate, 12–20 × 2–4.4 cm, very longly petiolate; inflorescence branched; branches suberect, 7–10-flowered; flowers with green sepals and petals, the lip white with yellow margins and red-veined side lobes; sepals slightly longer than the petals, 5–6 mm long; lip recurved, 4-lobed, 10 × 8 mm, papillose in throat, with 2 keels; spur straight, cylindrical, 2.5–3 mm long. Antsiranana, Mahajanga, Toliara; also in the Comoros. Seasonally dry, deciduous woodland and xerophytic scrub on sandstone and sand; up to 600 m. Fl. December–April.

52.20. **O. ambongensis** (Schltr.) Garay & P.Taylor

(syn. *Eulophidium ambongense* Schltr.; *Eulophia schlechteri* H.Perrier)

Plant 20–25 cm tall; pseudobulbs oblong, 1.5–2 cm long; leaf petiolate, broadly ovate, 10–13 × 3–3.5 cm; inflorescence simple; flowers 8–12; sepals 9 mm long; lip straight, 3-lobed, 8 × 7 mm; callus bilobed; spur subglobose, 2.5 mm long. Mahajanga. Seasonally dry, deciduous forest and woodland; up to 200 m. Fl. January.

52.21. **O. antsingyensis** G.Gerlach *(opposite, bottom right)*

Related to *O. rauhii* (p. 418) but differs by the broader, ovate, horizontal, variegated leaf, up to 3 × 1.5 cm, densely flowered inflorescence, 30–40 cm tall, larger flowers with rounded petals and sepals; sepals 12 mm long; lip 10 mm long and broad; spur clavate 3 mm long, and with acuminate calli. Mahajanga. Humus pockets in 'tsingy'; 60–100 m. Fl. February.

52.22. **O. boinensis** (Schltr.) Garay & P.Taylor *(p. 417, top left)*

(syn. *Eulophidium boinense* Schltr.; *Lissochilus boinensis* (Schltr.) H.Perrier)

Plant up to 40 cm tall; pseudobulbs ovoid, 2.5–3 cm long; roots warty; leaf more or less cordate at base, elliptic, 10–12 × 5 cm; inflorescence distantly few-flowered; flowers 6–10, 16–26 mm long, yellow or greenish-yellow with the lip with dark red markings; lip straight, 4-lobed, 6 mm long, almost orbicular; callus 2-ridged; spur clavate, 2.5–3 mm long, half length of lip; close to *O. maculata* (p. 418). Antsiranana, Mahajanga. Humid, evergreen, lowland forest; up to 500 m. Fl. January–March.

52.16

52.18

52.19

52.21

52.23. O. furcata Bosser & Morat

Plant up to 60 cm tall; pseudobulbs ovoid, 1–2 cm tall; leaves ovate, acute, 4–5.5 × 2–2.5 cm, shortly stalked; raceme simple, much longer than the leaves; flowers small, whitish yellow with purple veins on lip side lobes; sepals 7–8 mm long; lip 4-lobed, 9–10 × 9–10 cm; side lobes narrow, acute; midlobes spreading obovate, truncate, 6 × 3.5–4 mm; calli 2, small, triangular; spur recurved, S-shaped, 4–5 mm long, forked at the tip. Mahajanga. In sand; up to 100 m. Fl. April.

52.24. O. gracillima (Schltr.) Garay & P.Taylor   *(opposite, top right)*
(syn. *Eulophia gracillima* Schltr.; *Eulophidium gracillimum* (Schltr.) Schltr.; *Eulophidium roseovariegatum* Senghas; *Oeceoclades roseovariegata* (Senghas) Garay & P.Taylor)

Plant up to 50 cm tall; pseudobulbs ovoid; leaves 2, ovate-oblong, 3–7.5 × 1.7–3 cm, tessellated, margins corrugate-crisped; inflorescence broadly branching; flowers small, 1–1.2 cm long; sepals 4 mm long, a third longer than the petals; lip recurved, obscurely 4-lobed; callus 2-ridged; spur tapering, longer than lip. Antsiranana, Mahajanga. Seasonally dry, deciduous forest and woodland on limestone. Fl. March–June, November.

52.25. O. humbertii (H.Perrier) Bosser & Morat
(syn. *Lissochilus humbertii* H.Perrier; *Eulophia humbertii* (H.Perrier) Butzin)

Plant 35–40 cm tall; pseudobulbs conical-discoid, 2 cm across; leaves 3–4, lanceolate, cuspidate, 5–6 × 1 cm, green; inflorescence simple, 10–12-flowered; flowers green, purple veined on lip; sepals oblong-elliptic, 10 mm long, acute, similar to petals; lip recurved, 4-lobed, 7 × 8 mm; spur cylindrical, 1–2.5 mm long. Toliara. Xerophytic forest; up to 1000 m. Fl. December–January.

52.26. O. lanceata (H.Perrier) Garay & P.Taylor
(syn. *Eulophia lanceata* H.Perrier)

Pseudobulbs on a distinct rhizome, fusiform; leaves 2, ovate-elliptic, acute, long-petiolate, lightly mottled; petiole half the length of the leaf; inflorescence simple, many-flowered; flowers white with pink markings; sepals and petals subsimilar; lip straight; side lobes narrowly oblong; midlobe transversely oblong; spur clavate, 3–4 mm long; close to *O. pandurata* (p. 413), but the lip is distinct. Antananarivo. Open forest; 1500 m. Fl. December.

52.22

52.24

52.27

52.28

53.1

52.27. O. maculata (Lindl.) Lindl. (p. 417, middle left)

(syn. *Angraecum maculatum* Lindl.; *Eulophia monophylla* (A.Rich.) S.Moore; *Eulophidium maculatum* (Lindl.) Pfitz.; *Eulophia ledienii* N.E.Br.; *Eulophia mackeni* Rolfe ex Hemsl.; *Eulophidium monophyllum* (A.Rich) Schltr.; *Oeceoclades mackenii* (Rolfe ex Hemsl.) Garay & P.Taylor; *Oeceoclades monophylla* (A.Rich.) Garay & P.Taylor)

Pseudobulbs ovoid; leaf 1, ovate to oblong-ovate, mottled grey-green mottled with dark green; inflorescence simple or few-branched; branches erect; lip straight, 4-lobed, white with 2 purple-red blotches at the base of the midlobe, callus 2-ridged; spur clavate, slightly apically bilobed. Mahajanga, Toliara; also in the Mascarenes, tropical Africa and S and C America. Lowland and coastal, seasonally dry forest; up to 200 m. Fl. February–April.

52.28. O. peyrotii Bosser & Morat (p. 417, bottom left)

Plant 20–45 cm tall; pseudobulbs ovoid, 2–2.5 × 2 cm; leaf ovate, acute, 4–5 × 2–2.5 cm, tessellated; petiole c. 10 mm long; inflorescence simple or branched, much longer than the leaves; sepals and petals spatulate, 5–6 mm long; lip straight, 4-lobed, 6.5–8 × 7–10 mm; side lobes rounded; midlobes well-separated, obovate; the disk with an acute bilobed callus, the lobes obovate, the sinus broad; spur cylindrical-conical, 1.5–2 mm long. Toliara. Forest undergrowth in evergreen forest. Fl. December–February.

52.29. O. rauhii (Senghas) Garay & P.Taylor

(syn. *Eulophidium rauhii* Senghas)

Close to *O. boinensis* (p. 414) but sepals and petals linear-lanceolate; lip 4-lobed; spur 6–7 mm long. Antsiranana. Fl. July.

53. GRAPHORKIS

A small genus of 4 species in tropical Africa, Madagascar and the Mascarenes. Three species in Madagascar. Epiphytes with clustered ovoid pseudobulbs; roots both spreading and erect. Leaves deciduous, pleated, developing after flowering or at the same time. Inflorescences branched, many-flowered. Flowers yellow of green with dark spotting; lip white or yellow. Sepals and petals similar, spreading. Lip 3-lobed, with or without a spur; callus verrucose or keeled. Column short; pollinia 2 sessile on a large triangular viscidium.

Key to species of *Graphorkis*

53.1. **G. ecalcarata** (Schltr.) Summerh.

(p. 417, bottom right)

(syn. *Eulophiopsis ecalcarata* Schltr.; *Eulophia ecalcarata* (Schltr.) M.Lecoufle)

Pseudobulbs ovoid-elliptical; flowers c. 2 cm long; lip 7 mm long, without a spur. Toamasina. In humid, lowland and lower montane forest; up to 1000 m. Fl. October–January.

53.2. **G. concolor** (Lindl.) Kuntze var. **alphabetica** F.N.Rasm.

(syn. *Graphorkis scripta* (Thouars) Lindl.)   (p. 421, top)

Epiphyte; root-mass extensive, developing around the base of the plant, the erect roots sharply pointed; older pseudobulbs carrying teat at the apex; pseudobulbs 3–14 × 1–4 cm; leaves linear-lanceolate, 8–11 × 0.8–1.4 cm; inflorescences paniculate, many-flowered; produced before the leaves; flowers 15–25 mm in diameter, petals and sepals greenish yellow to yellow, spotted with red. Antsiranana, Fianarantsoa, Toamasina, Toliara; also in the Comoros, the Mascarenes and the Seychelles. Humid, lowland forest; littoral forest, on *Raphia* palms, on mango trees, on *Tiphonodorum lindleyanum*; up to 100 m; Fl. September–November. The typical variety, probably also found in Madagascar, lacks spotting on the sepals and petals and is very rare in nature.

53.3. **G. medemiae** (Schltr.) Summerh.

Pseudobulbs ovoid-oblong, 6 × 2 cm; leaves appearing after the flowers, grass-like, 6 mm broad; inflorescence slender, paniculate, c. 25 cm long; branches 3, many-flowered; flowers brown with a yellow lip; sepals and petals narrowly lanceolate, 9 mm long; lip 3-lobed near the middle, 7 x 3.5 mm, the midlobe larger than the side lobes, undulate crenulate on the margins; spur obscure. Mahajanga. On trunks of *Medemia nobilis*; up to 200 m. Fl. August–September.

54. CYMBIDIELLA

A genus of 3 species, endemic to Madagascar. Large terrestrial or epiphytic plants with stout elongate rhizomes and elongate roots. Pseudobulbs ovoid to fusiform, leafy in upper part. Leaves pleated, suberect to spreading. Inflorescences basal, erect or spreading, simple or branched, few–many-flowered. Flowers large, showy, yellow-green to green; lip green or red with black veins and spots. Sepals and petals subsimilar. Lip 3-lobed; side lobes erect; midlobe ovate or flabellate and emarginated. Column arcuate; pollinia 4, attached to a large viscidium.

Key to species of *Cymbidiella*

1. Plants terrestrial . . . . **54.1. C. flabellata**
1. Plants epiphytic 2
2. Lip pale greenish white heavily marked with black; lip midlobe ovate, acute, the margins very undulate . . . . **54.2. C. falcigera**
2. Lip red, spotted with black; lip midlobe flabellate, emarginated, the margins straight . . . . **54.3. C. pardalina**

54.1. **C. flabellata** (Thouars) Lindl.   *(opposite, bottom left and right)*
(syn. *Limodorum flabellatum* Thouars; *Cymbidiella perrieri* Schltr.)

A slender plant, 1–1.5 m tall, on a long thin rhizome; pseudobulbs 8–10 cm apart, leaves 6–8, ligulate, 20–50 × 1.7–2 cm; flowers somewhat similar in appearance to those of *C. pardalina* (p. 422) but more slender and much smaller; sepals and petals 1.5–2 cm long, yellow-green; lip obovate, 3-lobed, 1.5–2 × 1.4–1.8 cm, red with black spots. Antsiranana, Fianarantsoa, Mahajanga, Toamasina, Toliara. In coastal forest, on shaded and humid quartzite rocks; in *Philippia* scrub; in marshland in peaty sphagnum moss on coarse acid quartzite sand saturated with water; constantly humid and with black humus, often in the shade of *Ericaceous* scrub on laterite; up to 1500 m. Fl. September–February.

54.2. **C. falcigera** (Rchb.f.) Garay   *(p. 423, top left and right)*
(syn. *Grammangis falcigera* Rchb.f.; *Cymbidium humblotii* Rolfe; *Cymbidiella humblotii* (Rolfe) Rolfe)

Plant very large with a stout creeping rhizome; pseudobulbs up to 30 cm long; leaves 7–40, arranged in a fan, 25–60 × 2–2.5 cm; inflorescence branched; flowers apple-green to yellow-green with dark blackish spots on lip; sepals 40–45 mm long; lateral sepals dorsally keeled; lip recurved, 3-lobed, 3 cm long, the side lobes bigger than the middle one. Antsiranana, Fianarantsoa, Toamasina, Toliara. Possibly in the Comoros. On river margins, littoral forest, and marshland on *Raphia*, *Afzelia* sp., and *Vonitra thouarsiana* palms; up to 400 m. Fl. December–March.

53.2

54.1

54.1

54.3. C. pardalina (Rchb.f.) Garay   *(opposite, bottom left)*
(syn. *Grammangis pardalina* Rchb.f.; *Cymbidium rhodochilum* Rolfe; *Cymbidiella rhodochila* (Rolfe) Rolfe)

Plant large on a short stout rhizome; pseudobulbs 7.5–12 cm long; leaves 5–10, up to 65 cm long; inflorescence not branched; flowers large; tepals 35–45 mm long, yellowish-green, carrying dark spots; lip 4-lobed, 40 × 30 mm, green with black spots, the lobes red. Fianarantsoa, Toamasina. In humid, evergreen forest, growing in association with *Platycerium madagascariensis*, on *Albizzia fastigiata*; 500–2000 m. Fl. October–December.

55. EULOPHIELLA

A genus of 5 species, endemic to Madagascar. Large terrestrial or epiphytic plants with stout elongate rhizomes and roots. Pseudobulbs ovoid to fusiform, leafy in upper part. Leaves pleated, arching to spreading. Inflorescences basal, erect, simple, few–many-flowered. Flowers large, showy, white, yellowish or rose-purple. Sepals and petals subsimilar, elliptic, concave. Lip 3-lobed, lacking a spur; side lobes erect; midlobe smaller; callus of 2–3 basal ridges. Column arcuate; pollinia 4 attached to a large viscidium.

Key to species of *Eulophiella*

1. Plant terrestrial; leaves 1–2, not noticeable pleated, longly, slenderly petiolate . . . .
. . . .**55.1. E. ericophila**
1. Plant epiphytic; leaves usually 3 or more, pleated or not, tapering but not
obviously petiolate 2
2. Flowers yellowish or greenish yellow 3
2. Flowers white, pinkish-white or rose-purple 4
3. Leaves pleated, 50–70 cm long; lip midlobe glabrous, side lobes narrowly
oblong, obscure . . . . **55.2. E. capuroniana**
3. Leaves not pleated; lip midlobe papillose; side lobes very obliquely triangular,
acute . . . . **55.3. E. galbana**
4. Sepals and petals 20 mm long, pale pinkish to white, the petals with a pink
spot in the middle; lip 3-lobed, 14 × 13 mm, white with a yellow callus and
central area and red spotting around the callus . . . . **55.4. E. elisabethae**
4. Sepals and petals 35–40 mm long, rose-purple; lip suborbicular, 45 × 40 mm,
white with a yellow callus and rose-purple apex . . . . **55.5. E. roempleriana**

55.1. E. ericophila Bosser

A more slender plant than the other species of the genus, with smaller linear, slenderly petiolate leaves, 20–35 × 0.4–1.2 cm, and pale rose-coloured flowers with orange spots on the lip, and the shallowly 2-crested lip; sepals 10.5–12 mm long; lip 3-lobed, 7–8 × 6–7 mm; callus of 2 longitudinal tuberculate crests; column 5–6 mm long. Antsiranana. In montane, ericaceous scrub; 1300–1500 m. Fl. November.

54.2

54.2

54.3

55.4

55.2. **E. capuroniana** Bosser & Morat

Terrestrial similar to *E. elisabethae*, but it is more slender, with narrower plicate leaves, 50–70 × 1.5–3 cm; flowers 4–9, smaller, yellow and with the lip bearing crests of a different shape; sepals 14–18 mm long; lip obscurely 3-lobed, 10–12 × 7–8 mm; side lobes narrow; midlobe subcircular-ovate; callus of 2 keels in basal half of lip. Toamasina. In humid, lowland forest. Fl. October.

55.3. **E. galbana** (Ridl.) Bosser & Morat   *(opposite, top)*
(syn. *Eulophia galbana* Ridl.; *Graphorchis galbana* (Ridl.) Kuntze)

Small epiphyte with shiny ovoid pseudobulbs; leaves 2–3, oblanceolate, up to 18 × 1.3 cm; inflorescence up to 8-flowered, shorter than the leaves; flowers greenish yellow, tinged red, not opening widely; sepals and petals 9–11 mm long; lip 3-lobed, 15–16 × 10 mm; side lobes obliquely triangular, acute; midlobe elliptic, papillose; callus of 2 glabrous ridges to base of midlobe. Antananarivo, Fianarantsoa. In humid, medium-elevation forest. Fl. January.

55.4. **E. elisabethae** Linden & Rolfe   *(p. 423, bottom right)*
(syn. *Eulophiella perrieri* Schltr.)

Plant large; pseudobulbs oblong, 10–15 cm long; leaves narrowly lanceolate, plicate, 45–60 × 3.5–5 cm; flowers about 4 cm across; sepals and petals pale pinkish-white, 2 cm long; lip 3-lobed, 14 × 13 mm, white with a yellow callus and central area and red spotting around the callus. Antsiranana, Toamasina. In coastal forest on *Dypsis fibrosa*; up to 200 m. Fl. November–December.

55.5. **E. roempleriana** (Rchb.f.) Schltr.   *(opposite, bottom left and right)*
(syn. *Grammatophyllum roemplerianum* Rchb.f.; *Eulophiella peetersiana* Kraenzl.; *E. hamelini* Baill. ex Rolfe)

Plant very large; rhizome stout, creeping; pseudobulbs 8–28 cm long; leaves 4–8, large, lanceolate, plicate, 90–120 × 8–10 cm; inflorescence up to 120 cm tall; flowers 15–25, c. 9 cm, rose-purple with a white disc and yellow-tipped callus ridges; sepals 35–45 mm long; lip suborbicular, 45 × 40 mm. Antananarivo, Fianarantsoa, Toamasina. On *Pandanus* in coastal and mid-elevation humid, montane forest; up to 1300 m. Fl. December–January.

55.3

55.5

55.5

56. PARALOPHIA

A Madagascan endemic genus of 2 species. Epiphytes with short to long stout rhizomes and cylindrical stems or pseudobulbs. Leaves pleated. Inflorescences lateral, unbranched. Flowers showy, similar to those of *Eulophia*.

Key to species of *Paralophia*

1. Stems cylindrical, not pseudobulbous, many-leaved in apical part; lip callus 3-ridged, purple .... **56.1. P. epiphytica**
1. Stems pseudobulbous, 3–4-leaved; inflorescence longer than the leaves; lip callus 2-ridged .... **56.2. P. palmicola**

56.1. P. epiphytica (P.J.Cribb, DuPuy & Bosser) P.J.Cribb & Hermans
(opposite, top and bottom left)

(syn. *Eulophia epiphytica* P.J.Cribb, DuPuy & Bosser)

Plant large, epiphytic; rhizomes elongate, creeping; stems trailing, lacking pseudobulbs; leaves thin-textured; inflorescence shorter than the leaves; flowers c. 3.5 cm long; sepals and petals yellow-green; lip 3-lobed, white with 3 purple-marked callus ridges. Fianarantsoa, Toliara. On *Elaeis guineensis*, *Raphia farinifera* and probably also *Dypsis* palm trees, climbing amongst leaf bases on trunk of palm trees; sea level–600 m. Fl. December–February.

56.2. P. palmicola (H.Perrier) P.J.Cribb & Hermans
(opposite, middle right)

(syn. *Eulophia palmicola* H.Perrier; *Lissochilus palmicolus* (H.Perrier) H.Perrier)

Rhizomes short; pseudobulbs 3–4-leaved; inflorescence longer than the leaves, erect; flowers with apiculate sepals and petals; lip with a 2-ridged callus at the base, a shortly apiculate midlobe and a 4 mm long spur. Toliara. On the palm *Ravenia xerophila* in *Didiereaceae* forest. Fl. October.

56.1

56.1

56.2

57.2

57. GRAMMANGIS

A genus of 2 species endemic to Madagascar. Large epiphytes with clustered, ovoid, tetragonal pseudobulbs. Leaves 3–5, pleated. Inflorescence basal, unbranched, few-flowered. Flowers large, fleshy-waxy, yellow to orange, blotched with brown or purple on the sepals and petals; lip white marked with purple. Dorsal sepal and petals free, spreading but not widely; lateral sepals forming a distinct mentum with the column-foot. Petals porrect. Lip 3-lobed, lacking a spur; callus ridged or thorny-papillate. Column stout, with a distinct foot; pollinia sessile on a common viscidium.

Key to species of *Grammangis*

1. Sepals and petals yellow with brown base and apex and a transverse central brown bar in the middle of the sepals; sepals more than 3 cm long; lip midlobe deltoid, glabrous; callus of 2 longitudinal low ridges on disc of lip . . . . **57.1. G. ellisii**
1. Sepals and petals yellow with purple tips and central longitudinal purple vein; sepals 2.6–2.8 cm long; lip midlobe oblong, bearing a callus of 7 rows of thorn-like papillae . . . . **57.2. G. spectabilis**

57.1. G. ellisii (Lindl.) Rchb.f.   (*opposite*)
(syn. *Grammangis fallax* Schltr.)

Plant robust with large tetragonal pseudobulbs, 50–60 cm tall; pseudobulbs 8–10 cm long; leaves 3–5, oblong, 16–40 × 1.3–4 cm; flowers large, golden-brown, glossy; sepals 4 cm long; lip 3-lobed, 1.8 × 1.8 cm, with pronounced white keels. Antananarivo, Antsiranana, Fianarantsoa, Toamasina. Coastal forest; humid, evergreen forest; branches overhanging rivers; on *Raphia farinifera* and on *Pandanus*; up to 1300 m. Fl. November–January.

57.2. G. spectabilis Bosser & Morat   (*p. 427, bottom right*)

Pseudobulbs conical, up to 10 × 3–3.5 cm; leaves 2–5, loriform, up to 40 × 3 cm; inflorescence spreading, up to 50 cm long, 15–20-flowered; flowers orange-brown; sepals 2.6–2.8 cm long; lip 1.5–1.8 cm long; side lobes obliquely elliptic, obtuse; midlobe oblong, obtuse, bearing 7 barbed keels. More slender overall than *G. ellisii* and easily distinguished by floral characteristics, particularly the orange-brown flowers and the thorn-like appendages on the lip surface. Toliara. Very rare on large trees, such as figs, in seasonally dry, deciduous forest or woodland; 300–800 m. Fl. November.

57.1

Photography

With the exception of the species photographs listed below, all photographs in this book were taken by Johan Hermans.

Sven Buerki: *Bulbophyllum brevipetalum*,

Mark Clements: *Benthamia majoriflora, Didymoplexis madagascariensis, Gastrorchis simulans*

Philip Cribb: *Aerangis ellisii, Aeranthes parvula, Angraecum ferkoanum, Angraecum leonis, Angraecum setipes, Benthamia monophylla, Brachycorythis disoides, Brachycorythis pleistophylla, Bulbophyllum coriophorum, Bulbophyllum subclavatum, Cymbidiella falcigera, Cynorkis aurantiaca, Cynorkis cardiophylla, Cynorkis filiformis, Cynorkis gigas, Cynorkis stolonifera, Disa incarnata, Disperis erucifera, Disperis masoalensis, Eulophia hians, Eulophia livingstoniana, Eulophiella roempleriana, Graphorkis ecalcarata, Habenaria ambositrana, Habenaria bathiei, Habenaria cirrhata, Habenaria clareae, Habenaria incarnata, Habenaria praealta, Habenaria simplex, Habenaria tianae, Habenaria truncata, Oeceoclades longebracteata, Oeceoclades peyrotii, Polystachya aurantiaca, Polystachya monophylla, Satyrium rostratum, Tylostigma foliosum*

David DuPuy: *Jumellea alionae, Satyrium rostratum*

Eberhard Fischer: *Benthamia bathieana, Benthamia glaberrima, Tylostigma nigrescens*

Gunter Fischer: *Bulbophyllum afzelii, Bulbophyllum ambatoavense, Bulbophyllum debile, Bulbophyllum elliotii, Bulbophyllum henrici* var. *rectangulare, Bulbophyllum henrici* var. *rectangulare* yellow form, *Bulbophyllum hirsutiusculum, Bulbophyllum jackyi, Bulbophyllum lakatoense, Bulbophyllum ophiuchus*

Moritz Grubbeman: *Cynorkis melinantha, Oeceoclades humbertii*

Jean-Claude Hervouet: *Bulbophyllum acutispicatum, Bulbophyllum coccinatum, Bulbophyllum henrici, Bulbophyllum mangenotii, Bulbophyllum masoalanum, Bulbophyllum minutum, Bulbophyllum septatum, Cynorkis stolonifera, Cynorkis subtilis, Didymoplexis madagascariensis, Platylepis polyadenia, Vanilla planifolia*

Rudi Hromniak: *Bulbophyllum ambatoavense, Bulbophyllum edentatum, Bulbophyllum hapalanthos, Bulbophyllum liparidioides*

Dominique Karadjoff: *Angraecum litorale*

Jean-Noel Labat: *Microcoelia cornuta, Neobathiea spatulata*

Isobyl and Eric la Croix: *Corymborkis corymbis*

Marcel Lecoufle: *Bulbophyllum lecouflei, Paralophia palmicola*

Georges Morel: *Bulbophyllum brachystachyum*

Olaf Pronk: *Aeranthes crassifolia, Aeranthes polyanthemus, Angraecum curnowianum, Angraecum oblongifolium, Angraecum protensum, Bulbophyllum luteobracteatum, Bulbophyllum perpusillum, Disperis discifera* var. *borbonica, Eulophia mangenotiana, Habenaria arachnoides, Nervilia renschiana*

David Roberts: *Disa andringitrana, Habenaria hilsenbergii*
Walter Roosli: *Sobennikoffia poissoniana*
Karlheinz Senghas: *Aeranthes erectiflora, Angraecum mirabile, Angraecum triangulifolium, Nervilia affinis*
Anton Sieder: *Bulbophyllum bryophytoides* sp nov., *Bulbophyllum cirrhoglossum, Bulbophyllum debile, Bulbophyllum erectum, Bulbophyllum francoisii, Bulbophyllum oreodorum, Bulbophyllum pleiopterum, Bulbophyllum protectum, Bulbophyllum quadrialatum, Bulbophyllum toilliezae, Disa caffra, Platycoryne pervillei*
Eva Smrzova: *Angraecum filicornu*
Hendrik Venter: *Ambrella longituba*

Further Reading

A detailed annotated listing of the bibliography on Madagascan orchids is provided in Du Puy *et al.* (1999) and in Hermans *et al.* (2007).

Allorge-Boiteau, L. (2008). *Plantes de Madagascar: Atlas*. Ulmer Verlag, Paris.

Bradt, H. (1997). *Guide to Madagascar* 5th Ed. Bradt Publ., Chalfont St Peter, UK.

Du Puy, D., Bosser, J., Cribb, P.J. , Hermans, J. & C. (1999). *Orchids of Madagascar*. Royal Botanic Gardens, Kew.

Ecott, T. (2004). *Vanilla. Travels in Search of the Luscious Substance*. Michael Joseph & Penguin Books, London.

Goodman, S M. (ed) (2008). *Paysages Naturels et Biodiversite de Madagascar*. Museum d'Histoire Naturelle, Paris.

Goodman, S.Г. & Benstead, J.P. (eds) (2003). *The Natural History of Madagascar*. University of Chicago Press.

Govaerts, R. (2009). *World Checklist of Orchidaceae*. Royal Botanic Gardens, Kew. Published on the Internet; http://www.kew.org/wcsp/monocots.

Hermans, J. & C, Du Puy, D., Cribb, P.J. & Bosser, J. (2007). *Orchids of Madagascar* 2nd Ed. Royal Botanic Gardens, Kew.

Hillerman, F. (1986). *An Introduction to the Cultivated Angraecoid Orchids of Madagascar*. Timber Press, Portland, Oregon.

Moat, J. & Smith, P. (2007). *Atlas of the Vegetation of Madagascar*. Royal Botanic Gardens, Kew.

Perrier de la Bathie, H. (1939–1941). *Flore de Madagascar: Orchidées*. Tananarive: Imprimerie Officelle.

Schatz, G. E. (2001). *Generic Tree Flora of Madagascar*. Royal Botanic Gardens, Kew and Missouri Botanical Garden, St Louis.

Stewart, J. L. (2006). *Angraecoid Orchids: Species from the African Region*. Timber Press, Portland, Oregon.

Teissier, M. (2006). *Madagascar: A Paradise at Risk*. Cactus & Co., Tradate, Italy.

Tropicos.org. Missouri Botanical Garden. 2009; http://www.tropicos.org.

Glossary

Abaxial, the side away from the stem, normally the lower surface.

Abrupt, suddenly ending as though broken off.

Acuminate, having a gradually tapering point.

Acute, distinctly and sharply pointed, but not drawn out.

Adnate, attached to the whole length.

Adventitious, applied to roots arising from a node on the stem.

Alternate, placed on opposite sides of the stem on a different line.

Anther, that part of the stamen in which the pollen is produced.

Anther cap, hinged cover enclosing the pollinia and pollinial attachments.

Apical, applied to an inflorescence borne at the top of the pseudobulb.

Appressed, lying flat for the whole length of the organ.

Arching, curved like a bow.

Arcuate, curved like a bow.

Aristate, collection of awn-like bristle hairs.

Articulate, jointed; used to describe leaves or other parts that have an abscission layer or a joint at the base.

Auriculate, with small lobes or ear-like appendages.

Basal, applied to the inflorescence at the base of an organ or part such as the pseudobulb.

Bifid, divided by a deep cleft or notch into two parts.

Bract, a much reduced leaf-like organ bearing a flower, inflorescence or partial inflorescence in its axil.

Callus, a thickened area of the lip.

Calyx, the outermost layer of the floral envelope.

Cataphylls, the early leaf forms of a plant or shoot.

Caudicle, a slender, mealy or elastic extension of the pollinium, or a mealy portion at the end of the pollinium; the structure is part of the pollen mass, and is produced within the anther.

Caulescent, becoming stalked, where the stalk is clearly apparent.

Channelled, hollowed out like a gutter, as in a leaf-stalk.

Chataceous, resembling paper or parchment.

Clavate, club-shaped, thickened towards the apex.

Claw, the conspicuously narrowed and attenuated base of an organ.

Column, an advanced structure composed of a continuation of the flower-stalk, together with the upper part of the female reproductive organ (pistil) and the male reproductive organ (stamen).

Column foot, an extension at the base of the *column* to which the *labellum* is attached.

Concave, hollow, as the inside of a saucer.

Conduplicate, (of leaf-like organs) with a single median fold, with each half being flat.

Connate, joined to for entire length.

Convex, having a more or less rounded surface.

Convolute, when one part is wholly rolled up in another.

Cordate, heart-shaped.

Coriaceous, leathery.

Crenate, scalloped, toothed with crenatures.

Crenulate, crenate, but the teeth small.

Crisp, curled.

Crispate, curled.

Cristate, crested.

Crowded, closely pressed together or thickly set.

Cucullate, hooded or hood-shaped.

Cuspidate, tipped with a sharp rigid point.

Decumbent, lying along the ground, with extremity upturned.

Decurrent, extending down the stem below the point of attachment.

Deflexed, bent outwards, the opposite of inflexed.

Dentate, toothed.

Denticulate, minutely toothed.

Dilated, expanding into a blade.

Disc or **disk,** the face of any organ, used with reference to the *lip* in orchids and sometimes to the removable part of the ***rostellum*** projection. (In orchids, disk usually refers to the area between the side lobes in the basal half of the lip, the place where the callus is usually placed.)

Distichous, having leaves or other organs in two opposite rows, as opposed to a spiral or whorled arrangement.

Dorsal, referring to the back, or attached thereto.

Duplicate, refers to the folding of the leaves during development; folded once with each half flat.

Ellipsoid, an elliptic solid.

Elliptic, ellipse-shaped, oblong with regularly rounded ends.

Emarginate, notched, usually at the apex.

Embracing, clasping at the base.

Ensiform, sword-shaped.

Entire, without tooth or divisions.

Epichile, the terminal part of the lip when it is distant from the basal portion.

Epiphyte, a plant growing on another plant but not parasitic.

Erect, upright, perpendicular to the ground or its attachment.

Erose, bitten or gnawed.

Falcate, sickle-shaped.

Filiform, thread-like.

Fimbriate, fringed.

Flaccid, withered and limp, flabby.

Flexuose, bent alternatively in opposite directions, zig-zagged.

Free, not adhering.

Fusiform, spindle-shaped.

Genuflexed, bent.

Gibbous, more convex in one place than another.

Globose, nearly spherical.

Globular, spheroidal.

Heteromycotroph (adj. **heteromycotrophic**), a plant that is a *mycotroph* as part of its method of nutrition, usually with inadequate photosynthesis (hence often not green); a facultative mycotroph.

Holomycotroph (adj. **holomycotrophic**), a plant that is a *mycotroph* as its sole method of nutrition and hence without chlorophyll; an obligate mycotroph. This condition has often been erroneously referred to as *saprophytic*.

Hooded, a plane body, the apex or sides of which are curved inwards to resemble a hood.

Hypochile, the basal portion of the lip.

Imbricate, overlapping as the tiles of a roof.

Incised, cutting sharply into the margin.

Incurved, bending from without, bending inwards.

Inflorescence, the disposition of the flowers on the floral axis or the flowers, bracts and floral axis together.

Isthmus, the narrowed connection between parts of a flower.

Karst, limestone regions with underground drainage and many cavities and passages caused by the dissolution of the rock.

Keel, a ridge resembling the keel of a boat.

Knee, an abrupt bend in the stem or part of a flower.

Labellum, the enlarged, often highly modified, third petal of the orchid flower.

Lacerate, torn, with an irregular cleft.

Lacinate, slashed, cut into narrow lobes.

Lanceolate, narrow, tapering at each end, lance-shaped.

Lateral (inflorescence), borne on or near the side of the pseudobulb or stem, usually in the axils of the bracts or leaves.

Lax, loose, distant.

Linear, at least 12 times longer than broad, with the sides close to parallel.

Lip, the labellum of an orchid.

Lithophyte, a plant growing upon stones and rocks.

Lobe, any division of an organ or especially rounded division.

Lorate, thong- or strap-shaped.

Margin, the edge or boundary line of a body.

Median, belonging to the middle.

Membranaceous, thin or semi-transparent, like a thin membrane.

Mentum, a spur-like or chin-like extension of the flower composed of the variably united *column-foot*, *labellum*, and lateral *sepals*.

Mid-rib, the principle nerve of a leaf.

Monilform, necklace-shaped, like a string of beads.

Monopodial, referring to a growth habit in which new leaves develop from the same meristem or growing point as all previous leaves; cf. *sympodial*. (See Glossary Fig. 1)

Montane, pertaining to mountains, so a montane plant grows on mountains.

Morphology, the study of form, particularly external structure.

Mucro, a sharp terminal point.

Mucronate, possessing a mucro.

Nerve, prominent unbranched rib of a leaf, especially the midrib of the leaf.

Oblanceolate, tapering towards the base more than towards the apex.

Oblique, slanting or of unequal sides.

Oblong, much longer than broad, with nearly parallel sides.

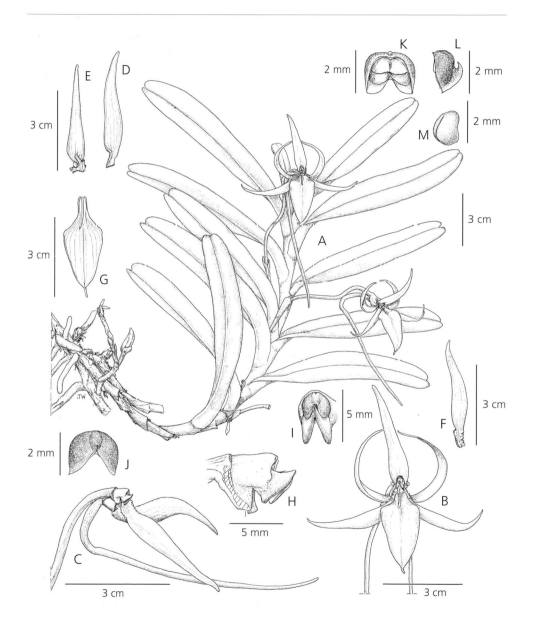

Glossary Fig. 1. Parts of a monopodial epiphytic orchid, *Jumellea alionae*. A, habit, showing roots, stem, leaves and inflorescence; B, flower, front view; C, flower, side view with lip, dorsal sepal, nearside lateral sepal and petal removed; D, dorsal sepal; E, lateral sepal; F, petal; G, lip; H, column, side view; I, rostellum lobes and anther cap from front; J, anther cap, front view, K, anther cap, back view; L, anther cap, side view; M, pollinium. All drawn by Juliet Beentje from the type collection. Scale bars as indicated.

Obovate, reversed ovate.

Obtuse, blunt or rounded at the end.

Obverse, the side facing, as opposed to reverse.

Operculate, lid-like.

Ovate, egg-shaped in longitudinal section, broader at the base.

Ovoid, egg-shaped solid.

Pandurate, fiddle-shaped.

Papillose, covered with soft superficial glands or protuberances, i.e. papillae.

Papyraceous, papery.

Patent, spreading, divergent from the axis at almost 90°.

Pectinate, combed.

Pedicel, the stalk of a single flower in an inflorescence.

Peduncle, the stalk bearing an inflorescence or solitary flower.

Peloric, an abnormality in which the flower in which it appears is more or less regular; for example, when lip has the form of a petal; a mutant form.

Petal, a flower leaf.

Petiole, the foot stalk of a leaf.

Plicate, leaves having several or many longitudinal veins and usually folded at each one, pleated.

Pollinarium (pl. **pollinaria**)**,** the functional unit of pollen transfer in orchid pollination, consisting of two or more *pollinia* (sometimes with *caudicles*), often a *stipes*, and a *viscidium*.

Pollinium (pl. **pollinia**)**,** a coherent mass of pollen grains.

Porrect, directed outwards and forward.

Pseudobulb, a thickened and a bulb-like stem in orchids.

Pubescent, softly hairy.

Pyriform, pear-shaped.

Rachis, the axis of an inflorescence to which the pedicels are attached, above the peduncle.

Radical, coming from the root or base of the plant stem.

Recurved, curved backwards and downwards.

Reflexed, abruptly bent downwards or backwards.

Resupinate, with the pedicel twisted so that the *labellum* is always in the same position (usually at the bottom of the flower) regardless of the position of the inflorescence.

Retuse, with a shallow notch at a rounded apex.

Rhizome, the indeterminate stem or system of stems of many plants, such as *sympodial* orchids, which successively give rise to new shoots and flowers, often horizontal or underground but sometimes appressed to branches or rocks.

Rostellum, that portion of the stigma that aids in gluing the pollinia to the pollinating agent; tissue that separates the anther from the fertile stigma; sometimes beak-like.

Saccate, a conspicuous hollow swelling.

Saprophyte (adj. **saprophytic**)**,** deriving its nourishment, in whole or part, from decaying organic matter. Often used incorrectly for a *heteromycotroph* or *holomycotroph* that lacks chlorophyll. Fungi are true saprophytes.

Scale, any thin scarious body, usually a degenerate leaf, sometimes of epidermal origin.

Scape, the peduncle and the rachis of the inflorescence.

Sclerophyllous, woody plants with tough and thick evergreen leaves that reduce water loss.

Sepal, term in universal use for each segment composing a calyx.

Serrate, sharp saw-like teeth.

Serrulate, finely serrate with a series of small notches.

Sessile, without a stalk.

Simultaneous, at the same time.

Spathulate, oblong and attenuated at the base, like a spatula.

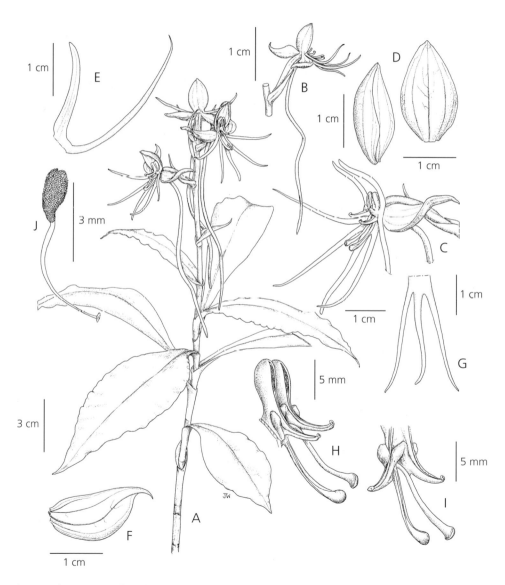

Glossary Fig. 2. Parts of a tuber-bearing terrestrial orchid, *Habenaria tianae*. A, habit, stem, leaves and inflorescence; B, C, flower, two views; D, dorsal sepal, two views; E, petal; F, lateral sepal; G, lip, flattened; H, column, oblique view; I, rostellar and stigma lobes; J, pollinarium. All drawn from the type by Juliet Beentje. Scale bars as indicated.

Spur, a long, usually nectar-containing, tubular projection of the lip.

Stelidia, column wings or teeth.

Stigma, the sticky receptive part of the pistil, produces a viscid, sugary material that receives the pollinia and permits the pollen grains to germinate.

Stipe (pl. **stipites**), a *pollinium* stalk derived from the *rostellum*.

Subtend, to extend under, or be opposite to.

Sulcate, grooved or furrowed.

Sympodium (pl. **sympodia**, adj. **sympodial**), a discontinuous main axis, where the stem is made up of a series of superposed branches, these imitating a single main axis, each new shoot developing from an axillary bud on the previous shoot unit; stem whose growth is continued not by the main stem but by lateral branches; prevalent in monocots; sympodial inflorescences include the dichasium, rhipidium, cincinnus, and false umbel.

Taxon (pl. **taxa**), a taxonomic group of any rank.

Tepal, a floral part, usually used to describe sepals and petals that are similar or indistinguishable.

Terete, circular in transverse section, cylindrical and usually tapering.

Terrestrial, growing in the earth.

Tuber, tuberoid, underground swollen storage organ. (See Glossary Fig. 2)

Undulate, having a wavy outline or appearance.

Unguiculate, contracted at the base into a claw. Also clawed.

Velamen, the outer layer or layers of the root, consisting of dead cells at maturity and bordered internally by an exodermis.

Ventral, on the lower side; belly.

Verrucose, covered with wart-like projections.

Villose, covered with soft, shaggy hairs.

Viscidium (pl. **viscidia**), the sticky portion of the *rostellum* that is often connected to *pollinia*.

Wing, a membraneous expansion attached to an organ.

Zig-zag, having short bends or angles from side to side, fractiflex.

Index

Accepted names are in roman text, synonyms in *italics*.

Local Names